I0038138

Cyber Security Using Modern Technologies

The main objective of this book is to introduce cyber security using modern technologies such as Artificial Intelligence, Quantum Cryptography, and Blockchain. This book provides in-depth coverage of important concepts related to cyber security. Beginning with an introduction to Quantum Computing, Post-Quantum Digital Signatures, and Artificial Intelligence for cyber security of modern networks and covering various cyber-attacks and the defense measures, strategies, and techniques that need to be followed to combat them, this book goes on to explore several crucial topics, such as security of advanced metering infrastructure in smart grids, key management protocols, network forensics, intrusion detection using machine learning, cloud computing security risk assessment models and frameworks, cyber-physical energy systems security, a biometric random key generator using deep neural network and encrypted network traffic classification. In addition, this book provides new techniques to handle modern threats with more intelligence. It also includes some modern techniques for cyber security, such as blockchain for modern security, quantum cryptography, and forensic tools. Also, it provides a comprehensive survey of cutting-edge research on the cyber security of modern networks, giving the reader a general overview of the field. It also provides interdisciplinary solutions to protect modern networks from any type of attack or manipulation. The new protocols discussed in this book thoroughly examine the constraints of networks, including computation, communication, and storage cost constraints, and verifies the protocols both theoretically and experimentally. Written in a clear and comprehensive manner, this book would prove extremely helpful to readers. This unique and comprehensive solution for the cyber security of modern networks will greatly benefit researchers, graduate students, and engineers in the fields of cryptography and network security.

Cyber Security Using Modern Technologies
Artificial Intelligence, Blockchain and Quantum Cryptography

Edited by
Om Pal
Vinod Kumar
Rijwan Khan
Bashir Alam
Mansaf Alam

CRC Press
Taylor & Francis Group
Boca Raton London New York

CRC Press is an imprint of the
Taylor & Francis Group, an **informa** business

First edition published 2024
by CRC Press
2385 Executive Center Drive, Suite 320, Boca Raton, FL 33431

and by CRC Press
4 Park Square, Milton Park, Abingdon, Oxon, OX14 4RN

© 2024 selection and editorial matter, Om Pal, Vinod Kumar, Rijwan Khan, Bashir Alam and Mansaf Alam; individual chapters, the contributors

CRC Press is an imprint of Taylor & Francis Group, LLC

Reasonable efforts have been made to publish reliable data and information, but the author and publisher cannot assume responsibility for the validity of all materials or the consequences of their use. The authors and publishers have attempted to trace the copyright holders of all material reproduced in this publication and apologize to copyright holders if permission to publish in this form has not been obtained. If any copyright material has not been acknowledged please write and let us know so we may rectify in any future reprint.

Except as permitted under U.S. Copyright Law, no part of this book may be reprinted, reproduced, transmitted, or utilized in any form by any electronic, mechanical, or other means, now known or hereafter invented, including photocopying, microfilming, and recording, or in any information storage or retrieval system, without written permission from the publishers.

For permission to photocopy or use material electronically from this work, access www.copyright.com or contact the Copyright Clearance Center, Inc. (CCC), 222 Rosewood Drive, Danvers, MA 01923, 978-750-8400. For works that are not available on CCC please contact mpkbookspermissions@tandf.co.uk

Trademark notice: Product or corporate names may be trademarks or registered trademarks and are used only for identification and explanation without intent to infringe.

Library of Congress Cataloging-in-Publication Data
Names: Pal, Om, editor. | Kumar, Vinod (Professor of electronics and communication), editor. |
Khan, Rijwan, 1981- editor. | Alam, Bashir, editor. | Alam, Mansaf, editor.
Title: Cyber security using modern technologies : artificial intelligence, blockchain and quantum cryptography / edited by Om Pal, Vinod Kumar, Rijwan Khan, Bashir Alam, Mansaf Alam.
Description: Boca Raton : CRC Press, 2023. | Includes bibliographical references and index.
Identifiers: LCCN 2023000887 (print) | LCCN 2023000888 (ebook) |
ISBN 9781032213194 (hardback) | ISBN 9781032213217 (paperback) | ISBN 9781003267812 (ebook)
Subjects: LCSH: Computer security–Technological innovations.
Classification: LCC QA76.9.A25 C919196 2023 (print) |
LCC QA76.9.A25 (ebook) | DDC 005.8–dc23/eng/20230201
LC record available at https://lccn.loc.gov/2023000887
LC ebook record available at https://lccn.loc.gov/2023000888

ISBN: 978-1-032-21319-4 (hbk)
ISBN: 978-1-032-21321-7 (pbk)
ISBN: 978-1-003-26781-2 (ebk)

DOI: 10.1201/9781003267812

Typeset in Times
by codeMantra

Contents

Preface

Cyber security is the protection of networks and network services from unauthorized access, alteration, and destruction, and there is no harmful effect even in critical situations. Modern networks are connected with heterogeneous devices, heterogeneous protocols, various network configurations, diverse network implementing methodologies, etc. Communication systems, critical infrastructure plants, defense network, social network, financial market, distributed workplace, etc., all are connected through a network. Therefore, maintaining the connectivity services for legitimate users is the prime concern for network administrators and policymakers.

Due to diversity and heterogeneity, it is a very tough and challenging task to provide security at each level. Day by day, new protocols, new technologies, and network methodologies are emerging. Hackers are also finding a noble way to breach the security of the network.

Therefore, there is a need to handle the modern threat with more intelligence. To handle this, it requires the assessment of modern networks in terms of possible security breach, protocols, present defense mechanism, etc. Also, it requires to suggest the interdisciplinary solutions for protecting the modern network from any kind of attack or disturbance in the network.

The objective of this book is to explore the concepts and applications related to the cyber security of modern network with the vision to identify and address existing challenges. Besides this, it shall also provide future research directions in this domain. This book is meant for students, practitioners, industry professionals, researchers, and faculty working in the field of cyber security.

This book introduces the latest weakness of the modern network systems, which includes various attacks, policies, defense mechanism, latest tools, and technologies. In addition, this book provides new techniques to handle modern threats with more intelligence. It also includes some modern techniques for cyber security such as cloud security best approaches, artificial intelligence and machine learning, security of IoT systems, use of blockchain for modern security, quantum cryptography, and forensic tools. It begins by providing a comprehensive survey of state-of-the-art research on cyber security of modern networks, giving the reader a general overview of the field. This book provides the interdisciplinary solutions for protecting the modern network from any kind of attack or disturbance. The new protocols discussed in the book thoroughly investigate the most important network devices' constraints including computation, communication, and storage cost constraints. It subsequently verifies the protocols both theoretically and experimentally. This unique and complete solution for cyber security of modern network will greatly benefit researchers, graduate students, and engineers in the fields of cryptography and network security.

Om Pal
New Delhi, India

Vinod Kumar
Prayagraj, India

Rijwan Khan
Ghaziabad, India

Bashir Alam
New Delhi, India

Mansaf Alam
New Delhi, India

Acknowledgments

The making of this book is a long journey that required a lot of hard work, patience, and persistence. We wish to express our heartfelt gratitude to our families, friends, colleagues, and well-wishers for their endless support throughout this journey. We would particularly like to express our gratitude to Prof. M.N. Doja, Director, IIIT Sonepat, Haryana for his constant encouragement. We would also like to thank Dr. Vinay Thakur, Ministry of Electronics & Information Technology (Government of India), New Delhi for his unconditional support. Besides, we owe a deep sense of appreciation to faculty members of Jamia Millia Islamia, New Delhi and of University of Allahabad, Prayagraj for their invaluable suggestions and critical review of the manuscript and for inspiring me to write this book.

How will we forget all our students Mr. Pradeep Kumar Tiwari, Mr. Animesh Tripathi, Ms. Khushboo Gupta, Mr. Arunesh Dutt, Mr. Neeraj Patel, Ms. Himanjali Singh, and others for their continuous support during the preparation of this book.

Finally, we wish to acknowledge and appreciate the Taylor & Francis team for their continuous support throughout the entire process of publication. Our gratitude is extended to the readers, who gave us their trust, and we hope this work guides and inspires them.

Om Pal

Vinod Kumar

Rijwan Khan

Bashir Alam

Mansaf Alam

Editors

Dr. Om Pal received his B.E. in Computer Science & Engineering from Dr. B. R. Ambedkar University, Agra, MBA(O&M) from IGNOU, MS (Research) in field of Cryptography from IIT Bombay and Ph.D. in field of Cryptography from Jamia Millia Islamia, New Delhi. Presently he is Pursuing "PG Diploma program (Part Time) in Cyber Law, Cyber Crime Investigation and Digital Forensics" from National Law Institute University, Bhopal. He has more than 20 years of Academic & Research experience in various areas of Computer Science. Presently he is working as an Associate Professor in Department of Computer Science, University of Delhi (Central University), North Campus, Delhi. Previously, he worked as a Scientist-D at Ministry of Electronics and Information Technology (MeitY), Government of India. He evolved and worked on research projects of national interest in various areas including Cryptography, Cyber Security, Digital Personal Data Protection (DPDP) Bill 2022, National Strategy on Blockchain, Post-Quantum Cryptography, Quantum Computing, Cyber Law etc. and drafting of R&D policies. Prior to this, he worked at Centre for Development of Advanced Computing (C-DAC) as a Sr. Technical Officer and as an IT Resource Person in National Thermal Power Corporation (NTPC). He has published many research articles in International Journals (Transaction/SCI/Scopus) and International Conferences of repute. He has filed patents on Conditional Access Systems and Cryptographic Systems. He has published a book: Unleashing the Potentials of Blockchain Technology for Healthcare Industries, by Elsevier. He has also served as a Guest Editor for special issues of International Journals. He has been nominated in the board of reviewers of various peer-reviewed and refereed Journals. He has delivered many talks on cyber security, Quantum Cryptography, Blockchain Technology and Cyber Law.

Dr. Vinod Kumar is presently working as an Assistant Professor with the Department of Electronics and Communication (Computer Science and Engineering), University of Allahabad, Prayagraj, Uttar Pradesh, India. Before joining the University of Allahabad, he worked as an Assistant Professor in the Department of Computer Science and Engineering, Rajkiya Engineering College (REC) Kannauj, UP, India. Prior to this, he worked as Assistant Professor in the School of Information Technology, Centre for Development of Advanced Computing (C-DAC), Noida, UP, India. He received his MCA from Uttar Pradesh Technical University (UPTU) Lucknow, UP, India in 2005; M.Tech in Computer Science and Engineering from Guru Gobind Singh Indraprastha University (GGSIPU) Delhi in 2011; and Ph.D. in the field of Cryptography from Jamia Millia Islamia (JMI), New Delhi. He has published various quality papers in reputed International Journals like Elsevier, Springer, Wiley, etc. He has also presented various research papers in reputed International Conferences. He has written two books for undergraduate and postgraduate students of computer science and engineering. He has a rich academics & research experience in various areas of computer science. His areas of research interest include Cryptographic Protocols, Key Management in Multicasting/Secure Group Communications, and Authentication Protocol and Security in WSNs.

Dr. Rijwan Khan received his B.Tech in Computer Science & Engineering from BIT, M.Tech in Computer Science & Engineering from IETE, and Ph.D. in Computer Engineering from Jamia Millia Islamia, New Delhi. He has 16 years of Academic & Research Experience in various areas of Computer Science. He is currently working as a Professor and Head of Department of Computer Science and Engineering at ABES Institute of Technology, Ghaziabad, U.P. He is the author of three subject books. He has published more than 50 research papers in different journals and conferences. He has been nominated to the board of reviewers of various peer-reviewed and refereed journals. He is the editor of three research books. He has chaired three international conferences and was a keynote speaker in some national and international conferences.

Prof. Bashir Alam received his B.Tech. (Computer Engineering) from Aligarh Muslim University (A Central University of Government of India), Aligarh, Uttar Pradesh, India; M.Tech. (IT) from Guru Gobind Singh Indraprastha University, Delhi, India; and Ph.D. from Jamia Millia Islamia (A Central University of Government of India), New Delhi, India. Currently, he is a Professor & Head of the Department, Department of Computer Engineering, Faculty of Engineering and Technology, Jamia Millia Islamia, New Delhi. He has published several research articles/papers in various reputed International Journals and Conference Proceedings published by reputed publishers like Elsevier, Springer, IEEE, etc. He has filed patents on Conditional Access System, Cryptographic systems. His areas of research interest include Blockchain Technology, Network Security, GPU Computing, Big Data, Parallel Computing, Soft Computing, Computer Network, Operating System, Distributed and Cloud Computing, Advanced Computer Architecture, and Intellectual Property Rights. He is also a reviewer for several reputed international journals and conferences. He is a Lifetime Member of ISTE (Indian Society for Technical Education) and other technical societies.

Prof. Mansaf Alam has been working as a Professor in the Department of Computer Science, Faculty of Natural Sciences, Jamia Millia Islamia, New Delhi 110025, Young Faculty Research Fellow, DeitY, Government of India & Editor-in-Chief, Journal of Applied Information Science. He has published several research articles in reputed International Journals and Proceedings at reputed International conferences published by IEEE, Springer, Elsevier Science, and ACM. His areas of research interest include Artificial Intelligence, Big Data Analytics, Machine Learning & Deep Learning, Cloud Computing, and Data Mining. He is a reviewer of various journals of international repute, such as *Information Science* published by Elsevier Science. He is also a member of the program committee of various reputed International conferences. He is on the Editorial Board of some reputed International Journals in Computer Sciences. He has published three books: *Digital Logic Design by PHI, Concepts of Multimedia* by Arihant; *Internet of Things: Concepts and Applications* by Springer; and *Big Data Analytics: Applications in Business and Marketing, Big Data Analytics: Digital Marketing and Decision Making* by Taylor & Francis. He recently received an International Patent (Australian) on An Artificial Intelligence-Based Smart Dustbin.

List of Contributors

Eshtiak Ahmed
Faculty of Information Technology and
    Communication Sciences
Tampere University
Tampere, Finland

Rezwan Ahmed
Department of Computer Science Engineering
United International University
Dhaka, Bangladesh

Sayada Sonia Akter
Department of Computer Science Engineering
United International University
Dhaka, Bangladesh

Farzana Anowar
Department of Computer Science
University of Regina
Regina, Canada

N. Sarat Chandra Babu
Society for Electronic Transactions and
    Security [SETS]
CIT Campus
Chennai, India

Mukesh Kumar Bhardwaj
Department of Computer Science
Harlal Institute of Management & Technology
Uttar Pradesh, India

M. Prem Laxman Das
Society for Electronic Transactions and
    Security [SETS]
CIT Campus
Chennai, India

Apurv Garg
ABES Institute of Technology
Uttar Pradesh, India

Subarna Ghosh
Maharishi International University
Fairfield, Iowa

Anmol Gupta
ABES Institute of Technology
Uttar Pradesh, India

Khushboo Gupta
Department of Electronics and
    Communication
University of Allahabad
Uttar Pradesh, India

Md Imran Hossen
School of Computing and Informatics
University of Louisiana at Lafayette
Lafayette, Louisiana

Ashraful Islam
School of Computing and Informatics
University of Louisiana at Lafayette
Lafayette, Louisiana

Dhaval S. Jha
Institute of Technology
Nirma University
Gujarat, India

Aashna Kapoor
ABES Institute of Technology
Uttar Pradesh, India

Sapna Katiyar
Impledge Technologies
Uttar Pradesh, India

Sandeep Kaur
Guru Nanak Dev University
Amritsar, India

Ferdous Hasan Khan
Department of Computer Science Engineering
United International University
Dhaka, Bangladesh

Pravin Kumar
Department of Electronics and Communication
University of Allahabad
Uttar Pradesh, India

Rajendra Kumar
Department of Computer Science
Jamia Millia Islamia
New Delhi, India

Nafees Mansoor
University of Liberal Arts (ULA)
Dhaka, Bangladesh

Jason Elroy Martis
NMAM Institute of Technology
Karnataka, India

Jyoti Mishra
Department of Electronics and
  Communications
University of Allahabad
Uttar Pradesh, India

Pawan Mishra
Department of Electronics and Communication
University of Allahabad
Uttar Pradesh, India

Abhyudaya Mittal
ABES Institute of Technology
Uttar Pradesh, India

Ravi Kamal Pandey
Department of Physics
University of Allahabad
Uttar Pradesh, India

Pooja
Department of Electronics and Communication
University of Allahabad
Uttar Pradesh, India

Shiv Prakash
Department of Electronics and Communication
University of Allahabad
Uttar Pradesh, India

Mohammed Masudur Rahman
Department of Computer Science and
  Engineering
University of Engineering and Technology
Dhaka, Bangladesh

Mohammad Shahriar Rahman
United International University (UIU)
Dhaka, Bangladesh

Manjit Sandhu
Guru Nanak Dev University
Amritsar, India

Sannidhan M. S.
NMAM Institute of Technology
Karnataka, India

Ravinder Singh Sawhney
Guru Nanak Dev University
Amritsar, India

Het Shah
Institute of Technology
Nirma University
Gujarat, India

Bhartendu Sharma
ABES Institute of Technology
Uttar Pradesh, India

Sandeep Sharma
Guru Nanak Dev University
Amritsar, India

Shiv Veer Singh
Department of Computer Science and
  Engineering
IIMT College of Engineering
Greater Noida, India

Sudeepa K. B.
NMAM Institute of Technology
Udupi, Karnataka

Vinay Thakur
Ministry of Electronics and Information
  Technology
Government of India
New Delhi, India

Aditi Tiwari
ABES Institute of Technology
Uttar Pradesh, India

Mahendra Tiwari
Department of Electronics and
    Communications
University of Allahabad
Uttar Pradesh, India

Pradeep Kumar Tiwari
Department of Electronics and Communication
University of Allahabad
Uttar Pradesh, India

Animesh Tripathi
Department of Electronics and Communication
University of Allahabad
Uttar Pradesh, India

Narendra Kumar Updhyay
Department of Computer Science
Harlal Institute of Management & Technology
Uttar Pradesh, India

Jai Prakash Verma
Institute of Technology
Nirma University
Gujarat, India

1 Quantum Computing
A Global Scenario

Om Pal
University of Delhi

Vinod Kumar
University of Allahabad

Vinay Thakur
Ministry of Electronics and Information Technology

Bashir Alam
Jamia Millia Islamia

CONTENTS

1.1 INTRODUCTION

Today, we are in the era of classical computing where information is stored in the form of 'on' or 'off' state of the transistor. Charged state of the transistor represents the classical bit 1 ('on state') and discharged state represents the classical bit 0 ('off state'). All classical models of computing work on a series of bits. To store a number of 'd' digits, a sequence of $(d-1)\log_2 10 + 1$ binary digits is needed. When a number of 'd' digits is used in any classical computing, then all binary digits $(d-1)\log_2 10 + 1$ of classical memory are affected. In the early 1980s, Physicist Paul Benioff presented a Quantum model of Turing machine. Physicist Benioff's idea of computation was based on the physical states of quantum particles [1]. This Quantum mechanical model opened the door

DOI: 10.1201/9781003267812-1

for new thinking in the field of information representation. In 1982, Richard Feynman designed an abstract model of quantum computing and suggested that principles of quantum physics could be used to build a quantum computer. He imagined a quantum machine which can do computation based on the principles of quantum mechanics. Feynman also observed that problems which require exponential growing time on a classical computer can be solved in polynomial time complexity on a quantum computer [2].

In the present scenario, services of quantum simulators are only available to a common researcher or to a general public. Today, quantum computers are existed in theories and only available in high-end laboratories as reality. As quantum field is an emerging field and a computing technology of tomorrow so, various countries are investing a good amount of money to promote the quantum research. The rest of this chapter is organized as follows. Section 1.2 covers the basic description of quantum computing and its theories. Section 1.3 emphasizes the need for quantum computing. Section 1.4 discusses various quantum computing approaches and the challenges associated with them. Section 1.5 covers the present status of quantum computing technology and its future market. Finally, in Section 1.6, the conclusion of this chapter and future directions are presented.

1.2 QUANTUM COMPUTING TERMINOLOGY

1.2.1 Qubit

In the field of classical computing, any memory bit represents the information as 0 or 1 at a time but not both simultaneously. If bit is 1, then the probability of charge state of transistor is 1, and if the state is in discharge mode, then the probability of 0 is 1. Quantum computers store information in the form of qubits. Qubit is the basic unit of information in the field of quantum computing. Qubit stores information as 0 or 1 or both 0 and 1 simultaneously.

In classical computing, the transistor's state represents the bit, but in quantum computing, the quantum particle's physical state represents the qubit. The physical state may be 'spin up'–'spin down' in the case of atom and 'Horizontal Polarization–Vertical Polarization' in the case of photon; two energy levels as 'excited state–ground state' can also be used for quantum particle. To represent information, the excited state can be represented as 1 and the ground state can be represented as 0 (Figure 1.1). To represent the quantum bit, Dirac notation $|0\geq0$ $|1\geq1$ is used [3].

1.2.2 Superposition

The interesting feature of quantum physics is that quantum particles may remain in the third state, which is neither the excited state nor the ground state. If the quantum particle is in this new state, then this state represents both 1 and 0 simultaneously. Let a, b are two complex numbers over a two-dimensional vector space of length 1, then the superposition state of the particle can be expressed as

$$|\psi\rangle = a|0\rangle + b|1\rangle$$

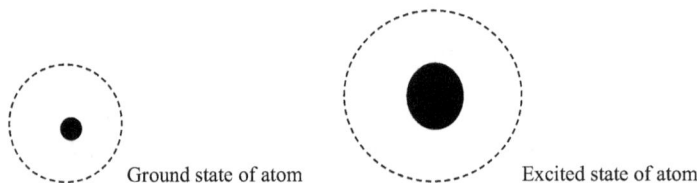

Ground state of atom   Excited state of atom

FIGURE 1.1 Qubit representation.

where a, b are complex numbers such that $|a|^2 + |b|^2 = 1$. Here, a represents the probabilistic amplitude of base state $|0>$ and b represents the probabilistic amplitude of base state $|1>$. Due to the property of superposition, a qubit can superimpose various values corresponding to different bits in classical computing. A single qubit with 2^n valid states contains 2^n values, which can be represented by n bits in classical computing.

1.2.3 PARALLELISM

Due to the superposition feature of quantum computing, single qubit represents the various values for which classical computing requires multiple bits. To get the benefit of superposition, maintaining coherence is a pre-requisite but maintaining coherence is a challenge in quantum computing. Another feature of quantum computing is the ability to perform multiple quantum operations on multiple qubits simultaneously. Quantum resources are independent of each other; therefore, quantum computing provides exponential speedup in solving tasks [4].

1.2.4 ENTANGLEMENT

Entanglement is the correlation or coupling of quantum states such that if there is a change in the property of a quantum state then there will be the opposite change in the property of the correlated state instantaneously. The entanglement feature of quantum particles always works regardless of the distance between the entangled particles. Let there be two qubits both are in superposition state $_j|0\rangle +_j|1\rangle$, $f(j)$ is a measurement function which returns value j on measurement of superposition state $_j|0\rangle +_j|1\rangle$ of first qubit, $f^{-1}(j)$ is an entangled function and $f^{-1}(j)$ instantaneously observed value j^{-1} on entangled second qubit when the measurement of superposition of first qubit is done. Values j and j^{-1} are opposite of each other [5].

1.3 QUANTUM GATES

Quantum gates are like classical gates, quantum gates take the inputs of qubits and produce the qubits with the same or different quantum states. Quantum gates are reversible; therefore, input states can be derived from the output states. In general, Quantum Gate operators are represented in 2×2 or 4×4 matrix form [3].

1.3.1 CONTROLLED NOT (CNOT) GATE

This gate operates on two qubits; one qubit is called controlled qubit and another is called target qubit. When the state of controlled qubit is 1, then target qubit flips from 1 to 0 and vice versa (Figure 1.2).

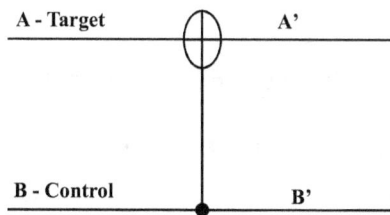

FIGURE 1.2 CNOT gate.

1.3.2 Hadamard Gate

This quantum gate is used to map a qubit from the basis state to the superposition state. This map $|0>$ to $\frac{|0\rangle + |1\rangle}{\sqrt{2}}$ and $|1>$ to $\frac{|0\rangle - |1\rangle}{\sqrt{2}}$. Hadamard matrix operator

$$H = \frac{1}{\sqrt{2}} \begin{bmatrix} 1 & 1 \\ 1 & -1 \end{bmatrix}$$

1.3.3 Pauli-X Gate

Pauli gate is equivalent to NOT gate of classical computing. It rotates the quantum state by 180°. This gate maps $|0>$ to $|1>$ and $|1>$ to $|0>$. Pauli matrix operator $X = \begin{bmatrix} 0 & 1 \\ 1 & 0 \end{bmatrix}$

1.3.4 Toffoli Gate

Toffoli gate is also called Controlled Not gate (CCNOT gate) universal reversible gate. It takes three qubits as input and inverts the third qubit if both first and second qubits are 1.

1.4 NEED OF QUANTUM COMPUTING

There are many problems that are impractical to solve by classical computing methodology or require exponential time growth to solve such problems. Utility of solution of any problem decreases with time; therefore, timely solution of any problem is a prime concern. To solve the complex problems like weather forecast, molecular comparison and simulation of chemical reactions for making medicines, to understand the behaviour of deadly viruses like COVID-19 using a large database and prediction of the nature of such virus for the next thousand years, optimization of transportation cost in the field of operation research, optimum assignment of jobs to a large population, optimization of inventory cost at the country level, key distribution in cryptography and to make a real-time decision in the financial sector, we are in need of a computing power that can solve these complex problem very quickly. NP-complete problems such as Travelling-Sales-Man Problem, Subset Sum Problem, Integer Factorization Problem, Set Cover Problem, etc. are not solvable in polynomial time using the classical computing model. It is expected that optimization will be the most used application of quantum computing in near future. Other applications such as machine learning, simulation and sampling will also use quantum computing power to a great extent in near future [6–10].

With the feature of superposition state and quantum parallelism, complex problems of classical computing models can be solved quickly using quantum computing. Few problems that are NP-Hard and NP-complete are solvable in polynomial time using the quantum computers. In 1994, Shore developed a quantum algorithm, which is capable to solve the problem of integer factorization in polynomial time [11]. In 1996, Grover proposed a quantum search algorithm, which can search an element from a large database with complexity of $\sqrt{N}$ only [12].

1.5 QUANTUM COMPUTING APPROACHES AND CHALLENGES

The major challenge in the field of quantum computing is the conversion of a theoretical model into reality. Qubits are highly sensitive to the surroundings; therefore, a constrained environment that can control the qubits efficiently is the pre-requisite for the development of a quantum computer. There are various quantum elements such as atoms, nuclear and electronic spins, ions and photons, which are being used by researchers for designing a Quantum system [13–15]. Various quantum computing implementing approaches that can be used to control the qubits to build a quantum computer are explained in the following sections.

i. **Nuclear Magnetic Resonance (NMR) Approach:**
    NMR is one of the implementing approaches of quantum computer that uses spin states of nuclei as qubits within the molecules. It is necessary that de-coherence time of any quantum state should be more than the time required to perform quantum operation by quantum logic gate. The NMR approach gives enough relaxation time for demonstration purposes and building of small qubits quantum computers. There are many nuclei candidates such as ^{1}H, ^{15}N, ^{13}C, ^{31}P, ^{19}F, ^{195}Pt which are suitable to be used for spin-1/2 nuclei in a magnetic field [13,16]. Rather than a single pure state, the NMR approach works on the molecule which is an example of an ensemble system.

ii. **Ion Trap Approach:**
    In this approach, ions or charged particles are confined in free space by applying the electromagnetic field. The electronic state of ions represents the qubit and a linear trap is used to hold the ions for representing the desired information. For coupling among the qubit's states of ions, laser beams are used [34]. Recently (March 2020), the president of Honeywell Quantum Solutions expected the release of the world's most powerful quantum computer from his research team, which will be based on ion trap methodology [17].

iii. **Neutral Atom Approaches:**
    The diameter of an atom is about 0.1 nm, and at room temperature, these atoms move at a speed of 1,000 km/h. Atoms are so small and dynamic in nature so that there is a need to cool down items and trap in a controlled environment. An array of neutral atoms cooled at micro-kelvin level temperature is used to encode the quantum information. Optical and microwave fields control the logic gate operations on the cooled atoms [18,19].

iv. **Optical Approach:**
    In the optical approach, photons are used to transfer the information from one place to another.

    The main advantage of the optical approach is the connectivity of quantum gates and memory devices through optical fibres or waveguides. Distribution of quantum key at a long distance needs optical fibre. However, heavy-cost infrastructure is involved in creating a quantum key distribution setup. Setup involves physical components such as photon transmitter, photon polarizer, beam splitter, quantum repeaters, avalanche photon-detector and various services such as methods of phase or polarization control and entanglement protocols. For long-distance communication, optical fibre is a preferable choice [20].

v. **Solid-State Approach:**
    Due to issues such as slowness of gate operation, scalability of qubit systems, efficient fabrication for close donor separation and maintaining low inter-donor separation time of various quantum computing approaches, it is desirable to consider alternative quantum computing approaches in the condensed phase. Solid-state approach is one of the possibilities that can provide the long decoherence time and scalability of system with high precision. However, the development of efficient solid-state devices for quantum computing is a challenge. Some of the solid-state approaches based on silicon are electron spin in semiconductor quantum dot, chains of ^{29}Si nuclei, donor electron spins in Si/Ge heterostructures, etc. [21,22].

vi. **Superconducting Approach:**
    In the superconducting approach, pair of electrons behaves as a charge carrier. To neutralize the repulsive force and resistance, pair of electrons is cooled down to near 1 K. Superconducting approach works at macroscopic quantum phenomena, where population of paired electrons produces a coherent quantum state. The superconducting approach is the leading implementing approach in the field of quantum computing. This approach gives

a fast gate operation time and a high scalability rate in comparison to others. Josephson junction-based quantum bit has the ability to maintain the desired coherence [23].

There are numerous challenges in designing a computing processor. Qubits are highly sensitive to surroundings; therefore, any interaction with surroundings, the superposition state leads to collapse. Another challenge is that there is no authentic mechanism for designing quantum gates and error correction. Due to high sensitivity, quantum computing implementing approaches are not easily scalable, therefore to propose a design that can accommodate a large number of qubits is a big challenge [14,24].

1.6 QUANTUM COMPUTING RESEARCH STATUS

Many quantum computing centres of applied as well as basic research have been established worldwide. Institute for Quantum Computing in Waterloo, Canada is the largest established centre in the field of quantum computing. China developed a strategic plan for quantum computing in the year 2006 and performed well in quantum technology, especially in the field of quantum communication and quantum sensors. Centre for Quantum Technology, Singapore was established in 2007. This centre actively works in the area of coherent control of photons, quantum cryptography and implementing technologies for cold atoms and molecules, optics, atom and ion trapping. Japan is a leading country in the field of space-based quantum communication and quantum key distribution. Defense Advanced Research Projects Agency (DARPA), National Aeronautics and Space Administration (NASA) and Intelligence Advanced Research Projects Activity (IARPA) establishments of USA are leading institutions in the field of quantum technologies. Paris Centre for Quantum Computing (PCQC) is working in the area of computing, communication and quantum security. Other countries Germany, Netherlands and Poland are also actively working in the field of quantum technologies. First, NMR technology-based quantum computer was demonstrated at the University of Oxford. The University of Oxford is one of the leading institutions in the field of quantum technology.

The strength of Massachusetts Institute of Technology (MIT) is theoretical physics and in the area of quantum computing, MIT deeply works in Quantum algorithms and complexity, measurement and control, applications of quantum computing and Quantum information theory. The National Institute of Standards and Technology (NIST) is working on standards and strategy building for preventing quantum attacks and exploitation of quantum systems [25]. D-Wave, one of the leading institutions, is using the technique of annealing for making the quantum computers. D-Wave has facilitated the open-source Ocean software development kit (SDK) for application development. This kit facilitates the development of quantum algorithms and development of new codes. In April 2020, UNSW, Sydney claimed a breakthrough in implementing the proof of concept using conventional silicon chip foundries. Researchers of UNSW published the results in Nature and claimed the scaling of quantum dots embedded in silicon. In June 2020, researchers [26] published an article in the field of spintronics, which gives hope for movement of electronics beyond Moore's law. Worldwide quantum computing growth for the last 20 years (1998–2019) is given in Figure 1.3.

In March 2018, Google announced the development of 72-qubit chip called 'Bristlecone'. Google claimed (on 23 October 2019) the quantum supremacy and announced the development of a 53-qubit quantum computer. Google group also claimed that a task that can be completed in 10,000 years on the fastest super computer of IBM has been completed by their 53-qubit quantum computer in just 200 seconds only. IBM has also announced the development of a 53-qubit quantum computer [27] and simulation of a 56-qubit design [28,29]. In 2018, Intel also announced the development of 49-qubit superconducting quantum chip called 'Tangle Lake' [30]. Intel is also working on the development of neuromorphic computing to study the working of biological brain. Microsoft is also a leading player in the field of quantum computing; it is working on a scalable quantum computer based on topological qubits. In topological quantum computing, Microsoft is using electron fabrication technology, which splits electrons into two parts for making redundant qubit states. Microsoft's aim is to create quantum states, which are less vulnerable to quantum noise or interference [31].

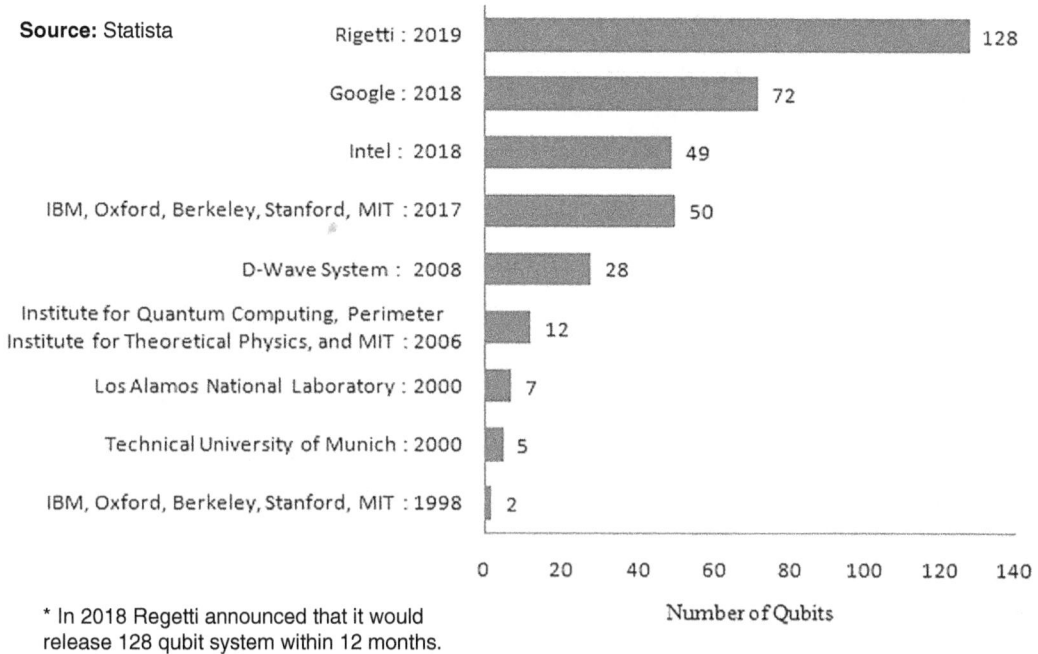

Source: Statista

Rigetti : 2019	128
Google : 2018	72
Intel : 2018	49
IBM, Oxford, Berkeley, Stanford, MIT : 2017	50
D-Wave System : 2008	28
Institute for Quantum Computing, Perimeter Institute for Theoretical Physics, and MIT : 2006	12
Los Alamos National Laboratory : 2000	7
Technical University of Munich : 2000	5
IBM, Oxford, Berkeley, Stanford, MIT : 1998	2

Number of Qubits

* In 2018 Regetti announced that it would
release 128 qubit system within 12 months.

FIGURE 1.3   Quantum computing growth.

Source: Statista.

Rigetti has also built up three quantum processors, namely, 16Q Aspen-1, 8Q Agave and 19Q Acorn [31]. Regetti is offering quantum computing services through cloud.

Since few years, there is a high-level impact on the growth of quantum computing market. Countries are investing heavily in the quantum computing sector. The market value of quantum computing in the year 2019 was witnessed to USD 101.12 Million [32]. Few companies have already started the commercialization of quantum computing services. It is anticipated that there will be US$23 Bn revenues in the field of quantum computing by the year 2025. Growth in the quantum computing market for the decade 2015–2025 is expected exponentially at a Compound Annual Growth Rate (CAGR) of 30.9% [33,34].

It is projected that the revenue for ten years (2019–2029) quantum computing market will be around US$2.54 billion [35]. Year-wise revenue detail is given in Figure 1.4, where the X-axis represents the year and the Y-axis represents the revenue in millions of US dollars.

Worldwide patenting detail on quantum computing technologies up to 2015 is given in Figures 1.5–1.8. United States is the leading country in patenting Quantum computing and Quantum sensor technologies, and China is leading in patenting Quantum Cryptographic technologies. As per the Quantum Computing Report, Microsoft Corporation is the leading organization in filing patent applications followed by Elwha LLC [36].

1.6.1   QUANTUM COMPUTING RESEARCH STATUS IN INDIA

In India, companies like IBM India, Qunu Labs, Automatski and Entanglement Partners are working in the area of quantum computing. In 2017, the Department of Science and Technology (DST) launched a quantum mission program on Quantum Science and Technology called Quantum Enabling Science and Technology (QuEST). In theoretic quantum research, institutions like IISc Bangalore, TIFR Mumbai, IISER Pune, RRI Bangalore and IIT Kanpur are leading institutions in India [37–39]. Indian Government has boosted the quantum research by announcing Rs. 8,000 crores in its 2020–2021 union budget for quantum computing over a period of five years. National

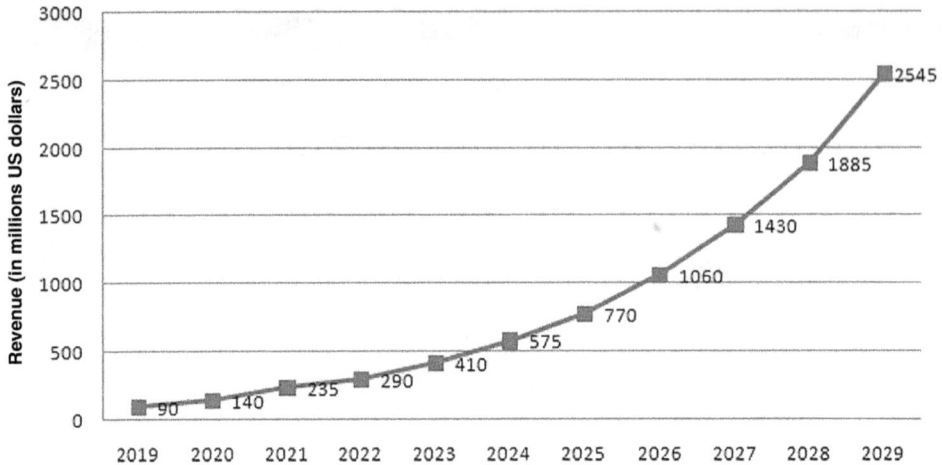

Source: Statista

FIGURE 1.4   Year-wise revenue.

Source: Statista.

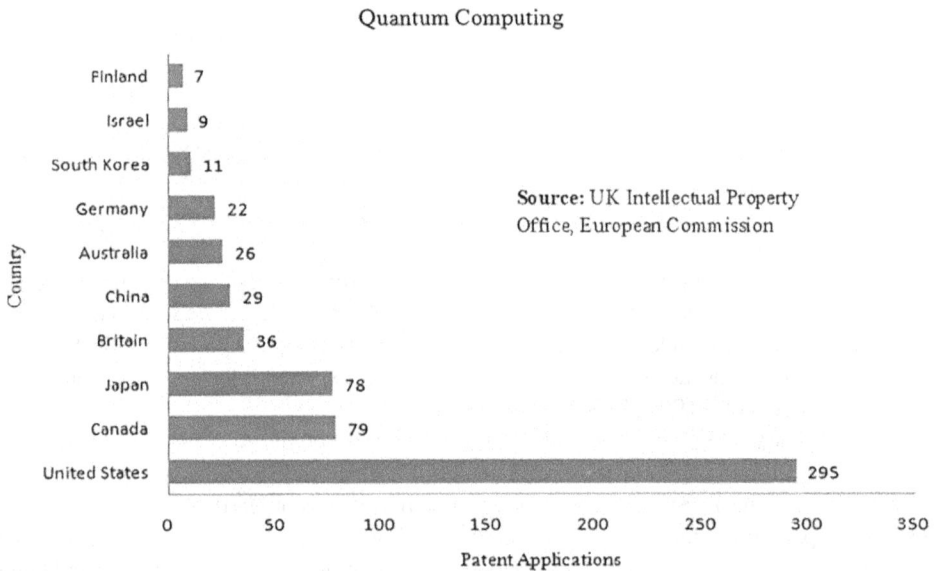

FIGURE 1.5   Quantum computing patent applications.

Mission on Quantum Technologies (NM-QTA) is the ambitious mission of Government of India. The major objectives of this mission are to create skilled manpower in the country in the field of quantum technologies, motivation for quantum devices and equipment's manufacturing, formulation of quantum strategy roadmap, eco-system development for quantum research, development of quantum legal framework, etc.

1.7   CONCLUSION AND FUTURE DIRECTIONS

Quantum technology is the technology of future. This technology has the ability to reshape the industries such as Medical Industry, Financial sector, Energy sector, Aerospace and various other

Quantum cryptography

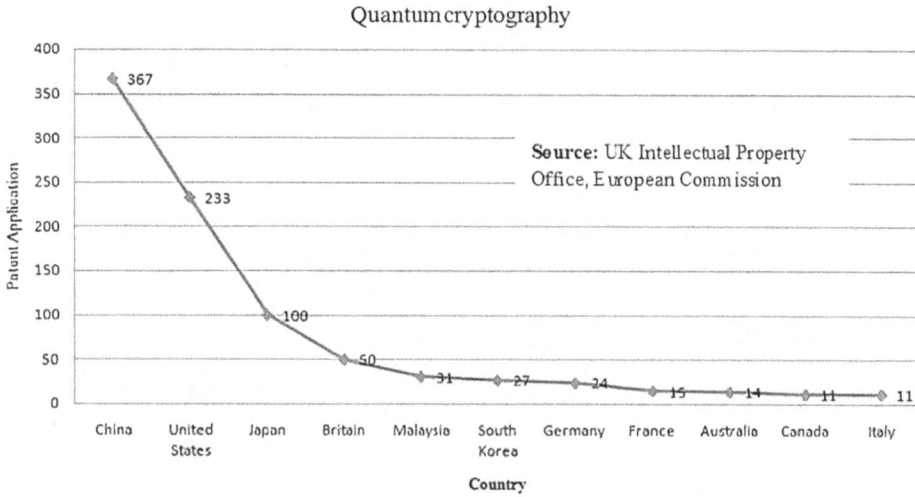

FIGURE 1.6 Quantum cryptography patent applications.

Quantum sensors

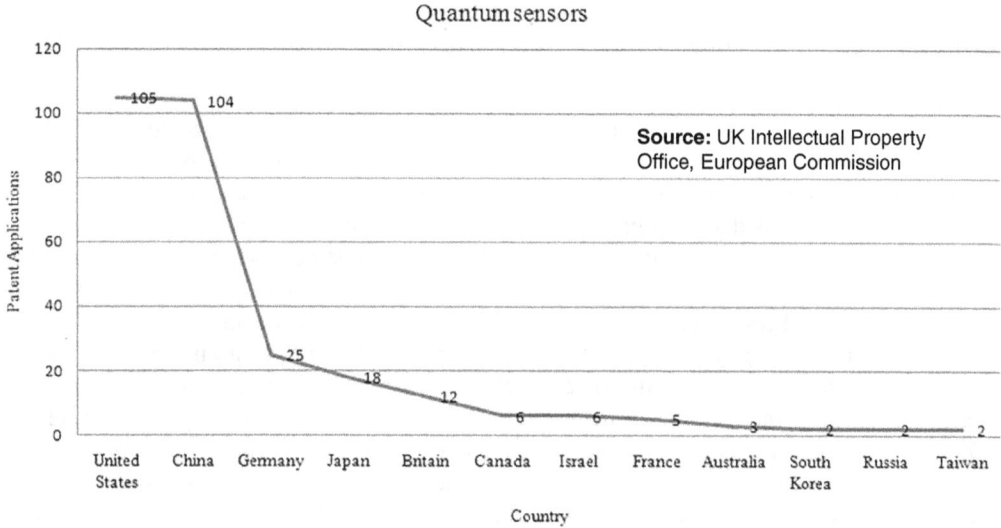

FIGURE 1.7 Quantum sensors patent applications.

Source: UK Intellectual Property Office, European Commission.

sectors. This has various applications in the field of algorithms optimization, artificial intelligence, cyber security, pharmaceuticals research & development, oil well optimization, risk analysis, trading strategies, and many more.

Quantum technology is an emerging technology that has the ability to transform the present computing power to the next level. Using the quantum computing power, problems of exponential complexities such as integer factorization are solvable in polynomial time complexity. Worldwide, a lot of research has been already going on in various dimensions of quantum technology. The real challenge in quantum technology is the practical implementation of theoretical concepts of this technology. There are many approaches to implement the quantum computing concepts but to find the stable and scalable implementing approach is still a challenge. Countries are investing a good amount in quantum computing research; therefore, it is expected that there will be a high-level impact on the growth of quantum computing market.

Quantum-key distribution

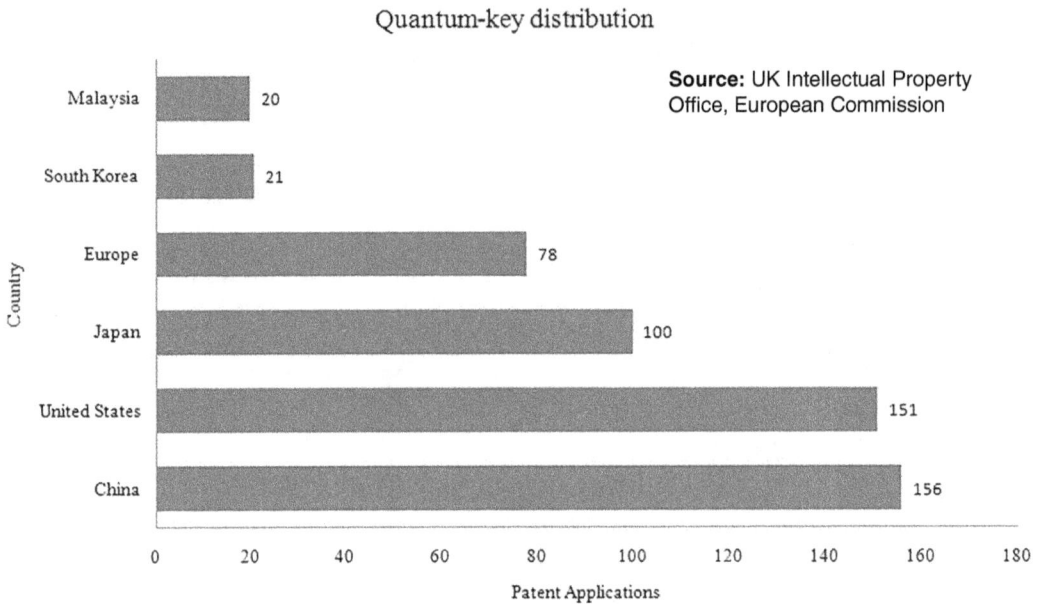

FIGURE 1.8   Quantum key distribution patent applications.

Following are some future steps and possible plans for boosting up the quantum research: (a) First, explore the technologies for qubits realization, (b) development of quantum computer test beds and their architectural optimization, (c) development of quantum computing algorithms and demonstration of developed algorithms on quantum computing test beds and ensure the superiority of results over the classical computing approach, (d) resolve the issues like speedup, optimization, quantum error corrections, quantum communication and quantum storage, (e) collaboration among the theoretical researchers, industry partners, mathematicians, etc., (f) conduction of regular workshops and conferences, (g) encourage doctorate research in the field of quantum computing, (h) create a quantum computing eco-system, (i) design elective courses on quantum computing for master and graduate students and (j) basic awareness certificate courses on quantum technologies, etc.

Due to the advancement in computing power, quantum technology can be utilized unethically in the form of cyber-attacks, financial frauds, social security breaches, etc. Considering the possible risks of this technology, there is a need to plan the appropriate quantum computing standards, quantum-safe crypto solutions. World realized the threat of this upcoming technology and accordingly Accredited Standards Committee X9 Inc has started work of creating appropriate standards against quantum attacks.

REFERENCES

1. P. Benioff, "The computer as a physical system: A microscopic quantum mechanical Hamiltonian model of computers as represented by Turing machines," *J. Stat. Phys. Vol.*, vol. 22, pp. 563–591, 1980.
2. R. P. Feynman, "Simulating physics with computers," *Int. J. Theor. Phys.*, vol. 21, pp. 467–488, 1982.
3. P. S. Menon and R. M., "A Comprehensive But Not Complicated Survey on Quantum Computing," in *International Conference on Future Information Engineering*, 2014, pp. 144–152.
4. N. Margolus, "*Parallel Quantum Computation*," Cambridge, MA, USA, 1990.
5. P. Jorrand, "A programmer's survey of the quantum computing paradigm," *Int. J. Nucl. Quantum Eng.*, vol. 1, no. 8, pp. 409–415, 2007.
6. M. M̈oller and C. Vuik, "On the impact of quantum computing technology on future developments in high-performance scientific computing," *Ethics Inf. Technol.*, vol. 19, pp. 253–269, 2017.
7. "Quantum computing: An emerging ecosystem and industry use cases," 2021.

8. H. Soeparno and A. S. Perbangsa, "Cloud Quantum Computing Concept and Development: A Systematic Literature Review," in *5th International Conference on Computer Science and Computational Intelligence*, 2021, pp. 944–954.

9. J. D. Whitfield, J. Yang, W. Wang, J. T. Heath, and B. Harrison, "*Quantum Computing 2022*," 2022.

10. O. Ayoade, P. Rivas, and J. Orduz, "*Artificial Intelligence Computing at the Quantum Level*," 2022.

11. P. W. Shor, "Algorithms for Quantum Computation: Discrete Logarithms and Factoring," in *35th Annual Symposium on Foundations of Computer Science*, 1994, pp. 124–134.

12. L. Grover, "*A Fast Quantum Mechanical Algorithm for Database Search*," New Jersey, 1996.

13. Y. Kanamori, S.-M. Yoo, W. D. Pan, and F. T. Sheldon, "A Short Survey On Quantum Computers," *Int. J. Comput. Appl.*, vol. 28, no. 3, pp. 227–233, 2006.

14. K. Kumar et al., "A Survey on Quantum Computing with Main Focus on the Methods of Implementation and Commercialization Gaps," in *2nd Asia-Pacific World Congress on Computer Science and Engineering (APWC on CSE)*, 2015.

15. M. M. Savchuk and A. V. Fesenko, "Quantum computing: Survey and analysis," *Cybern. Syst. Anal.*, vol. 55, no. 1, pp. 10–21, 2019.

16. J. A. Jones, "Quantum computing and nuclear magnetic resonance," *PhysChemComm*, no. 4, pp. 49–56, 2001.

17. S. K. Moore, "Honeywell's Ion Trap Quantum Computer Makes Big Leap," 2020. [Online]. Available: https://spectrum.ieee.org/tech-talk/computing/hardware/honeywells-ion-trap-quantum-computer-makes-big-leap. [Accessed: 11-Jun-2020].

18. M. Saffman, "Quantum computing with atomic qubits and Rydberg interactions: Progress and challenges," *J. Phys. B At. Mol. Opt. Phys.*, vol. 49, pp. 1–27, 2016.

19. T. Mukai, "Making a Quantum Computer Using Neutral Atoms," NTT Basic Research Laboratories. [Online]. Available: https://www.ntt-review.jp/archive/ntttechnical.php?contents=ntr200801sp7.html. [Accessed: 15-Jun-2020].

20. H. Krovi, "Models of optical quantum computing," *Nanophotonics*, vol. 6, no. 3, pp. 531–541, 2017.

21. B. Golding and M. I. Dykman, "*Acceptor-Based Silicon Quantum Computing*," Michigan State University, 2004. [Online]. Available: https://arxiv.org/ftp/cond-mat/papers/0309/0309147.pdf. [Accessed: 20-Jun-2020].

22. A. Gaita-Ariño, F. Luis, S. Hill, and E. Coronado, "Molecular spins for quantum computation," *Nat. Chem.*, vol. 11, pp. 301–309, 2019.

23. M. Kjaergaard et al., "Superconducting qubits: Current state of play," *Annu. Rev. Condens. Matter Phys.*, vol. 11, pp. 369–395, 2020.

24. M. J. Kumar, "Quantum computing in India: An opportunity that should not be missed," *IETE Tech. Rev.*, vol. 31, no. 3, pp. 187–189, 2014.

25. L. Chen et al., "NIST: Report on Post-Quantum Cryptography," NIST, Tech. Rep, 2016.

26. A. Avsar, H. Ochoa, F. Guinea, B. Özyilmaz, B. J. van Wees, and I. J. Vera-Marun, "Colloquium: Spintronics in graphene and other two-dimensional materials," *Rev. Mod. Phys.*, vol. 19, no. 2, 2020.

27. A. Cho, "IBM Casts Doubt on Google's Claims of Quantum Supremacy," 2019. [Online]. Available: https://www.sciencemag.org/news/2019/10/ibm-casts-doubt-googles-claims-quantum-supremacy. [Accessed: 25-Jun-2020].

28. E. Pednault, "Quantum Computing: Breaking through the 49 Qubit Simulation Barrier," 2017. [Online]. Available: https://www.ibm.com/blogs/research/2017/10/quantum-computing-barrier/. [Accessed: 30-Jun-2020].

29. E. Pednault, J. Gunnels, D. Maslov, and J. Gambetta, "*On 'Quantum Supremacy'*," IBM, 2019. [Online]. Available: https://www.ibm.com/blogs/research/2019/10/on-quantum-supremacy/. [Accessed: 02-Jul-2020].

30. J. Hsu, "*CES 2018: Intel's 49-Qubit Chip Shoots for Quantum Supremacy*," Intel, 2018. [Online]. Available: https://spectrum.ieee.org/tech-talk/computing/hardware/intels-49qubit-chip-aims-for-quantum-supremacy. [Accessed: 04-Jul-2020].

31. J. Ridley, "Quantum Computing: The Key Players Leading Us into the Quantum Age," 2019. [Online]. Available: https://www.pcgamesn.com/quantum-computing-researchers-development-where-are-we-now. [Accessed: 06-Jul-2020].

32. "*Global Quantum Computing Market (2020 to 2025) – Investment in R&D of Technology and Development is Strategically Important*," Businesswire, 2020. [Online]. Available: https://www.businesswire.com/news/home/20200317005476/en/Global-Quantum-Computing-Market-2020-2025. [Accessed: 07-Jul-2020].

33. D. S. Abdou, A. G. Salem, and M. Samir, "A battle for the future: Quantum computing chips markets analysis in the light of game theory," *Am. J. Mod. Phys. Appl.*, vol. 5, no. 2, pp. 30–36, 2018.

34. *"Global Quantum Computing Market is Expected to Reach US$ 23 Bn by 2025- PMR's Study,"* Persistence Market Research Pvt. Ltd., 2017. [Online]. Available: https://www.prnewswire.com/news-releases/global-quantum-computing-market-is-expected-to-reach-us-23-bn-by-2025---pmrs-study-636708523.html. [Accessed: 08-Jul-2020].

35. T. Alsop, *"Forecast Size of the Quantum Computing Market Worldwide from 2019 to 2029 (in Million U.S. Dollars),"* Statista, 2020. [Online]. Available: https://www.statista.com/statistics/1067216/global-quantum-computing-revenues/#:~:text=Global quantum computing revenues 2019–2029&text=The revenues for the global, million U.S. dollars in 2019. [Accessed: 11-Jul-2020].

36. "Quantum Computing Patent Analysis," Quantum Computing Report, 2020. [Online]. Available: https://quantumcomputingreport.com/quantum-computing-patent-analysis/. [Accessed: 11-Jul-2020].

37. "Union Budget 2020–21: Quantum Computing Gets Funds," *DownToEarth*, 2020. [Online]. Available: https://www.downtoearth.org.in/news/science-technology/union-budget-2020-21-quantum-computing-gets-funds-69117. [Accessed: 11-Jul-2020].

38. "FM's Rs 8,000 Crore Boost Will Help India Bridge Gap in Quantum Computing with US, China," *The Economic Times*, 2020. [Online]. Available: https://economictimes.indiatimes.com/tech/hardware/fms-rs-8000-crore-boost-will-help-india-bridge-gap-in-quantum-computing-with-us-china/articleshow/73835819.cms?from=mdr. [Accessed: 11-Jul-2020].

39. "Why India Is Falling behind in the Y2Q race," *LiveMint*, 2020. [Online]. Available: https://www.livemint.com/technology/tech-news/why-india-is-falling-behind-in-the-y2q-race-11579018170008.html. [Accessed: 11-Jul-2020].

2 Post-Quantum Digital Signatures

N. Sarat Chandra Babu and M. Prem Laxman Das
Society for Electronic Transactions and Security [SETS]

CONTENTS

DOI: 10.1201/9781003267812-2

2.1 INTRODUCTION

Any digital transaction which one carries out daily, like checking emails, internet banking and connecting to a remote server, uses network security protocols. These protocols use cryptographic schemes to provide functionalities like confidentiality and authenticity. While symmetric key ciphers can be used to provide these functionalities, they depend on the assumption that the peers have a common key. Hence, public key cryptosystems such as Rivest-Shamir-Adleman (RSA) cryptosystem (Rivest, Shamir and Adleman 1978) and Diffie-Hellman (DH) cryptosystem (Diffie and Hellman 1976) are ubiquitously used for providing key agreement and authentication. The security of these cryptosystems depends, respectively, on the computational hardness of integer factorization problem and that of solving the so-called DH problem (or the discrete logarithm problem over a finite field). Both these problems are very hard to solve on a classical computer, the Number Field Sieve (Pomerance 1996) can be used which has sub-exponential complexity in general. The exact complexities of these problems are not known.

Digital signatures are public key primitives, which are used to provide authentication (Katz and Lindell 2014, Chapter 12). The sender appends a signature to the message; these together are verified by the receiver for authenticity. The security of the primitive is captured by unforgeability, where a forger can access the signing oracle and has to output a valid message-signature pair. The Digital Signature Algorithm[1] is used in most cases. This chapter talks only about the functionality of authentication and public key signatures, which are used to achieve this.

Quantum computing provides another model of computing, which uses quantum mechanical phenomena like superposition and entanglement. Shor's algorithm (Shor 1994) gives a polynomial time method for integer factorization. Thus, a practical quantum computer would render RSA cryptosystem, and hence any internet protocol which uses this cryptosystem, insecure. "I estimate a one in five chance that within 10 years RSA-2048 will be broken using quantum computers", wrote Michele Mosca of University of Waterloo and President of evolutionQ. With companies like IBM and Google racing to build a larger quantum computer, the research community has begun to analyse and remedy the threat of quantum computers to security of digital communications. The quantum key distribution (QKD) and post-quantum cryptography (PQC) are two approaches for addressing this security issue.

QKD, implemented using the Bennet and Brassard protocol (Bennett and Brassard 2014) for example, is a method for key agreement over insecure channels. This method provides unconditional security, where the guarantee is a result of physical laws. Thus, QKD for key distribution and fast symmetric key ciphers could be used for providing confidentiality and authentication. This is due to the fact that symmetric key ciphers are not amenable to fast quantum attacks. Efforts are on to make the QKD technology viable. Thus, the above method will be secure even if the adversary has access to a large quantum computer.

QKD requires the usage of additional devices and infrastructure. So, one may ask whether it would be feasible to ensure quantum-resilient communication on the existing network infrastructure. Obviously, cryptosystems like RSA and DH need to be replaced. Classical symmetric key ciphers are largely immune to quantum attacks, but need to be tweaked to ensure quantum resilience (Perlner and Cooper 2009), thus providing the requisite functionalities. The study of design and analysis of public key ciphers, which resist attacks by quantum-enabled adversaries (and classical ones), is termed "Post-Quantum Cryptography" (PQC).

Computational problems that are hard even for the quantum computers to solve, security models and proofs incorporating the actions of a quantum-enabled adversary and security of classical cryptographic transformations are the main objects of study under PQC. Such cryptosystems can be classified, based on the computational problem they use for ensuring security, as lattice-based (Hoffstein et al. 2003; Gentry, Peikert and Vaikuntanathan 2008), multivariate quadratic (MQ)-based (Patarin 1996; Ding and Schmidt 2005) or as those based on miscellaneous techniques (Chase et al. 2017; Bernstein et al. 2015). While the techniques used are varied and literature vast, we restrict our attention to specific design strategies listed above. These are relevant in view of the NIST Post-Quantum Competition. However, though not relevant to NIST competition (from signature perspective), we discuss signatures constructed using supersingular isogenies in a separate section.

NIST had announced[2] a competition for arriving at a portfolio of Key Encapsulation Mechanism/Public Key Encryption Schemes and Digital Signature Schemes. A scalable quantum computer can break cryptosystems like RSA, DH and ECC, which are essential for providing authentication and confidentiality over networks. While post-quantum cryptosystems have been designed and studied by many researchers, there is a need for further investigation into security and performance aspects of these systems. NIST "Post-Quantum Cryptography Standardization" competition mentions advancements in building quantum computers and the complexity involved in migration to post-quantum candidates as reasons behind such a call. The proposals are evaluated for their performance and security guarantee. The competition is in its third round now, in the signature category, there are three finalists and three alternate candidates in the fray. While Falcon, CRYSTALS-DILITHIUM and Rainbow are the finalists, GeMSS, SPHINCS+ and Picnic are the alternates. It is expected that the future systems which use signatures will use these constructions.

This work surveys the construction aspects, security and most-recent implementation benchmarks of these signature schemes. As we had mentioned, there is extensive literature on any particular idea mentioned above. The article discusses the NIST PQC competition candidates, mainly. We focus on giving the major ingredients for describing the construction aspects of the above signature candidates. Wherever possible, complete design details are given. In some cases, only an outline of the scheme is presented. We have given the implementation benchmark obtained for the three finalists on an ARM platform. This would give the reader some idea of which design is best among the three. But such a concrete comparison is far from complete. The use cases mentioned here are only representative and not exhaustive. The article also touches upon some global initiatives on the migration plan.

2.1.1 SECTION-WISE PLAN

In Section 2.2, we discuss some preliminaries required, such as definition of signature and its security, Fiat-Shamir transform and Unruh transform. In Section 2.3, we describe the NIST PQC competition process. With respect to signatures, we discuss the evaluation criteria and the current status of the competition. In Sections 2.4–2.6, we discuss signatures built, respectively, using lattice, multivariate and symmetric key cryptography techniques. In Section 2.7, we discuss signature construction using super singular isogenies. In Section 2.8, we discuss some interesting use cases for these signatures. The section also discusses some migration plans, which are being drafted by bodies. These include NIST and ETSI's plan for migration to PQC. Finally, we have concluded our work in Section 2.9.

2.1.2 BACKGROUND EXPECTED OF THE READER

The subject of post-quantum signatures uses techniques from computer science, cryptography, discrete mathematics and quantum computation. This article, as mentioned earlier, has the modest goal of presenting the basic aspects of signatures, which are candidates in the NIST competition. We wish to collate some references for topics dealt with here. Katz and Lindell (2014) is an excellent reference for public key cryptography, digital signatures and random oracle methodology. For an excellent summary of sigma protocol and identification schemes, the reader may refer to Galbraith, Petit and Silva (2020). De Feo (2017) is an excellent article for learning the mathematical aspects of isogenies. Ding, Petzoldt and Schmidt (2020) is an excellent book for multivariate cryptosystems. It describes, in particular, the Unbalance Oil and Vinegar (UOV) and Hidden Field Equation (HFE) paradigms and attacks. For basics on lattice-based cryptography, the reader may refer to Peikert (2015).

2.2 PRELIMINARIES AND NOTATIONS

We discuss some basic terminology and techniques that are used in this article. We recall the definitions related to signatures and their security. We discuss identification schemes and their properties. We state the essential definitions and properties of identification schemes. We discuss Fiat-Shamir transform, which gives secure signatures starting from identification schemes, in the classical Random Oracle Model. We discuss the Unruh transform, which is the quantum analogue of the Fiat-Shamir transform. We discuss some results that provide security guarantees of signatures based on the properties of the underlying identification schemes.

2.2.1 DIGITAL SIGNATURES AND THEIR SECURITY

In this section, we discuss the definition and security of digital signatures. These are used for providing the functionality of authentication. For more details on the cryptographic aspects of digital signature, the reader may refer to Katz and Lindell (2014).

Definition 2.1

(Digital Signature) A digital signature scheme is a triple of probabilistic polynomial time (PPT) algorithms, namely, KeyGen, Sign and Verify:

KeyGen.
   Given a security parameter λ, this module outputs a key pair (pk, sk) (where sk is the secret key used for signing and pk is the public key used for verifying)

Sign.
   Given a message m and the secret key sk, this outputs a signature σ

Verify.
   Given a message m, public key pk and a signature σ_0, this module outputs accept or reject (checks whether σ_0 is a valid signature on the message)
   Correctness. The digital signature scheme is said to be correct, if, for security parameter λ and message m

$$\Pr\left[Verify(pk, m, \sigma) = 1 : pk, sk \leftarrow KeyGen(1^{\lambda}); \sigma \leftarrow Sign(m, sk) \right] \geq 1 - negl(\lambda).$$

That is, the verification module should output 1 on a message, whenever the correct signature and public key are supplied to it.

Next, we recall an important notion for discussing security of a signature scheme. The notion is described as game between an adversary and a challenger, where the adversary can ask signature queries and has to come up with a forgery. A digital signature scheme is said to be secure if it is *existentially unforgeable under chosen message attack* (EUF-CMA). The definition is given below.

Definition 2.2

(EUF-CMA) Let $\mathcal{A}$ be an adversary playing the following game $Game_{\mathcal{A}}^{\lambda}$:

KeyGen Phase.
The challenger generates a pair of keys $pk, sk \leftarrow KeyGen\left(1^{\lambda}\right)$ and $\mathcal{A}$ is given pk

Query Phase.
The adversary makes polynomial-many (in λ) signing queries m_i and the challenger responds with $\sigma_i \leftarrow Sign\left(m_i, sk\right)$

Challenge Phase.
The adversary outputs $\left(m^*, \sigma^*\right)$ at any stage.
    The adversary $\mathcal{A}$ wins if $\left(m^*, \sigma^*\right) \neq \left(m_i, \sigma_i\right)$ for all i and $Verify\left(pk, m, \sigma\right) = 1$. The signature scheme is said to be $\left(t, q, \varepsilon\right)-$ secure if for any $\mathcal{A}$, running in time at most $t = Poly\left(\lambda\right)$ and making at most $q = Poly\left(\lambda\right)$ signing queries

$$\Pr\left[Game_{\mathcal{A}}^{\lambda} \text{ outputs } 1\right] \leq \varepsilon\left(\lambda\right).$$

2.2.2 Secure Signatures in ROM

One way of obtaining provably EUF-CMA signature schemes is to do the following. One starts with a (interactive) canonical identification scheme and combines it with a hash function using the Fiat-Shamir transform. EUF-CMA in the Random Oracle Model can be proven based on the properties of the underlying identification scheme. We give an outline of the techniques here.

Random Oracles. We discuss the notion of a (quantum) random oracle model first. This a heuristic framework in which cryptographic security proofs are given. In this model, a random function H is chosen and it is assumed that all the parties involved, as well as the honest algorithms in the protocol, can access the oracle classically. In the quantum random oracle model (QROM), the access to H is quantum. That is, the random function is modelled as a unitary map, which can be evaluated at any valid state.

Sigma Protocols. We now discuss the definition and security of identification schemes.

Definition 2.3

Let X and Y be sets, whose sizes depend on the security parameter λ. A relation is a subset of $R \subseteq Y \times X$. This defines a language $L = \left\{y \in Y : \exists x \in X, R\left(y, x\right) = 1\right\}$. A sigma protocol for R is a three-round interactive protocol between two PPT algorithms prover $\mathcal{P}$ and verifier $\mathcal{V}$. The prover holds a witness x for $y \in L$ and the verifier is given y. The protocol proceeds as follows. In the commitment phase, the prover sends a value α, called the commitment, to the verifier. In the challenge phase, the verifier submits a challenge, β, drawn from a suitable set. In the response phase, the prover responds with γ. The verifier verifies and outputs one or zero. The triple (α, β, γ) is called a transcript.

- The sigma protocol is said to be complete if the verifier outputs 1 with probability 1. In such a case, a transcript on which the verifier outputs 1 is called a valid transcript.
- A (witness) extractor algorithm takes as input two transcripts (α, β, γ) and $(\alpha, \beta', \gamma')$, with the same first message but with $\beta \neq \beta'$, and outputs a valid witness for x. A sigma protocol is said to be 2 -special sound if it has an extractor.
- A sigma protocol is honest verifier zero knowledge (HVZK) if there is a simulator that, given $y \in L$, can generate valid transcripts (α, β, γ), which are identically distributed to those of the real protocol. In other words, a PPT adversary cannot distinguish whether it is interacting with the simulator or the real protocol.

An identification scheme is an example of a sigma protocol, where the prover holds a secret key corresponding to a public key.

Definition 2.4

A canonical identification scheme is $ID = (K, P, V, c)$, where

- K is the key generation algorithm which outputs (PK, SK) on the input of security parameter
- The two phases of the prover are described by the PPT algorithm P, in the first phase it computes commitment and in the second phase it provides response.
- The verifier submits the challenge and, finally, accepts or rejects. The PPT algorithms modelling these actions are denoted by V.
- c is the length of the challenge.

A transcript of an honest execution of the scheme is the sequence $CMT \leftarrow P(SK, r)$, $CH \leftarrow \{0,1\}^c$ and $RSP \leftarrow P(SK, r, CMT, CH)$, with the condition that

$$V(PK, CMT, CH, RSP) = 1.$$

An impersonator simulates a dishonest prover. It receives the public key and transcripts of honest executions. It outputs the commitment, receives the challenge and outputs the response. We say that the impersonator wins, if the verification goes through for this set of input. The advantage of the impersonator is $\left| \Pr[V(\cdot) = 1] - 1/2^c \right|$.

- The identification scheme is said to be non-trivial if $c \geq \lambda$.
- The identification scheme is said to be recoverable if there is a deterministic algorithm Rec, which outputs the commitment, $Rec(PK, CH, RSP) = CMT$.

Fiat-Shamir Transform. Fiat-Shamir transform converts a non-tirivial canonical identification scheme into a signature scheme. Suppose the protocol has to be repeated t times in parallel for it to be non-trivial. Let H be a hash function with digest length ct.

KeyGen.
The key generation of the signature scheme is the same as that of ID.

Sign.

Sign takes as input the secret key and message. Compute t commitments $CMT_i = P(SK,r_i)$ for $1 \leq i \leq t$. Let $h = H(m,CMT_1,\ldots,CMT_t)$. Parse h as t values $CH_i \in \{0,1\}^c$. Compute $RSP_i = P(SK,r_i,CMT_i,CH_i), 1 \leq i \leq t$. The output signature is

$$\sigma = (CMT_1,\ldots,CMT_t,RSP_1,\ldots,RSP_t).$$

Verify.

Verify takes as input the message, the public key and signature. First the h value is computed as above and parsed as t challenges. Accept if $V(PK,CMT_i,CH_i,RSP_i) = 1, \forall 1 \leq i \leq t$.

Remark 2.1

If the identification scheme is recoverable, a more compact signature scheme can be given. It is enough to give $(h,RSP_1,\ldots,RSP_t)$ as signature. After parsing and obtaining the challenges, the verifier can use Rec to obtain the commitments and verify.

Security of the signature scheme thus obtained is proven using the following general result (Abdalla et al. 2002).

Theorem 2.1

Let ID be a non-trivial canonical identification protocol that is secure against impersonation under passive attacks. Let S be the signature scheme derived from ID using the Fiat-Shamir transform. Then, S is secure against chosen-message attacks in the random oracle model.

2.2.3 Modelling Quantum Adversary

It is logical to extend the above notion of security and construction in the presence of an adversary who is equipped with a quantum computer. Notice that the hash function used in the above transformation can be analysed offline. Superposition queries can be used. Thus, the QROM was introduced.

Let (P,V) be a sigma protocol for the relation, with $\{0,1\}^c$ being the set of challenges (polynomial in security parameter). Suppose the protocol is complete, n-special sound and HVZK. Let t be such that 2^{ct} is exponential in security parameter and $H : \{0,1\}^* \to \{0,1\}^{tc}$ a hash function modelled as a random oracle. Let Γ be the set γ (or RSP) of responses. The signature scheme is described below.

KeyGen.

Upon input the security parameter λ, $(pk,sk) \leftarrow^R K(1^\lambda)$

Sign.

Compute the commitments $CMT_i \leftarrow P(sk,pk)$, $1 \leq i \leq t$. Let $CH_{i,j}, 0 \leq j < 2^c$ be the set of all challenges (bit representation of j). Compute $RSP_{i,j} = P(pk,sk,CMT_i,CH_{i,j})$ and $g_{i,j} = G(RSP_{i,j})$, where $G: \Gamma \to \Gamma$ is a quantum random oracle with polynomially many preimages. Let

$$T = \left(CMT_1,\ldots,CMT_t,CH_{1,0},\ldots,CH_{t,2^c-1},g_{1,0},\ldots,g_{t,2^c}\right).$$

Let $h = H(pk, m, T) = CH_1 \| \ldots \| CH_t$, where $CH_i \in \{0,1\}^c$. Write $CH_i = CH_{i,J_i}$. Signature $\sigma = (T, RSP_{1,J_1}, \ldots, RSP_{t,J_t})$.

Verify.

Let $h = H(pk, m, T) = J_1 \| \ldots \| J_t$. Check that challenges are correctly formed, $g_{i,J_i} = G(RSP_i, J_i)$ and $V(pk, CMT_i, CH_{i,J_i}, RSP_{i,J_i}) = 1, \forall 1 \le i \le t$. If all checks are correct accept signature.

The following theorem summarizes the security claim.

Theorem 2.2

If the sigma protocol is complete, n-special sound and HVZK, the above transform gives a EUF-CMA signature scheme in QROM.

2.3   NIST PQC STANDARDIZATION COMPETITION

NIST had announced a "Call for Proposals" for solicitation, evaluation and standardization of one or most post-quantum cryptographic algorithms, recognizing the threat posed by quantum computers. Such computers can solve certain mathematical problems, efficiently, which classical computers cannot. We focus mainly on aspects that are relevant to digital signatures. A document[3] brought out by NIST describes the submission requirements and evaluation criteria. Key-Establishment/Public Key Encryption (Key Encapsulation Mechanism) and Digital Signatures are covered under this call. The developed standards will be quantum-resilient counterparts for digital signature schemes specified in Federal Information Processing Standards (FIPS) Publication 186[4] and key establishment schemes specified in NIST Special Publications 800-56 A[5] and B.[6]

The NIST document describes the security criteria which the submissions are expected to possess. With respect to digital signatures, the submissions are expected to be existentially unforgeable with respect to an adaptive chosen message attack (proof was not mandatory). For estimating the signature strength, it may be assumed that the forger has access to no more than 2^{64} message-signature pairs. The submission was required to satisfy a certain security level. Five security levels were defined, in increasing order of strength, which depends on the computational effort required to find a key for a 128-bit block cipher encryption or that required to find a collision in a 256-bit hash function.

Resistance against side-channel attacks, multi-key attacks and misuse were also listed as additional security requirements. The cost of the scheme, measured in terms of public key and signature sizes and signing and verification time, needs to be low. The schemes are expected to be portable on a variety of platforms and are expected to be simple in terms of mathematical description. Among the 82 submissions received by NIST, 20 signature candidates and 49 PKE/KEM candidates were deemed to be conforming to the submission requirements. After the first round of evaluation, 9 digital signature candidates and 17 PKE/KEM candidates remained.

2.3.1   ROUND THREE CANDIDATES

Two rounds of evaluation, of the submitted candidates, have been completed. In the third round, the following digital signature candidates are being evaluated given in Table 2.1.

Alagic et al. (2020) describe the evaluation and selection process followed during the second round. The document summarizes the advantages and disadvantages of all the six candidates. We summarize them below:

TABLE 2.1

NIST Round Three Signature Candidates

Type	Name	Classification
Finalist	CRYSTALS-DILITHIUM	Lattice-based
	Rainbow	MQ-based
	Falcon	Lattice-based
Alternative Candidates	GeMSS	MQ-based
	SPHINCS+	Based on symmetric key primitives
	Picnic	Based on symmetric key primitives

1. DILITHIUM depends on the hardness of Module Learning With Errors (MLWE) and Module Short Integer Solution (MSIS) problems and follows Fiat-Shamir with aborts paradigm. Only uniform distribution sampling is used here. A QROM proof can be provided for this candidate. The scheme has balanced performance in terms of key sizes and signing/verification times and hence is applicable to a wide variety of applications.

2. Falcon depends on the hardness of Shortest Integer Solution (SIS) problem over NTRU lattices and follows "hash and sign" paradigm. A security proof in both classical and quantum random oracle models is available. Complicated tree data structure manipulation, sampling from discrete Gaussian distribution and several floating-point operations are the major drawbacks of this scheme. However, this candidate offers the smallest bandwidth in terms of communication complexity.

3. Rainbow is a multi-layer unbalanced oil-vinegar type signature. It offers small signature size and short signing/verification times, but has a large public key. Thus, this is suitable only for those applications which do not require frequent communication of public keys.

4. GeMSS is based on the HFEv- construction, following the "big field" paradigm. The EUF-CMA security in classical ROM can be argued using the unforgeability of the underlying HFEv- primitive. This offers a very small signature size and reasonably fast verification. On the other hand, the scheme suffers from large public key size and slow signing times. This makes the scheme unsuitable for low-end devices.

5. Picnic signature scheme depends on the random oracle assumption on the underlying hash function and security of LowMC block cipher. This scheme is a non-interactive proof of knowledge of the secret key. The construction uses an identification scheme (incorporating zero-knowledge proofs and circuit decomposition) and either Fiat-Shamir or Unruh transform. It has large signature size and large signing times. A naive implementation suffers from side-channel attacks.

6. SPHINCS+ is based on secure hash families. It is robust against cryptanalytic attacks but is vulnerable to side-channel analysis. There is a proof of security in classical ROM. The standard model proof has also been given, based on some plausible but nonstandard assumptions on tweakable hash functions.

2.4 LATTICE-BASED SIGNATURES

Digital signatures based on lattices have been proven to be hard to construct. The first such construction was probably due to Goldreich, Goldwasser, and Halevi (1997), which was cryptanalysed by Nguyen (1999). NTRUSign (Hoffstein et al. 2003) was broken by Nguyen and Regev (2006). The literature in this area is vast. We focus on essential ingredients for describing the Falcon and DILITHIUM signatures. The focus is on giving an intuitive description of the schemes and not on discussing security aspects and cryptanalytic attacks.

Two techniques are most prominent when it comes to signature construction. The first one is the hash-and-sign method. NTRUSign (Hoffstein et al. 2003), then its provable secure versions (Gentry, Peikert and Vaikuntanathan 2008; Stehlé and Steinfeld 2013) use this mechanism. The other class of signatures uses Fiat-Shamir transform on an identification scheme. Lyubashevsky (2009) describes an identification scheme with aborts and the converted signature scheme. Aborting the protocol becomes essential to ensure the security of the secret key. We discuss the two paradigms. Falcon follows the former and DILITHIUM the latter for their construction.

2.4.1 HARD PROBLEMS

We recall some relevant preliminaries on lattices and hard problems, which are relevant to the description of NIST round two candidates. We will follow Gentry, Peikert and Vaikuntanathan (2008).

Let $B = \{b_1,\ldots,b_n\} \subset \mathbb{R}^n$ be a collection of n linearly independent vectors. The n-dimensional lattice Λ generated by the basis B is

$$\Lambda = \mathcal{L}(B) = \left\{ Bc = \sum_{i\in[n]} c_i \cdot b_i, c \in \mathbb{Z}^n \right\}.$$

The minimum distance $\lambda_1(\Lambda)$ and the i-th successive minimum distance $\lambda_i(\Lambda)$ of the lattice are defined w.r.t. the ℓ_2-norm in the usual way. A lattice is an additive subgroup of the Euclidean space. Hence, cosets can be defined in the usual way. The Gram-Schmidt orthogonalization gives an orthonormal basis for the lattice, starting from a set of linearly independent vectors. The dual of a lattice can be defined in the usual way via the inner product.

We now recall some central hard problems from lattices.

Definition 2.5

(Shortest Vector Problem, SVP) Upon input the basis of a full-rank n-dimensional lattice, decide whether $\lambda_1\big(\mathcal{L}(B)\big) \leq 1$ or $\lambda_1\big(\mathcal{L}(B)\big) > \gamma(n)$.

Definition 2.6

The SIS problem (in the ℓ_2-norm) is defined as follows: given an integer q and a matrix $A \in \mathbb{Z}_q^{n\times m}$ and a real β, find a non-zero integer vector $e \in \mathbb{Z}^m$ such that $Ae = 0 \bmod q$ and $\| e \|_2 \leq \beta$.

We now recall ring learning with errors (R-LWE) problem from Lyubashevsky, Peikert and Regev (2010). Let K be a number field and $R = \mathcal{O}_K$ be its ring of integers. Let $R^\vee$ be its dual ideal. Let $q \geq 2$ be an integer modulus. Let $R_q = R/qR$ and $R_q^\vee = R^\vee/qR^\vee$. Let $\mathbb{T} = K_{\mathbb{R}}/R^\vee$. For $s \in R_q^\vee$ and a distribution ψ over $K_{\mathbb{R}}$, the distribution $A_{s,\psi}$ on $R_q \times \mathbb{T}$ is generated by choosing $a \in_R R_q$ uniformly, choosing e according to ψ and outputting $(a,(a.s)/q + e)$, where the addition is in $\mathbb{T}$.

Definition 2.7

(Ring Learning with Errors Problem, R-LWE) Let K be a number field and $R = \mathcal{O}_K$ be its ring of integers. Let $q \geq 2$ be a rational integer and let Ψ be a family of distributions over $K_{\mathbb{R}}$. The R-LWE problem in $R = \mathcal{O}_K$ is defined as follows: given access to arbitrarily many independent samples from $A_{s,\psi}$ for some $s \in R_q^\vee$ and $\psi \in \Psi$, find s.

2.4.2 NTRUSign

Hoffstein et al. (2003) introduce a lattice-based signature on NTRU-type lattices, based on the security of approximate CVP. This scheme was cryptanalysed by Nguyen and Regev (2006). Gentry, Peikert and Vaikuntanathan (2008) (GPV) corrected the flaw which led to the cryptanalysis and made the scheme relevant. Falcon uses GPV techniques in its construction. We recall the NTRUSign scheme here because the mathematical structure is essentially the same for Falcon. We give an informal, but complete description of the scheme.

The basic mathematical object where all the operations take place is the ring $R = \mathbb{Z}[X]/(X^N - 1)$. A parameter q is fixed. The first observation is that the set

$$M_{h,q} = \left\{ (u,v) \in R^2 \mid v \equiv u * h \bmod q \right\}$$

is an R-module of rank 2. The element h is chosen in such a way that $h = f^{-1} * g \bmod q$ for $(f,g) \in M_{h,q}$ satisfying $\|f\| = \sqrt{d_f(1 - d_f/N)}$, $\|g\| = \sqrt{d_g(1 - d_g/N)}$, and $\|(f,g)\| = \sqrt{|f|^2 + |g|^2}$, where f and g have only 1 or 0 as coefficients. For $r = \sum_{i=0}^{N-1} r_i X^i$, the centred Euclidean norm $\sum_{i=0}^{N-1} r_i^2 - (1/N)\left(\sum_{i=0}^{N-1} r_i\right)^2$ is used.

We recall the key generation process. The pair $\{(1,h),(0,q)\}$ is a basis for $M_{h,q}$. Since $(f,g) \in M_{h,q}$, under certain conditions on f and g, it is possible to extend (f,g) to a basis $\{(f,g),(F,G)\}$, such that $f * G - g * F = q$. If $f = \sum_{i=0}^{N} f_i X^i$ and $g = \sum_{i=0}^{N} g_i X^i$, one can write down the matrix

$$A = \begin{pmatrix}
f_0 & f_1 & \cdots & f_{N-1} & g_0 & g_1 & \cdots & g_{N-1} \\
f_{N-1} & f_0 & \cdots & f_{N-1} & g_{N-1} & g_0 & \cdots & g_{N-2} \\
\vdots & & \vdots & & \vdots & & & \vdots \\
f_1 & f_2 & \cdots & f_0 & g_1 & g_2 & \cdots & g_0 \\
F_0 & F_1 & \cdots & F_{N-1} & G_0 & G_1 & \cdots & G_{N-1} \\
F_{N-1} & F_0 & \cdots & F_{N-1} & G_{N-1} & G_0 & \cdots & G_{N-2} \\
\vdots & & \vdots & & \vdots & & & \vdots \\
F_1 & F_2 & \cdots & F_0 & G_1 & G_2 & \cdots & G_0
\end{pmatrix}$$

One can associate a lattice L generated by this matrix. With this background, the signature is described as follows.

KeyGen.
The element h is the secret key and $\{(f,g),(F,G)\}$ is the public key, which are generated as described above.

Sign.
The signing proceeds as follows. A public hash function is used to obtain a digest which is parsed as $m = (m_1, m_2) \in \{0,\dots,q-1\}^N \times \{0,\dots,q-1\}^N$. A closest vector (s,t) to (m_1,m_2) is found using Babai's rounding algorithm (Babai 1986). The signature is simply s.

Verify.

To verify, the t such that $(s,t) \in L$ is recovered using the public key and then it is verified whether (s,t) is sufficiently close to (m_1, m_2).

2.4.3   GPV FRAMEWORK

Gentry, Peikert and Vaikuntanathan (2008) gave a framework for obtaining many secure primitives including signatures. The essential features are described below. Note that the description is in generic setting. One key ingredient of Falcon is doing GPV over NTRU lattices.

GPV begins by fixing a q-ary lattice Λ and a generator matrix $A \in \mathbb{Z}_q^{n \times m}, (m > n)$. Let

$$\Lambda_q^{\perp} = \left\{ x \in \mathbb{Z}^m \,|\, Ax \equiv 0 \bmod q \right\}$$

be the orthogonal lattice. Let $B \in \mathbb{Z}_q^{m \times m}$ be a generator for the orthogonal lattice. The two matrices satisfy $B \times A^t = 0$. The key generation module obtains such matrices.

KeyGen.

Obtain matrices as described above. The matrix A will be the public key and B the secret key.

Sign.

For signing, a public hash function $H : \{0,1\}^* \to \mathbb{Z}_q^n$ is first used. The secret key is used to obtain a short $s \in \mathbb{Z}_q^m$ such that $sA^t = H(m)$. This is done in two steps. First, a preimage $c_0 \in \mathbb{Z}_q^m$ (not necessarily short) satisfying $c_0 A^t = H(m)$ is computed. Then, B is used to obtain a v in the orthogonal lattice which is close to c_0. Then, $s = c_0 - v$ is a valid signature. If c_0 and v are near, then s will be short.

Verify.

Verification is simple. Given a s, one needs to verify that s is short and that $sA^t = H(m)$ holds.

Remark 2.2

Stehlé and Steinfeld (2013) use the NTRU lattice setting to instantiate GPV. This resulted in a provably secure NTRUSign.

2.4.3.1   Falcon Signature Scheme

Falcon[7] is a NIST round three finalist. It follows a hash-and-sign paradigm, where the hash function used is modelled as a random oracle. It uses the GPV framework on NTRU lattices. We discuss the salient features of the scheme here.

The first deviation from NTRUSign is the choice of the modulus polynomial. Here, $\phi = x^n + 1, n = 2^\kappa$ is chosen. A positive integer q is fixed. As usual, the secret basis and public basis are chosen, which are represented as block matrices as follows:

$$\begin{pmatrix} f & g \\ F & G \end{pmatrix} \quad \text{and} \quad \begin{pmatrix} 1 & h \\ 0 & q \end{pmatrix}$$

respectively, with the understanding that h stands for the block matrix of cyclic shifts similar to the A matrix in NTRUSign. If f and g are chosen randomly, then h would look uniformly random (Stehlé and Steinfeld 2013).

' Additional techniques are required to instantiate GPV on NTRU lattice. The orthogonal lattice is considered with A and B as basis matrices, which satisfy an orthogonality condition. For signing, a pair (s_1, s_2) is obtained such that $s_1 + s_2 h = H(r \| m)$, where r is a random salt. The tuple (r, s_2) is used as a signature.

The next ingredient in Falcon is the "fast Fourier sampling". Given a public matrix A, a trapdoor T and a target c, the sampling technique outputs a short vector s such that $s^t A = c \bmod q$. Ducas and Prest (2016) techniques are used.

We now give an outline of the Falcon signature. For an exact description, the reader may refer to the NIST submission.

KeyGen.

The first step is to sample f and g from the lattice according to discrete Gaussian sampling (Zhao, Steinfeld and Sakzad 2020). A precomputed cumulative density function table is used. Then, F and G are obtained using techniques from Pornin and Prest (2019). The next step is to construct a Falcon tree for the LDL* decomposition of the matrix $G = BB^*$, where

$$B = \begin{pmatrix} g & -f \\ G & -F \end{pmatrix}.$$

The public key is $h = gf^{-1} \bmod q$. The secret key is a matrix, $\hat{B}$, containing Fast Fourier Transform (FFT) of entries of B and the tree T.

Sign.

For the hash digest $c \in \mathbb{Z}_q[x]/(\phi)$ of a message m and a random salt r, the sign module computes short values s_1 and s_2 such that $s_1 + s_2 h = c \bmod q$. The sampling is done in such a way that successive computation of signatures does not reveal the secret key. Fast Fourier sampling is used. Then, the s_2 is compressed to a bitstring s. The pair (r, s) is the signature.

Verify.

The signature is first parsed as (r, s). From the salt r and the message m, the digest c is computed. Then, s is decompressed as $s_2 \in \mathbb{Z}_q[x]/(\phi)$. The value $s_1 = c - s_2 h \bmod q$ is computed. It is verified whether $\|(s_1, s_2)\|^2 \leq |\beta^2|$ is satisfied.

Remark 2.3

1. Note that both floating point and finite field arithmetic are involved. These operations are carried out using FFT and number theoretic transform (NTT), respectively. Another observation is that most of the operations are carried out over the Fourier transform domain. This warrants the use of FFT. Some structures are maintained over the FFT domain.
2. Sampling from the special discrete Gaussian distribution is a costly operation.
3. It has also been suggested that the Falcon tree part of the secret key can be generated on the go to save memory.
4. The hash function used is based on SHAKE-256.

TABLE 2.2

Falcon Parameter Sets and Timing

Sec. Level	Name	PK Size bytes	Sig Size bytes	Key Gen ms	Sign per sec	Verify per sec
I	Falcon-512	897	666	8.64	5948.1	27933
V	Falcon-1024	1793	1280	27.45	2913.0	13650.0

Parameters and Benchmarking. Falcon has been benchmarked on an Intel(R) Core(R) i5-8259U CPU ("Coffee Lake" core, clocked at 2.3 GHz) processor. The parameter sizes and benchmarking are summarized in Table 2.2. These figures are as reported in the NIST submission document.

Pornin (2019) reports the implementation of Falcon on the ARM Cortex M4, STM32F4 "discovery" board. Key generation, signing and verification modules take $171M$, $21M$ and $0.5M$ cycles, respectively.

2.4.4 FIAT-SHAMIR WITH ABORTS

As we have already seen, one route for constructing signatures is by first constructing an identification scheme and then using Fiat-Shamir transform. The challenge in using lattices for designing an IDS is that the response phase should not reveal information about the secret key. Lyubashevsky (2009) suggests a mechanism for constructing lattice-based IDS with this property. The basic algebraic structure considered here is the ring $R = \mathbb{Z}_p[x]/(x^n + 1)$. The lattices associated with this ring are precisely the ideals. The lattices will be cyclic, i.e., if $(v_o, v_1, \ldots, v_{n-1})$ is an element of the lattice, then, so is $(-v_{n-1}, v_0, \ldots, v_{n-2})$. The identification scheme also depends on the following hash family defined over lattices.

For integer m, the family $\mathcal{H}(R, D, m)$ mapping D^m to R is defined as

$$\mathcal{H}(R, D, m) = \left\{ h_{\hat{a}} : \hat{a} = (a_1, \ldots, a_m) \in R^m \right\},$$

where for any $\hat{z} = (z_1, \ldots, z_m) \in D^m$, $h_{\hat{a}}(\hat{z}) = \hat{a} \cdot \hat{z} = a_1 z_1 + \ldots + a_m z_m$.

The parameters m, n, σ, κ, p are chosen according to the required security level. The following sets are defined for describing the identification scheme:

$$D = \left\{ g \in R : \|g\|_\infty \leq mn\sigma\kappa \right\}$$

$$D_s = \left\{ g \in R : \|g\|_\infty \leq \sigma \right\}$$

$$D_c = \left\{ g \in R : \|g\|_1 \leq \kappa \right\}$$

$$D_y = \left\{ g \in R : \|g\|_\infty \leq mn\sigma\kappa \right\}$$

$$G = \left\{ g \in R : \| g \|_\infty \leq mn\sigma\kappa - \sigma\kappa \right\}.$$

The identification scheme is described as follows. The prover has $\hat{s}$, a set of m secret polynomials chosen from D_s. He wants to prove his knowledge of this key. The public key is the hash value $S = h(\hat{s})$, where $h \in \mathcal{H}(R, D, m)$ is randomly chosen from the family. The interactive protocol proceeds as follows:

Commit.

The prover chooses a random $\hat{y} \in D_y^m$, computes $Y = h(\hat{y})$ and sends Y to the verifier.

Challenge.

The verifier chooses a random $c \in D_c$ and sends it to prover.

Response.

The prover computes $\hat{z} = \hat{s}c + \hat{y}$. If $\hat{z} \notin G^m$, he aborts. Otherwise, the prover sends $\hat{z}$ to verifier.

Accept.

The verifier checks whether $\hat{z} \in G^m$ and $h(\hat{z}) = Sc + Y$. If such is the case, he accepts, otherwise rejects.

The scheme is perfectly witness-indistinguishable, complete and sound. The security is argued based on the SVP on ideal lattices. The protocol can be executed in parallel to increase the acceptance probability.

Converting the protocol to a signature scheme is easy. With the secret and the public keys as before, the signing algorithm uses a hash function $H: \{0,1\}^* \rightarrow D_c$. For signing a message μ, the signer picks $\hat{y} \in D_y^m$, computes $e = H(h(\hat{y}), \mu)$, $\hat{z} = \hat{s}e + \hat{y}$ and checks whether $\hat{z} \in G^m$. If such is the case, he sets $(\hat{z}, e)$ as signature, otherwise repeats the whole process. The signature is accepted if $\hat{z} \in G^m$ and $e = H(h(\hat{y}), \mu)$.

2.4.4.1 CRYSTALS-DILITHIUM Signature Scheme

CRYSTALS-DILITHIUM[8] is based on the Fiat-Shamir with aborts paradigm and builds upon (Bai and Galbraith 2014). The structure of the hash family, which acts as an inner product of any input, ensures that the keys can be setup in the following way. Let $R_q = \mathbb{Z}_q[x]/(x^n + 1)$. First, a matrix $A \in R_q^{k \times l}$ is chosen and then $(s_1, s_2) \in S_\eta^l \times S_\eta^k$ is sampled. Then, $t = As_1 + s_2$ is computed. The public key is (A, t) and the secret key is (s_1, s_2, A, t). DILITHIUM uses a decomposition of $w \in R_q^k$ as $w = w_1 \cdot 2\gamma_2 + w_0$, where coefficients of w_0 lie in $\{-\gamma_2, \ldots, \gamma_2\}$. The HighBits of w is w_1 and LowBits of w is w_2. Here, γ_2 is a suitably chosen threshold.

We give a high-level description of the signature scheme here. The reader may obtain the documentation from the NIST site and refer to it for more details.

KeyGen.

A seed ρ is chosen and expanded. Shake-256 is used. The part ρ is used to generate the matrix A. Then, ς is used to generate s_1, s_2. Then, $t = As_1 + s_2$ is computed, which is decomposed into t_1 and t_0 which are the high and low bits. The public key is (ρ, t_1) and the secret key is $(\rho, K, tr, s_1, s_2, t_0)$. Here, K is an additional seed to be used for obtaining a salted signing algorithm.

Sign.

The seeds are expanded and the public and secret keys are parsed. A $y \in R_q^l$ with coefficients in the range $-\gamma_1, \ldots, \gamma_1$ is sampled. Then, w_1 is computed as the HighBits of Ay of parameter γ_2. A hash function $\mathcal{H}$ takes $M \| w_1$ and computes $c \in B_h$, where B_h is the set of all polynomials from R_q which have h coefficients from the set $\{-1, 1\}$ and the rest are all zero. Then, $z = y + cs_1$ is computed. If $\|z\|_\infty < \gamma_1 - \beta$ and $\text{LowBits}(Az - ct) < \gamma_2 - \beta$, then (z, c) is returned as signature.

Verify.

The public key and signature are parsed, then A is computed. Then $w_1' = \text{HighBits}(Az - ct, 2\gamma_2)$ is computed. If $c = H(M \| w_1')$ and $\|z\|_\infty < \gamma_1 - \beta$, the signature is accepted.

TABLE 2.3
DILITHIUM Parameter Sets and Timing

Sec. Level	(q, d, τ)	PK Size bytes	Sig Size bytes	Key Gen	Sign	Verify
II	(8380417,13,39)	1312	2420	300,751	1,081,174	327,362
III	(8380417,13,49)	1952	3293	544,232	1,713,783	522,267
V	(8380417,13,60)	2592	4595	819,475	2,383,399	871,609

Remark 2.4

1. Signing can be deterministic or random.
2. The higher order bits of $Az - ct$ can be computed using the hint h provided by the signer.
3. Most of the operations take place over the NTT domain.

Parameters and Benchmarking. DILITHIUM has been benchmarked by the proposers on an Intel Core i7-6600U CPU that runs at a base clock frequency of up to 2.6 GHz. The parameter sizes and benchmarking are summarized in Table 2.3. The key generation, signing and verification modules' performance is measured in median cycles. These figures are as reported in the NIST submission document.

Greconici et al. (2021) discuss the implementation of DILITHIUM on ARM Cortex M4 board. The KeyGen, Sign and Verify take, respectively, $6M$, $6M$ and $2.7M$ cycles.

2.5 MQ-BASED SIGNATURES

Multivariate polynomial system solving over finite fields is known to be a computationally hard problem, even for quantum computers. Hence, it becomes relevant to the construction of post-quantum cryptosystems. The challenge, from a cryptographic construction point of view, is designing mechanisms for constructing polynomial systems, which are hard to invert unless trapdoor information is known about them. Otherwise, the polynomial system should be indistinguishable from a uniformly random choice.

It has been argued that quadratic polynomials are optimal for the design of an efficient and secure scheme. Most of the MQ-based schemes that are studied today use oil-vinegar (Kipnis, Patarin, and Goubin 1999) or HFE (Patarin 1996) paradigm. Rainbow and GeMSS, which are contenders in the NIST PQC competition, are of the above type, respectively. Another mechanism for the design of MQ-based signatures uses standard transformation on an identification scheme. MQDSS is a design that follows this paradigm, based on a 5-pass IDS (Chen et al. 2016).

Provable security of signature schemes and attacks on cryptosystems are vital to comprehend the usability of the scheme. These are essential for making parameter size recommendations. We focus mainly on the design and usage part of the schemes. Attacks and security proofs are not discussed here. However, we outline parameter selection rationale for Rainbow, just as an example.

It can be seen that forging a signature in this setting would be accomplished if a solution for a MQ system could be obtained. Gröbner basis techniques have been employed to mount attack on such cryptosystems. In particular, Faugere's F4 (Faugére 1999) and F5 (Faugère 2002) algorithms are most relevant in this scenario.

We discuss the oil-vinegar paradigms and describe Rainbow and GeMSS signature schemes. We mention the rationale behind Lifted Unbalanced Oil and Vinegar (LUOV) construction. We also mention the rationale used by the designers of Rainbow for making parameter choices.

2.5.1 MQ-Based Hard Problems

Let $\mathbb{F}_q$ be a finite field. The problem of solving a system of high-degree polynomials over this field has been proven to be NP-hard. An explicit reduction from 3-SAT to this problem has been given. This problem forms the basis of many public key cryptosystems.

Definition 2.8

MQ Problem Input: A MQ polynomial system $p_1(x_1,\ldots,x_n),\ldots,p_m(x_1,\ldots,x_n)$ over $\mathbb{F}_q$ Output: A vector $(z_1,\ldots,z_n)$ over $\mathbb{F}_q$, if one such exists, such that

$$p_1(z_1,\ldots,z_n) = 0,\ldots,p_m(z_1,\ldots,z_n) = 0$$

Another problem was introduced by Buss et al. (1999) called the MinRank problem. The authors also prove its NP-hardness. Though it appears like a problem from linear algebra, it forms the basis for security arguments for many multivariate cryptosystems. We recall its definition below.

Definition 2.9

MinRank Problem Input: A set of $k+1$ matrices $M_0,\ldots,M_k \in \mathcal{M}_{N \times n}(\mathbb{F}_q)$ and an integer $r > 0$. Output: A k-tuple, if any, $(\lambda_1,\ldots,\lambda_k) \in \mathbb{F}_q^k$ such that

$$rank\left(\sum_{i=1}^{k} \lambda_i M_i - M_0\right) \leq r.$$

2.5.2 Oil-Vinegar Signatures

The oil-vinegar signature scheme[9] was originally proposed by Patarin. An attack on this scheme was presented by Kipnis and Shamir (1998). Kipnis, Patarin and Goubin (1999), then modified the earlier scheme so that this attack could be thwarted. The unbalanced version of oil-vinegar signature scheme has been a building block for many other schemes like Rainbow (Ding and Schmidt 2005) and LUOV (Beullens and Preneel 2017). In this section, we discuss some prominent signature schemes built using the oil-vinegar paradigm and their security. Security aspects of these schemes against quantum adversaries are also discussed.

2.5.2.1 Unbalanced Oil-Vinegar Signature

All the operations take place over a finite field $\mathbb{F}$. This field will have size q, which is a prime power. The description of the schemes given here will not depend of the field size, hence we do not include the field size as a subscript. Let $\{X_1,\ldots,X_n\}$ be indeterminates over $\mathbb{F}$. The first $n - m$ variables will be called *vinegar variables* and the remaining *oil variables*. The goal here is to describe a MQ system, which can be solved using a trapdoor information, but comes with a computational hardness guarantee otherwise. This is achievd by the use of oil-vinegar polynomials. *The oil-vinegar type polynomial* is a MQ polynomial over $\mathbb{F}$ in the variables described above, but without any quadratic terms involving only the oil variables. In other words, an oil variable is not allowed to mix with another oil variable in such a polynomial. The general form of such a polynomial is as follows:

$$\phi(X_1, \ldots, X_n) = \sum_{j=1}^{n-m}\sum_{k=1}^{n-m} \gamma_{jk} X_j X_k + \sum_{j=n-m+1}^{n} \sum_{k=1}^{n-m} \lambda_{jk} X_j X_k + \sum_{j=1}^{n} \mu_j X_j + \delta.$$

As part of the trapdoor description m such polynomials are chosen and this system will be denoted by $\mathcal{F}$.

Recall that there are n variables. Let $T : \mathbb{F}^n \to \mathbb{F}^n$ be an affine map. Then, $\mathcal{P} = \mathcal{F} \circ T$ is computed. In other words, the affine transformation is applied on each polynomial in the system $\mathcal{F}$ and the resulting polynomials are collected into a system $\mathcal{P}$. For a given polynomial of the form Equation [eqn:provUOV-OVpolynomial], if values are assigned to all vinegar variables, the resulting polynomial is linear in oil variables. This feature is the central theme of the trapdoor. Thus, to invert a system $\mathcal{P}(X_1,\ldots,X_n) = h$, first one substitutes random values to all the vinegar variables and solves the resulting linear system with m equations in m unknowns. The linear system involving the oil variables may not be consistent for a particular vinegar assignment. If a solution is obtained for oil variables, then T^{-1} is applied to the complete vector of assignments. A simple signature scheme that uses this trapdoor is easily described.

Remark 2.5

We discuss the terminology "unbalanced". It becomes relevant to investigate security of the scheme with respect to choices for parameters m and n. Initially, Patarin had proposed his scheme with the choice $m = n$. Kipnis and Shamir (1998) gave an attack for this parameter choice. The unbalanced version was derived to resist such an attack. Thomae and Wolf (2012) study extension of Kipnis-Shamir ideas to underdetermined systems. Such attacks and many others are considered to fix parameter sizes.

In Sakumoto, Shirai and Hiwatari (2011), the authors discuss provable security aspects of the UOV signatures. The authors outline a method for obtaining a provably secure UOV signature. They study

1. how close to uniform, the distribution of $\mathcal{P}(\alpha)$ is, for uniform choice of α and
2. how close to uniform, the distribution of the signature is.

They introduce a random salt[10] in the UOV signature to obtain an EUF-CMA secure scheme. The notation $\mathcal{F}(x'_v, X_m)$ is used to denote the linear system in oil variables, which is obtained after the vinegar variables have been specialized to the vector x'_v. The scheme is described as follows:

KeyGen.
This module outputs the secret and the public keys for a given security parameter. The secret key comprises $\mathcal{F}$ and T, which are sampled as described above. There is a public hash function H and a salt space $\{0,1\}^\ell$. The public key is the system $\mathcal{P}$.

Sign.
This module uses the secret key to sign a message. First, a random salt r is sampled uniformly at random from the salt space $\{0,1\}^\ell$. The inputs to the signing module are $\mathcal{F}, T$ and the message μ. The output is a signature on M such that $\mathcal{P} = H(\mu \| r)$. The signing module is described below:
1. Sample $x'_v \xleftarrow{U} \mathbb{F}^v$
2. repeat
    a. Sample salt $r \xleftarrow{U} \{0,1\}^\ell$
    b. Compute $y = H(\mu \| r)$

3. until $\left\{ x_m \in \mathbb{F}^m : \mathcal{F}\left(x'_v, x_m \right) = y \right\}$ is non-empty

4. Sample $x'_m \xleftarrow{U} \left\{ x_m \in \mathbb{F}^m : \mathcal{F}\left(x'_v, x_m \right) = y \right\}$

5. Compute $x = T^{-1}\left(x'_v, x'_m \right)$

6. Output (x, r) as signature

Verify.

Verify takes the message, signature and the public key to output accept/reject. The signature is first parsed as (x, r). Then, $y = H(\mu \| r)$ is computed. Then, it is verified if $\mathcal{P}(x) = y$ is satisfied or not.

It is easy to verify that the signature scheme is correct.

2.5.2.2 Rainbow Signature Scheme

The Rainbow signature scheme (Ding and Schmidt 2005) is a multi-layer UOV. This is one of the three finalists in round three of the NIST PQC competition.[11] We describe the scheme here.

Let $\{X_1, \ldots, X_n\}$ be indeterminates over $\mathbb{F}$. Rainbow considers these variables in two layers as follows:

$$\text{First Layer}: \left\{ X_1, \ldots, X_{v_1} \right\}(\text{Vinegar}); \left\{ X_{v_1+1}, \ldots, X_{o_1} \right\}(\text{oil})$$

$$\text{Second Layer}: \left\{ X_1, \ldots, X_{o_1} \right\}(\text{Vinegar}); \left\{ X_{o_1+1}, \ldots, X_n \right\}(\text{oil})$$

Public system is obtained from the oil-vinegar type system in two stages. Apart from the transformation for mixing variables, another transformation that acts via linear combination of the polynomials is also used.

The complete signature is described below. The following convention is followed for the NIST submission. Finite Field $\mathbb{F}_q$; integers $0 < v_1 < \ldots v_u < v_{u+1} = n$; index sets $V_i = \{1, \ldots, v_i\}, O_i = \{v_i, 1, \ldots, v_{i+1}\}(i = 1, \ldots, u)$, such that each $k \in \{v_1 + 1, \ldots, n\}$ is contained in exactly one O_i; set $o_i = |O_i|$; number of equations $m = n - v_1$; number of variables n

KeyGen.

The secret key consists of two invertible affine maps $\mathcal{S} : \mathbb{F}_q^m \to \mathbb{F}_q^m$, $\mathcal{T} : \mathbb{F}_q^n \to \mathbb{F}_q^n$ and the central quadratic map $\mathcal{F} : \mathbb{F}_q^n \to \mathbb{F}_q^n$, consisting of m multivariate polynomials $f^{(v_1+1)}, \ldots, f^{(n)}$. The polynomials are of the form

$$f^{(k)}\left(x_1, \ldots, x_n \right) = \sum_{i, j \in V_l, i \leq j} \alpha_{ij}^{(k)} X_i X_j + \sum_{i \in V_l, j \in O_l} \beta_{ij}^{(k)} X_i X_j + \sum_{i \in V_l \cup O_l} \gamma_i^{(k)} X_i + \delta^{(k)},$$

where $l \in \{1, \ldots, u\}$ is the only integer such that $k \in O_l$. The coefficients $\alpha_{ij}, \beta_{ij}, \gamma_i, \delta_i$ all come from the base finite field. The public key is the composite map

$$\mathcal{P} = \mathcal{S} \circ \mathcal{F} \circ \mathcal{T} : \mathbb{F}_q \to \mathbb{F}_q,$$

consisting of m quadratic polynomials.

Sign.

Let d be the message to be signed. First compute $h = \mathcal{H}(d)$, where $\mathcal{H} : \{0,1\}^* \to \mathbb{F}_q^m$ is a hash function. Then compute $x = \mathcal{S}^{-1}(h) \in \mathbb{F}_q^m$. Then compute $y = \mathcal{F}^{-1}(x)$. The signature $z = \mathcal{T}^{-1}(y)$ is then computed.

Verify.

Let d be the message to be signed. First compute $h = \mathcal{H}(d)$. Then compute $h' = \mathcal{P}(z)$. If $h = h'$ accept.

Remark 2.6

1. The scheme described above is not provably EUF-CMA. To get EUF-CMA, the authors suggest the usual recipe. Instead of generating signature for $\mathcal{H}(d)$, generate it for $\mathcal{H}(\mathcal{H}(d) \| r)$, where r is a random salt. The signature is of the form (z, r), where z is the standard Rainbow signature. However, no formal reduction is given by the proposers.
2. The description is complete in terms of matrices rather than polynomials.

Parameter Sets and Benchmarking. We now summarize various attacks that have been mounted on Rainbow. More details can be obtained from documentation. The first attack tries to find a hash collision to obtain a forgery. Rainbow does not make any recommendations regarding the specific hash functions to be used. Some Gröbner basis algorithms (Faugére 1999) could be used directly to obtain a forgery. Algebraic algorithms like eXtended Linear (XL) and other variants can also be used. Some algebraic techniques are used to bind the complexity of such attacks. But, generically, solving polynomial systems is hard even in quantum computing setting. The MinRank attack considers the homogenous part of the public polynomials and the matrices associated with them. The MinRank problem asks for a linear combination of these matrices such that the resulting matrix has rank less than a threshold. Techniques developed by Kipnis and Shamir (1999) are used for solving the MinRank problem. From such a linear combination, an equivalent (giving the same public key) secret key can be recovered. The high rank attack uses techniques from Coppersmith, Stern and Vaudenay (1997) to find variables which appear the least number of times, which correspond to oil variables in the last layer. From this information, part of the secret polynomials can be found out. Reconciliation and intersection attacks try to find vectors in the oil subspace. Beullens (2021) discusses its impact on Rainbow and its security. The Rainbow band separation attack (Ding et al. 2008; Perlner and Smith-Tone 2020) tries to find the S and T transformations. Quantum attacks on the MQ problem are discussed in Bernstein and Yang (2018). Complexities of these attacks are considered for estimating parameter sizes.

We summarize the performance benchmark of the standard Rainbow signature scheme in Table 2.4. These figures are as reported in the NIST submission document. The signature size includes a 128-bit random salt, which is used to obtain EUF-CMA security. The benchmarking was carried out on a computer with Intel(R) Xeon(R) CPU E3-1275 v5 @ 3.60GHz (Skylake), 32GB of memory and running Linux OS.

Chou, Kannwischer and Yang (2021) discuss such an implementation of Rainbow. Key generation, signing and verification modules take 98431K, 957K and 239K cycles, respectively.

2.5.2.3 LUOV Signature Scheme

LUOV (Beullens and Preneel 2017) is another MQ signature scheme derived from UOV. It uses a PRNG to derive a large part of the public key and finite field lifting in UOV paradigm to get a new signature scheme. LUOV differs from standard UOV in three ways, namely,

TABLE 2.4

Rainbow Parameter Sets and Timing

Param Set	Parameters $(\mathbb{F}, v_1, o_1, o_2)$	PK Size kB	Sig Size bits	Key Gen Time (cycles)	Sign Time (cycles)	Verify Time (cycles)
I	$(\mathbb{F}_{16}, 36, 32, 32)$	157.8	528	$32M$	$319k$	$41k$
III	$(\mathbb{F}_{256}, 68, 32, 36)$	861.4	1,312	$197M$	$1.47M$	$203k$
V	$(\mathbb{F}_{256}, 96, 36, 64)$	1,885.4	1,696	$436M$	$2.48M$	$362k$

1. One can choose a large part of the public key as the output of a PRNG, and this part can be represented with the seed of the PRNG
2. LUOV gets its name from lifted UOV. The public key $\mathcal{P} : \mathbb{F}_q^n \to \mathbb{F}_q^m$ is lifted to the extension field as $\mathcal{P} : \mathbb{F}_{q^r}^n \to \mathbb{F}_{q^r}^m$
3. The secret map $\mathcal{T}$ is chosen to have a matrix representation of the form

$$\begin{pmatrix} 1_v & T \\ 0 & 1_m \end{pmatrix}$$

4. where T is $v \times m$ matrix

We do not describe the scheme completely. Many attacks have been presented on this scheme, resulting in it being eliminated in the second round of NIST competition.

2.5.3 HFE SIGNATURES

Patarin (1996) proposed another trapdoor based on the MQ problem. As usual, the description begins with a choice of a finite field $\mathbb{F}_q$. HFE paradigm uses a field extension of degree n over $\mathbb{F}_q$. The general idea of this trapdoor is that while solving a univariate high-degree polynomial over a finite field is easy (Cantor and Zassenhaus 1981), it is hard to solve a multivariate system even when the degree is as low as two. The Dembrowski-Ostrom polynomial plays a central role in HFE. It is defined as follows.

Definition 2.10

Let $D > 0$ be an integer. A polynomial $F(X) \in \mathbb{F}_{q^n}[X]$ is said to have HFE shape if it is of the form:

$$F(X) = \sum_{0 \le i \le j < n,\, q^i + q^j \le D} A_{i,j} X^{q^i + q^j} + \sum_{0 \le i < n,\, q^i \le D} B_i X^{q^i} + C.$$

There is the usual identification of $\mathbb{F}_{q^n}$ as a $\mathbb{F}_q$ vector space of dimension n, which we denote by φ. Let $(\theta_1, \ldots, \theta_n) \in (\mathbb{F}_{q^n})^n$ be a basis for $\mathbb{F}_{q^n}/\mathbb{F}_q$. Let

$$M_n = \begin{pmatrix} \theta_1 & \theta_1^q & \cdots & \theta_1^{q^{n-1}} \\ \theta_2 & \theta_2^q & \cdots & \theta_2^{q^{n-1}} \\ \vdots & \vdots & \vdots & \vdots \\ \theta_n & \theta_n^q & \cdots & \theta_n^{q^{n-1}} \end{pmatrix} \in GL_n\left(\mathbb{F}_{q^n}\right).$$

The structure of the polynomial ensures that the components of this polynomial over $\mathbb{F}_q$ are quadratic. We summarize the fact in the following lemma.

Lemma 2.1

Let $(\theta_1,\ldots,\theta_n)$ be a vector space basis for $\mathbb{F}_{q^n}/\mathbb{F}_q$. Define a system of polynomials

$$\left(f_1(x_1,\ldots,x_n),\ldots,f_n(x_1,\ldots,x_n)\right)$$

over $\mathbb{F}_q$ satisfying

$$F\left(\sum_{i=1}^{n}\theta_i x_i\right) = \sum_{i=1}^{n}\theta_i f_i,$$

for the HFE polynomial. Then, f_i are quadratic.

The key generation module chooses the univariate polynomial F and sets up the basis for extension field arithmetic. The S and the T matrices which mix the equations and the variables, respectively, are chosen. The triple (F,S,T) is the secret key. The public key is

$$\left(f_1\left((x_1,\ldots,x_n)T\right),\ldots,f_n\left((x_1,\ldots,x_n)T\right)\right)S.$$

When the signer signs the message, the preimage of the hash digest under S^{-1}, which is a n-tuple of elements from the base field, is identified with an element in the extension field. The public key system reconstructs to a univariate polynomial over an extension field. This can be solved in polynomial time and any zero of this polynomial is used to obtain a signature after an application of T^{-1}. The verification is, as usual, evaluation of the public key system at the signature coordinates and checking for equality with hash digest. It is now trivial to write down the whole signature scheme.

2.5.3.1 GeMSS

GeMSS[12] stands for great multivariate short signature. This is an alternate candidate in the round three of the NIST competition. It is a signature scheme in the HFE paradigm, in particular, its "v-minus variant". The GeMSS signature scheme has the following parameters:

degree of secret polynomial $= D = 2^i$ or 2^{i+j}

\# vinegar variables $= v$

\# equations in public key $= m$

size of hash digest $= K$

degree of extension of the finite field $= n$

\# minus $= \Delta = n - m$

\# iterations in the verification and signature processes $= \text{nb\_ite} > 0$

We now explain the vinegar variant of HFE. The secret polynomial $F \in \mathbb{F}_{2^n}[X, v_1, \ldots, v_v]$ has the structure

$$\sum_{0 \le j < i < n, 2^i + 2^j \le D} A_{ij} X^{2^i + 2^j} + \sum_{0 \le i \le n, 2^i \le D} \beta_i(v_1, \ldots, v_v) + \gamma(v_1, \ldots, v_v),$$

where $A_{ij} \in \mathbb{F}_{2^n}$, each $\beta_i: \mathbb{F}_2^v \to \mathbb{F}_{2^n}$ is linear and $\gamma: \mathbb{F}_2^v \to \mathbb{F}_{2^n}$ is quadratic. The variables $v_1, \ldots, v_v$ are called *vinegar variables*. Such a polynomial is said to be of *HFEv-shape*. It is easy to see that any specialization of F to vinegar values yields a one variable polynomial of HFE type over $\mathbb{F}_{2^n}$. Two matrices $(S, T) \in GL_{n+v}(\mathbb{F}_2) \times GL_n(\mathbb{F}_2)$ are sampled at random.

The mechanism for deriving the public key is similar to the HFE type schemes. Choose a vector space basis $(\theta_1, \ldots, \theta_n) \in (\mathbb{F}_{2^n})^n$ be a basis for $\mathbb{F}_{2^n}/\mathbb{F}_2$. Let $\phi: \sum_{k=1}^n e_k \theta_k \mapsto (e_1, \ldots, e_n)$ be the basis map. Let $\mathbf{f} = (f_1, \ldots, f_n) \in \mathbb{F}_2[x_1, \ldots, x_{n+v}]^n$ derived from F by $F\left(\sum_{k=1}^n \theta_k x_k, v_1, \ldots, v_v\right) = \sum_{k=1}^n \theta_k f_k$. We now rename the vinegar variables $(v_1, \ldots, v_v)$ as $(x_{n+1}, \ldots, x_{n+v})$. The public key $\mathbf{p} = (p_1, \ldots, p_m) \in \mathbb{F}_2[x_1, \ldots, x_{n+v}]^m$, where $m = n - \Delta$, where the component polynomials are the first m polynomials of

$$\left(f_1((x_1, \ldots, x_{n+v})S), \ldots, f_n((x_1, \ldots, x_{n+v})S)\right)T$$

reduced modulo $\left| x_1^2 - x_1, \ldots, x_{n+v}^2 - x_{n+v} \right|$.

The signing module uses an inversion submodule. The public key system $p_1(x_1, \ldots, x_{n+v}) - d_1 = 0, \ldots, p_m(x_1, \ldots, x_{n+v}) - d_1 = 0$ needs to be solved for $\mathbf{d} = (d_1, \ldots, d_{n+v})$. The output of the Inv module could be used as the signature. But, as noted in Gui and QUARTZ, this could allow the adversary to mount birthday type attacks. The Feistel-Patarin strategy has been proposed to handle this issue. The idea is to iterate Inv several times. The nb_ite, the parameter, can be chosen to be 4.

KeyGen.

The input is the security parameter λ. The secret key is (F, S, T) and the public key is $\mathbf{p}$.

Sign.

The signing module takes the message and secret key as inputs. It uses SHA3 for computing the digest. The complete signing module is described below:

Inversion for GeMSS	Sign for GeMSS
Repeat	$H = \text{SHA3}(M)$
$\mathbf{r} \in_R \mathbb{F}_2^{n-m}$	$\mathbf{S}_0 = 0 \in \mathbb{F}_2^m$
$\mathbf{d}' = \mathbf{d} \parallel \mathbf{r}$	for $i = 1$ to nb_ite do
$D' = \phi^{-1}(\mathbf{d}' \times T^{-1})$	Set $\mathbf{D}_i$ to be first m bits of H
$\mathbf{v} \in_R \mathbb{F}_2^v$	$(\mathbf{S}_i, \mathbf{X}_i) = \text{Inv}(\mathbf{D}_i \oplus \mathbf{S}_{i-1})$
Roots $=$ zeroes of $F_{D'}$	$H = \text{SHA3}(H)$
until Roots $\ne \varnothing$	return $S_{nb\_ite}, X_{nb\_ite}, \ldots, X_1$
$Z \in_R$ Roots	
return $(\phi(Z), \mathbf{v}) \times S^{-1} \in \mathbb{F}_2^{n+v}$	

Verify.

This module takes as input the message, the public key of the sender and the signature to return accept or reject. The process is described in the table below.

$$H = \text{SHA3}(M)$$

Parse signature as $S_{nb\_ite}, X_{nb\_ite}, \ldots, X_1$

for $i = 1$ to nb_ite do

Set $\mathbf{D}_i$ to be first m bits of H

$$H = \text{SHA3}(H)$$

for nb_ite down to $i = 1$ do

$$\mathbf{S}_i = \mathbf{p}(\mathbf{S}_{i+1}, \mathbf{X}_{i+1}) \oplus \mathbf{D}_{i+1}$$

return accept if $\mathbf{S}_0 = 0$, reject otherwise

Remark 2.7

1. Many variants of GeMSS have been proposed. Refer to NIST submission for more details.
2. Obtaining the public MQ system from the univariate private polynomial is a computationally intensive task. The submitters have used a package called MQsoft[13] for the software implementation of GeMSS.

Parameter Sets and Benchmarking. We discuss parameter sets and benchmarking of GeMSS. The proposal has many variants of the signature scheme. The platform used is a laptop running on Intel(R) Core (TM) i7-6600U CPU @3.40 GHz and 32GB RAM. We tabulate the performance of the plain GeMSS variant only. The performance is summarized in Table 2.5.

2.6   SIGNATURES BASED ON SYMMETRIC KEY TECHNIQUES

Symmetric key primitives could be used for signature construction. Only Grover and Simon attacks become relevant in this case due to the lack of structure. Picnic uses the trapdoor information of encryption key used for obtaining a ciphertext for a plaintext using a block cipher like LowMC. SPHINCS+ is obtained by making a hash-based signature stateless. We discuss some ingredients of these constructions in this section.

2.6.1   Picnic

Picnic signature scheme (Chase et al. 2017) uses only symmetric key primitives in its construction. It starts with the construction of an identification scheme and then uses Fiat-Shamir or Unruh

---

TABLE 2.5

GeMSS Parameter Sets and Timing

Param Set	Parameters $(\lambda, D, n, \Delta, v, \text{nb}\_\text{ite})$	PK Size kB	Sig Size bits	Key Gen Time (cycles)	Sign Time (cycles)	Verify Time (cycles)
GeMSS128	$(128, 513, 174, 12, 12, 4)$	352.19	258	$140M$	$2420M$	$211k$
GeMSS192	$(192, 513, 265, 22, 20, 4)$	1237.96	411	$600M$	$6310M$	$591k$
GeMSS256	$(256, 513, 354, 30, 33, 4)$	3040.70	576	$1660M$	$10600M$	$1140k$

transform for obtaining a signature. The identification scheme proves the knowledge of a secret key, which is used for encrypting a message. The message and ciphertexts are public. ZKBoo (Giacomelli, Madsen and Orlandi 2016), ZKB++ (Chase et al. 2017) and Katz, Kolesnikov and Wang (2018) methods use circuit decomposition (or MPC-on-the-head) to obtain an IDS. We discuss the outline of the Picnic construction and refer the reader to the scheme documentation[14] and the preceding papers.

2.6.1.1 ZKBoo

ZKBoo (Giacomelli, Madsen and Orlandi 2016) gives zero-knowledge proofs for Boolean circuits. Consider the case where one wishes to setup a trapdoor using only symmetric key primitives. Suppose one uses a secure hash function, H, to compute a digest as $d = H(M)$. He then publishes d and wishes to prove the knowledge of M satisfying the above relation. ZKBoo give an identification scheme for such a setting, using the notion of circuit decomposition. We discuss the salient features of ZKBoo.

The construction works for any relation $R \subseteq \{0,1\}^* \times \{0,1\}^*$ ($R = \{(d,M) : d = \text{SHA} - 256(M)\}$, for example). Let $L = \{y : \exists x \text{ s.t } R(y,x) = 1\}$. The construction depends on a $(2,3)$-decomposition (Giacomelli, Madsen and Orlandi 2016, Definition 3) of a circuit. Intuitively, this shares the input and output into three shares such that reconstruction (using Rec function) of the final output is possible, given the output shares. The decomposition satisfies a notion of privacy, if two of the three shares are revealed, the third share cannot be distinguished from uniform distribution. A mechanism called "linear decomposition" is described for any circuit. This is constructed by finding views $\{w_i\}$ (using Output function) iteratively. This is nothing but a mechanism for evaluating XOR and AND gates after decomposing the input into shares. The reader may consult (Giacomelli, Madsen and Orlandi 2016, Picture 5) for more details.

Given Q and y with the trapdoor information x such that $(y,x) \in L_\phi$, the identification scheme can be described as follows. The scheme uses the linear decomposition of ϕ.

Commit.
The prover chooses random values k_1, k_2, k_3, gets the views w_1, w_2, w_3 and the output shares y_1, y_2, y_3. He then computes the commitments $c_i = \text{Com}(k_i, w_i), i = 1,2,3$. He then sends $a = (y_1, y_2, y_3, c_1, c_2, c_3)$.

Challenge.
The verifier chooses an index $i \in [3]$ and sends it to the prover.

Prove.
The prover opens c_i, c_{i+1} and sends $(k_i, k_{i+1}, w_i, w_{i+1})$ (indices are added modulo 3).

Verify.
The verifier checks if $y = \text{Rec}(y_1, y_2, y_3)$, checks $y_j = \text{Output}(w_j), j = i, i+1$ and that the view w_i has been correctly computed from the entire chain of iterations. If these verify, he accepts.

In Giacomelli, Madsen and Orlandi (2016), a complete protocol for SHA-256 is described and the implementation details have been presented.

2.6.1.2 ZKB++ and Picnic

Chase et al. (2017) give an optimized version of the ZKBoo protocol. Some values are generated in a pseudorandom fashion, thereby resulting in reduction of communication complexity and proof sizes. A major improvement is obtained due to the usage of LowMC block cipher (Albrecht et al. 2015). This cipher has minimal usage of AND gates, which gives improvement in proof sizes over ZKBoo. The optimizations are described in Chase et al. (2017, Section 3.2).

The Picnic signature scheme is obtained by using Unruh transform on the identification scheme described using ZKB++. The identification scheme is run t times in parallel. The scheme has 3-special soundness property. This gives the security of the signature scheme.

Picnic submission to NIST competition uses both Fiat-Shamir and the Unruh transform. A version uses Katz, Kolesnikov and Wang (2018) techniques too. The proposers give parameters for security levels 1, 3 and 5. For more details, the reader may refer to the NIST documentation.

2.6.2   SPHINCS+

SPHINCS+[15] is a stateless, hash-based signature scheme. We outline the building blocks of this proposal and ask the reader to refer to the scheme documentation for full details.

Use of hash functions for constructing signatures began with the work of Lamport (1979), where a one-time signature (OTS) was given. The key pair can be used for signing only one message.

Merkle (1990) extends this to a few-time signature. Merkle used a construction of a hash tree, where the leaves are the Lamport hash values and the inner node is constructed as a hash of the concatenation of the two child nodes. This tree later came to be known as Merkle tree. The root of the Merkle tree is the public key and the OTS keys are the secret keys. The OTS keys can be generated pseudo-randomly using a seed. The OTS key pairs cannot be reused. So, a STATE is used to keep track of the used key pairs, leftmost keys first. The signature, apart from containing the OTS obtained using the i-th key pair, also contains the so-called authentication path. This path consists of all the sibling nodes of those nodes lying in the path from ith leaf to the root. This information is used for computing the root value in signature verification.

Buchmann, Dahmen and Hülsing (2011) proposed XMSS, which is based on Winternitz OTS (WOTS). WOTS achieves reduction in signature size; the public key is computed during the signature verification. XMSS uses bitmasks to mask the inputs before hashing, which results in weaker assumptions on the hash function.

SPHINCS+ uses a variant WOTS+. Using this XMSS, the many-time signature (MTS) is obtained. A hypertree, which is a tree of MTS, is constructed. The Forest Of Random Subsets (FORS) is defined, which is the FTS used in SPHINCS+ construction. The final signature scheme obtained is stateless.

SPHINCS+ is a variant of SPHINCS (Bernstein et al. 2015). The reader may refer to the design documentation for full details.

2.7   SIGNATURES BASED ON SUPERSINGULAR ISOGENIES

The discrete logarithm problem on elliptic curve groups is solvable in polynomial time on a quantum computer. But supersingular elliptic curves can still be used to construct quantum-resilient cryptographic schemes. These schemes rely on the hardness of isogeny computation. Isogenies are maps between elliptic curves. This section outlines signature construction using supersingular isogenies.

2.7.1   PRELIMINARIES ON ELLIPTIC CURVES

We recall some basic notions on elliptic curves, which are relevant to our study at hand. The reader may refer to Silverman (2009) for more details. Let E be an elliptic curve over a finite field $\mathbb{F}_q$. The curve is usually described using the Weierstraßmodel. For our discussion here, only curves over large characteristic fields are considered. The j-invariant and the discriminant of the curve are easily defined. The group law, making the set of rational points (including the point at infinity) into an abelian group, is well-known. The ℓ-torsion subgroup is isomorphic to $\mathbb{Z}/\ell\mathbb{Z} \oplus \mathbb{Z}/\ell\mathbb{Z}$.

An *isogeny* $\phi : E \to E'$ is a surjective rational map, preserving the points at infinity. An isogeny is a homomorphism between the groups of rational points. The *degree* of an isogeny is the degree of the underlying rational map. An isogeny induces an injective map between the function fields of the corresponding curves. The isogeny is said to be *separable* if this field extension is. For separable extensions, the degree of the isogeny is equal to the size of the kernel. An *isomorphism* is an isogeny with trivial kernel.

An isogeny from an elliptic curve to itself is called an *endomorphism*. The set of all endomorphisms of a curve form a ring $End(E)$. If, for an elliptic curve, the endomorphism ring is isomorphic to an order in the quaternion algebra, the curve is said to be *supersingular*.

2.7.2   YOO ET AL. SCHEME

Yoo et al. (2017) gave the first signature scheme based on supersingular isogenies. The authors construct an identification scheme and then use Unruh techniques (refer to Section [sec:prelims]) to obtain a signature. We describe the construction here.

A prime p of the form $\ell_A^{e_A} \ell_B^{e_B} f \pm 1$ is chosen, such that ℓ_A and ℓ_B are small primes, $\ell_A^{e_A} \approx \ell_B^{e_B}$ and f is a small cofactor. A supersingular elliptic curve E with cardinality $\left(\ell_A^{e_A} \ell_B^{e_B} f\right)^2$ is chosen. Then, E has two torsion subgroups $E\left[\ell_A^{e_A}\right]$ and $E\left[\ell_B^{e_B}\right]$. Let $\{P_A, Q_A\}$ and $\{P_B, Q_B\}$ be the generating sets of these torsion subgroups, respectively. In this setting, the Computational SuperSingular Isogeny (CSSI) problem, which is central to Yoo et al. type schemes, is described as follows.

Definition 2.11

(CSSI Problem) Let $\phi_A : E \to E_A$ be an isogeny with (cyclic) kernel $\langle R_A \rangle$, where R_A is a random point of order $\ell_A^{e_A}$. Given $\left(E_A, \phi_A(P_B), \phi_A(Q_B)\right)$, find a generator for $\langle R_A \rangle$.

The identification protocol is described as follows. As usual, the protocol is between the prover P and verifier V. The prover chooses as its secret key a subgroup $\langle S \rangle$, generated by S of order $\ell_A^{e_A}$, and publishes the triple $\left(E/\langle S \rangle, \phi(P_B), \phi(Q_B)\right)$ as its public key.

Commit.

The prover computes a random isogeny $\psi : E \to E/\langle R \rangle$, where R is of order $\ell_B^{e_B}$. The prover then computes $\phi' : E/\langle R \rangle \to E/\langle R, S \rangle$ and $\psi' : E/\langle S \rangle \to E/\langle R, S \rangle$, so that the commutative diagram

$$
\begin{array}{ccc}
E & \xrightarrow{\phi} & E/\langle S \rangle \\
\psi \downarrow & & \downarrow \psi' \\
E/\langle R \rangle & \xrightarrow{\phi'} & E/\langle R, S \rangle
\end{array}
$$

is completed. Let $E_1 = E/\langle R \rangle$ and $E_2 = \langle R, S \rangle$. The prover sends $com = (E_1, E_2)$ to the verifier.

Challenge.

The verifier sends a random bit b as challenge.

Response.

The prover sends as response $\psi(S)$ if $b = 1$ and if $b = 0$.

Verify.

The verifier, for the case $b = 0$, checks that R and $\phi(R)$ are of order $\ell_B^{e_B}$, and that these points generate the kernels of isogenies $E \to E_1$ and $E/\langle S \rangle \to E_2$, respectively. If the bit $b = 1$, the verifier checks that $\psi(S)$ is of order $\ell_A^{e_A}$ and that it generates the kernel of the isogeny $E_1 \to E_2$.

The authors show that (Yoo et al. 2017, Section 5.1) this scheme satisfies completeness, special soundness and HVZK. The proof is based on the hardness of the CSSI problem. The signature scheme constructed using Unruh techniques is EUF-CMA. For more details on construction and security of the signature scheme, refer to Yoo et al. (2017).

2.7.3 A Discussion on Other Isogeny-Based Signatures

Recall that the endomorphism ring of a supersingular curve is an order in the quaternion algebra. The Galbraith, Petit and Silva (2020) constructions of signatures depend on the assumption that computing the endomorphism ring of an elliptic curve is hard. The authors first describe an identification scheme. The prover chooses a nice initial curve, for which the endomorphism ring is well-known, and performs an isogeny walk to reach another curve. This information about the isogenies is sufficient to compute the endomorphism ring of the final curve. On the other hand, just the knowledge of the final curve is not enough for efficient computation of its endomorphism ring. The major challenge is in computing the commitment, so that the secret, namely the knowledge of the isogeny, is not leaked. The authors then use Unruh's techniques to derive a signature scheme and argue its security. The authors also discuss the algorithmic efficiency of the signature.

If E is supersingular, the $\mathbb{F}_p$-rational endomorphism ring is an order in an imaginary quadratic field. The class group of this order can be defined via the invertible fractional ideals. Castryck et al. (2018) used the splitting behaviour of the invertible integral ideals to define a class group action. While this action on ordinary curves was used to construct earlier schemes, supersingular case was considered only recently. SeaSign by De Feo and Galbraith (2019) and CSI-FiSh by Beullens, Kleinjung and Vercauteren (2019) are signature schemes, obtained based on an identification scheme which uses this class group action.

2.8 SOME INTERESTING USE CASES

Digital signatures are used for providing authentication. While post-quantum signatures can potentially be used in all places where the traditional ones are being used, some candidates are ideally suited for certain specific use cases. Tan, Szalachowski and Zhou (2019) surveyed various applications which would benefit from post-quantum signatures. We discuss some places where these signatures can be used and the rationale behind the choice. It is interesting to note that the proposers of NIST candidate signatures make recommendations regarding ideal use cases for their design. While discussing the use cases, we also mention such claims.

2.8.1 Certification Authority and Authentication in TLS

Any trusted communication must include an authentication phase. Such is the case even in Transport Layer Security protocol. Digital Certificate is issued by Certification Authority (CA), where the CA is an authority trusted by both the owner of the certificate and the parties who rely on the validity of the certificate. Digital Certificate is obtained using signatures and these follow the standard specified in X.509 (Hesse et al. 2005; Boeyen et al. 2008). There are various types of certificates, including TLS Client/Server Certificate, Email Certificate and Code-Signing Certificate.

The current infrastructure uses RSA- and DH-based systems. These need to be replaced to get quantum resilience.

Some applications have been demonstrated to be efficient using post-quantum signatures. Pradel and Mitchell (2020) use post-quantum signatures for providing digital passports. TLS protocol has an authentication phase, where CA infrastructure is needed. LIBOQS[16] has successfully integrated post-quantum signatures in this phase. The proposers of Rainbow mention that their design is ideal for such an application. They also claim that Certificate Transparency is easy to verify due to a small signature size.

2.8.2 SECURE AND VERIFIED BOOT

Secure boot is a service where the code used for booting is authentic. This is done by verifying a signature on the operating system code. Xilinx provides secure boot,[17] which uses digital signatures. Proposers of Rainbow suggest that such services can be derived using their proposal.

Android's verified boot ensures that all data and code are verified before the booting process begins. Again, Rainbow has been recommended for such a service.

2.8.3 MISCELLANEOUS APPLICATIONS

Rainbow has been recommended for usage in Internet of Thing (IoT) scenarios due to its small signature size. Falcon has been recommended for use in CA, blockchain and firmware update. DILITHIUM can be used for building zero-knowledge proofs, which has many applications.

2.8.4 CHALLENGES IN STANDARDIZATION, MIGRATION, AND UBIQUITOUS USAGE OF SUCH SCHEMES

This section discusses some efforts being made to plan for migration to quantum-resilient cryptography. Bodies like NIST and ETSI have brought out a document outlining a tentative plan. Some companies like ISARA already provide quantum-resilient cryptography products and services.

Chen (2017) challenges in migrating to PQC. The work lists three uncertainties. The first one is regarding the timeframe. Considering progress in building quantum computers, whether there is an immediate threat to the cryptography in-use needs careful evaluation. The second is the cost involved in deploying the new schemes. The third concern is that whether the progress in quantum algorithms could threaten these new schemes. The author mentions the argument that there is a fair chance of a quantum computer breaking today's cryptography by 2031, so migration should be imminent. The author foresees that new schemes may not prove to be drop-in replacements for the existing ones.

NIST Whitepaper,[18] authored by Barker et al., lays the foundation for US Government's plan for migration. NIST has also invited organizations to participate in the migration. The first step would be for the standard's development bodies to identify critical applications which would warrant migration. Some examples cited are Internet Engineering Task Force, International Organization for Standardization/International Electrotechnical Commission, American National Standards Institute/International Committee for Information Technology Standards and Trusted Computing Group. The plan hopes to identify all FIPS and NIST Special Publication 800-series documents which have to be updated. This would assist various organizations to identify the exact place where they use public key cryptography. The document envisages sector-wise plan for such a migration. Inter-operability has been identified as a major challenge in the plan.

A similar document brought out by ETSI[19] identifies inventory preparation, plan for migration and execution as the three stages. ISARA corporation has brought out quantum-resilient crypto-agile[20] solutions.

2.9 CONCLUSIONS

This article presented tools and techniques for constructing post-quantum digital signatures. Authentication is very vital in communication over insecure channels. The constructions discussed here can be used for providing authentication, that too in a quantum-resilient way. We discussed some techniques for proving security of such signatures. We presented some interesting use cases for these schemes. We restricted out discussion to tools and techniques useful for understanding NIST round three candidates. Tasks like understanding their quantum resilience, efficient implementation and comparison remain. It is hoped that future internet protocols would use these schemes after they have been analysed and standardized.

ACKNOWLEDGEMENTS

This study was carried out as part of a project funded by MEITY, Govt. of India. The second author acknowledges the inputs from Santhosh, Vyshna, Samyuktha and Nalini on technical and editorial aspects.

NOTES

1 https://nvlpubs.nist.gov/nistpubs/FIPS/NIST.FIPS.186-4.pdf.
2 https://csrc.nist.gov/CSRC/media/Projects/Post-Quantum-Cryptography/documents/call-for-proposals-final-dec-2016.pdf.
3 https://csrc.nist.gov/CSRC/media/Projects/Post-Quantum-Cryptography/documents/call-for-proposals-final-dec-2016.pdf.
4 https://nvlpubs.nist.gov/nistpubs/FIPS/NIST.FIPS.186-4.pdf.
5 https://nvlpubs.nist.gov/nistpubs/SpecialPublications/NIST.SP.800–56Br2.pdf.
6 https://nvlpubs.nist.gov/nistpubs/SpecialPublications/NIST.SP.800–56Br2.pdf.
7 https://csrc.nist.gov/projects/post-quantum-cryptography/round-3-submissions.
8 https://csrc.nist.gov/projects/post-quantum-cryptography/round-3-submissions.
9 The terminology, probably, originates from salad dressing. There is not much significance to this terminology. The philosophy is that an oil variable cannot coexist with another oil variable.
10 Salt is a random string used in order to bring randomness in signature even when the message is the same. This is useful while arguing security of signatures.
11 https://csrc.nist.gov/projects/post-quantum-cryptography/round-3-submissions.
12 https://csrc.nist.gov/projects/post-quantum-cryptography/round-3-submissions.
13 https://www-polsys.lip6.fr/Links/NIST/MQsoft.html.
14 https://csrc.nist.gov/projects/post-quantum-cryptography/round-3-submissions.
15 https://csrc.nist.gov/projects/post-quantum-cryptography/round-3-submissions.
16 https://github.com/open-quantum-safe/liboqs.
17 https://www.xilinx.com/support/documentation/application_notes/xapp1175_zynq_secure_boot.pdf.
18 https://nvlpubs.nist.gov/nistpubs/CSWP/NIST.CSWP.04282021.pdf.
19 https://www.etsi.org/deliver/etsi_tr/103600_103699/103619/01.01.01_60/tr_103619v010101p.pdf.
20 https://www.isara.com/products/isara-radiate.html.

REFERENCES

Abdalla, Michel, Jee Hea An, Mihir Bellare, and Chanathip Namprempre. 2002. "From Identification to Signatures via the Fiat-Shamir Transform: Minimizing Assumptions for Security and Forward-Security." In *Advances in Cryptology — Eurocrypt 2002*, edited by Lars R. Knudsen, pp. 418–33. Berlin, Heidelberg: Springer Berlin Heidelberg.

Alagic, Gorjan, Jacob Alperin-Sheriff, Daniel Apon, David Cooper, Quynh Dang, John Kelsey, Yi-Kai Liu, et al. 2020. "Status Report on the Second Round of the NIST Post-Quantum Cryptography Standardization Process." NISTIR 8309. NIST.

Albrecht, Martin R., Christian Rechberger, Thomas Schneider, Tyge Tiessen, and Michael Zohner. 2015. "Ciphers for Mpc and Fhe." In *Advances in Cryptology – Eurocrypt 2015*, edited by Elisabeth Oswald and Marc Fischlin, pp. 430–54. Berlin, Heidelberg: Springer Berlin Heidelberg.

Babai, László. 1986. "On Lovász' Lattice Reduction and the Nearest Lattice Point Problem." *Combinatorica* 6: 1–13.

Bai, Shi, and Steven D. Galbraith. 2014. "An Improved Compression Technique for Signatures Based on Learning with Errors." In *Topics in Cryptology – Ct-Rsa 2014*, edited by Josh Benaloh, pp. 28–47. Cham: Springer International Publishing.

Bennett, Charles H., and Gilles Brassard. 2014. "Quantum Cryptography: Public Key Distribution and Coin Tossing." *Theoretical Computer Science* 560: 7–11. doi:10.1016/j.tcs.2014.05.025.

Bernstein, Daniel J., Daira Hopwood, Andreas Hülsing, Tanja Lange, Ruben Niederhagen, Louiza Papachristodoulou, Michael Schneider, Peter Schwabe, and Zooko Wilcox-O'Hearn. 2015. "SPHINCS: Practical Stateless Hash-Based Signatures." In *Advances in Cryptology – Eurocrypt 2015*, edited by Elisabeth Oswald and Marc Fischlin, pp. 368–97. Berlin, Heidelberg: Springer Berlin Heidelberg.

Bernstein, Daniel J., and Bo-Yin Yang. 2018. "Asymptotically Faster Quantum Algorithms to Solve Multivariate Quadratic Equations." In *Post-Quantum Cryptography -9th International Conference*, Pqcrypto 2018, Fort Lauderdale, FL, USA, April 9–11, 2018, Proceedings, edited by Tanja Lange and Rainer Steinwandt, 10786:487–506 Lecture Notes in Computer Science. Springer. doi:10.1007/978-3-319-79063-3_23.

Beullens, Ward. 2021. "Improved Cryptanalysis of UOV and Rainbow." In Advances in Cryptology - EUROCRYPT 2021–40th Annual International Conference on the Theory and Applications of Cryptographic Techniques, Zagreb, Croatia, October 17–21, 2021, Proceedings, Part I, edited by Anne Canteaut and François-Xavier Standaert, 12696:348–73 Lecture Notes in Computer Science. Springer. doi:10.1007/978-3-030-77870-5_13.

Beullens, Ward, Thorsten Kleinjung, and Frederik Vercauteren. 2019. "CSI-Fish: Efficient Isogeny Based Signatures through Class Group Computations." In *Advances in Cryptology – Asiacrypt 2019*, edited by Steven D. Galbraith and Shiho Moriai, pp. 227–47. Cham: Springer International Publishing.

Beullens, Ward, and Bart Preneel. 2017. "Field Lifting for Smaller UOV Public Keys." In *Progress in Cryptology - INDOCRYPT 2017–18th International Conference on Cryptology* in India, Chennai, India, December 10–13, 2017, Proceedings, edited by Arpita Patra and Nigel P. Smart, 10698:227–46 Lecture Notes in Computer Science. Springer. doi:10.1007/978-3-319-71667-1_12.

Boeyen, Sharon, Stefan Santesson, Tim Polk, Russ Housley, Stephen Farrell, and David Cooper. 2008. "Internet X.509 Public Key Infrastructure Certificate and Certificate Revocation List (CRL) Profile." Request for Comments. RFC 5280; RFC Editor. doi:10.17487/RFC5280.

Buchmann, Johannes, Erik Dahmen, and Andreas Hülsing. 2011. "XMSS - a Practical Forward Secure Signature Scheme Based on Minimal Security Assumptions." In *Post-Quantum Cryptography*, edited by Bo-Yin Yang, pp. 117–29. Berlin, Heidelberg: Springer Berlin Heidelberg.

Buss, Jonathan F., Gudmund Skovbjerg Frandsen, and Jeffrey Shallit. 1999. "The Computational Complexity of Some Problems of Linear Algebra." *Journal of Computer and System Sciences* 58 (3): 572–96. doi:10.1006/jcss.1998.1608.

Cantor, David G., and Hans Zassenhaus. 1981. "A New Algorithm for Factoring Polynomials over Finite Fields." *Mathematics of Computation. JSTOR*, 36: 587–92.

Castryck, Wouter, Tanja Lange, Chloe Martindale, Lorenz Panny, and Joost Renes. 2018. "CSIDH: An Efficient Post-Quantum Commutative Group Action." In *Advances in Cryptology – Asiacrypt 2018*, edited by Thomas Peyrin and Steven Galbraith, pp. 395–427. Cham: Springer International Publishing.

Chase, Melissa, David Derler, Steven Goldfeder, Claudio Orlandi, Sebastian Ramacher, Christian Rechberger, Daniel Slamanig, and Greg Zaverucha. 2017. "Post-Quantum Zero-Knowledge and Signatures from Symmetric-Key Primitives." In *Proceedings of the 2017 ACM SIGSAC Conference on Computer and Communications Security, CCS 2017*, Dallas, TX, USA, October 30-November 03, 2017, edited by Bhavani M. Thuraisingham, David Evans, Tal Malkin, and Dongyan Xu, pp. 1825–42. ACM. doi:10.1145/3133956.3133997.

Chen, Lidong. 2017. "Cryptography Standards in Quantum Time: New Wine in an Old Wineskin?" *IEEE Security Privacy* 15 (4): 51–57. doi:10.1109/MSP.2017.3151339.

Chen, Ming-Shing, Andreas Hülsing, Joost Rijneveld, Simona Samardjiska, and Peter Schwabe. 2016. "From 5-Pass Mq -Based Identification to Mq -Based Signatures." In *Advances in Cryptology - ASIACRYPT 2016–22nd International Conference on the Theory and Application of Cryptology and Information Security*, Hanoi, Vietnam, December 4–8, 2016, Proceedings, Part II, edited by Jung Hee Cheon and Tsuyoshi Takagi, 10032:135–65 Lecture Notes in Computer Science. doi:10.1007/978-3-662-53890-6_5.

Chou, Tung, Matthias J. Kannwischer, and Bo-Yin Yang. 2021. "Rainbow on Cortex-M4." *IACR Transactions on Cryptographic Hardware and Embedded Systems* 2021 (4): 650–75. doi:10.46586/tches.v2021. i4.650-675.

Coppersmith, Don, Jacques Stern, and Serge Vaudenay. 1997. "The Security of the Birational Permutation Signature Schemes." *Journal of Cryptology* 10 (3): 207–21. doi:10.1007/s001459900028.

De Feo, Luca 2017. "Mathematics of Isogeny Based Cryptography." *CoRR* abs/1711.04062. http://arxiv.org/abs/1711.04062.

De Feo, Luca, and Steven D. Galbraith. 2019. "SeaSign: Compact Isogeny Signatures from Class Group Actions." In *Advances in Cryptology – Eurocrypt 2019*, edited by Yuval Ishai and Vincent Rijmen, pp. 759–89. Cham: Springer International Publishing.

Diffie, Whitfield, and Martin E. Hellman. 1976. "New Directions in Cryptography." *IEEE Transactions on Information Theory* 22 (6): pp. 644–54. doi:10.1109/TIT.1976.1055638.

Ding, Jintai, Albrecht Petzoldt, and Dieter S. Schmidt. 2020. *Multivariate Public Key Cryptosystems*, Second Edition. Vol. 80. Advances in Information Security. Springer. doi:10.1007/978-1-0716-0987-3.

Ding, Jintai, and Dieter Schmidt. 2005. "Rainbow, a New Multivariable Polynomial Signature Scheme." In *Applied Cryptography and Network Security, Third International Conference, ACNS 2005*, New York, USA, June 7–10, 2005, Proceedings, edited by John Ioannidis, Angelos D. Keromytis, and Moti Yung, 3531:164–75 Lecture Notes in Computer Science. doi:10.1007/11496137_12.

Ding, Jintai, Bo-Yin Yang, Chia-Hsin Owen Chen, Ming-Shing Chen, and Chen-Mou Cheng. 2008. "New Differential-Algebraic Attacks and Reparametrization of Rainbow." In *Applied Cryptography and Network Security, 6th International Conference, ACNS 2008*, New York, USA, June 3–6, 2008. Proceedings, edited by Steven M. Bellovin, Rosario Gennaro, Angelos D. Keromytis, and Moti Yung, 5037:242–57 Lecture Notes in Computer Science. doi:10.1007/978-3-540-68914-0_15.

Ducas, Léo, and Thomas Prest. 2016. "Fast Fourier Orthogonalization." In *Proceedings of the ACM on International Symposium on Symbolic and Algebraic Computation*, pp. 191–98. ISSAC '16. New York, USA: Association for Computing Machinery. doi:10.1145/2930889.2930923.

Faugére, Jean-Charles. 1999. "A New Efficient Algorithm for Computing Gröbner Bases (F4)." *Journal of Pure and Applied Algebra* 139 (1): 61–88. doi:10.1016/S0022-4049(99)00005-5.

Faugère, Jean Charles. 2002. "A New Efficient Algorithm for Computing Gröbner Bases Without Reduction to Zero (F5)." In *ISSAC '02*. New York, USA: Association for Computing Machinery. doi:10.1145/780506.780516.

Galbraith, Steven D., Christophe Petit, and Javier Silva. 2020. "Identification Protocols and Signature Schemes Based on Supersingular Isogeny Problems." *Journal of Cryptology* 33 (1): 130–75. doi:10.1007/s00145-019-09316-0.

Gentry, Craig, Chris Peikert, and Vinod Vaikuntanathan. 2008. "Trapdoors for Hard Lattices and New Cryptographic Constructions." In *Proceedings of the 40th Annual ACM Symposium on Theory of Computing*, Victoria, British Columbia, Canada, May 17–20, 2008, edited by Cynthia Dwork, 197–206. ACM. doi:10.1145/1374376.1374407.

Giacomelli, Irene, Jesper Madsen, and Claudio Orlandi. 2016. "ZKBoo: Faster Zero-Knowledge for Boolean Circuits." In *25th USENIX Security Symposium, USENIX Security 16*, Austin, TX, USA, August 10–12, 2016, edited by Thorsten Holz and Stefan Savage, pp. 1069–83. USENIX Association. https://www.usenix.org/conference/usenixsecurity16/technical-sessions/presentation/giacomelli.

Goldreich, Oded, Shafi Goldwasser, and Shai Halevi. 1997. "Public-Key Cryptosystems from Lattice Reduction Problems." In *17th Annual International Cryptology Conference*, pp. 112–31. Santa Barbara, CA: Springer.

Greconici, Denisa O. C., Matthias J. Kannwischer, and Daan Sprenkels. 2021. "Compact Dilithium Implementations on Cortex-M3 and Cortex-M4." *IACR Transactions on Cryptographic Hardware and Embedded Systems* 2021 (1): 1–24. doi:10.46586/tches.v2021.i1.1-24.

Hesse, Peter, Matt Cooper, Yuriy A. Dzambasow, Susan Joseph, and Richard Nicholas. 2005. "Internet X.509 Public Key Infrastructure: Certification Path Building." Request for Comments. RFC 4158; RFC Editor. doi:10.17487/RFC4158.

Hoffstein, Jeffrey, Nick Howgrave-Graham, Jill Pipher, Joseph H. Silverman, and William Whyte. 2003. "NTRUSign: Digital Signatures Using the Ntru Lattice." In *Topics in Cryptology — Ct-Rsa 2003*, edited by Marc Joye, pp. 122–40. Berlin, Heidelberg: Springer Berlin Heidelberg.

Katz, Jonathan, Vladimir Kolesnikov, and Xiao Wang. 2018. "Improved Non-Interactive Zero Knowledge with Applications to Post-Quantum Signatures." In *Proceedings of the 2018 ACM SIGSAC Conference on Computer and Communications Security*, pp. 525–37. CCS '18. New York, USA: Association for Computing Machinery. doi:10.1145/3243734.3243805.

Katz, Jonathan, and Yehuda Lindell. 2014. *Introduction to Modern Cryptography*, Second Edition. Boca Raton: Chapman; Hall/CRC.

Kipnis, Aviad, Jacques Patarin, and Louis Goubin. 1999. "Unbalanced Oil and Vinegar Signature Schemes." In *Advances in Cryptology - EUROCRYPT '99, International Conference on the Theory and Application of Cryptographic Techniques*, Prague, Czech Republic, May 2–6, 1999, Proceeding, edited by Jacques Stern, 1592:206–22 Lecture Notes in Computer Science. Springer. doi:10.1007/3-540-48910-X_15.

Kipnis, Aviad, and Adi Shamir. 1998. "Cryptanalysis of the Oil & Vinegar Signature Scheme." In *Advances in Cryptology - CRYPTO '98, 18th Annual International Cryptology Conference*, Santa Barbara, California, USA, August 23–27, 1998, Proceedings, edited by Hugo Krawczyk, 1462:pp. 257–66 Lecture Notes in Computer Science. Springer. doi:10.1007/BFb0055733.

Kipnis, Aviad, and Adi Shamir. 1999. "Cryptanalysis of the HFE Public Key Cryptosystem by Relinearization." In *Advances in Cryptology - CRYPTO '99, 19th Annual International Cryptology Conference*, Santa Barbara, California, Usa, August 15–19, 1999, Proceedings, edited by Michael J. Wiener, 1666:19–30 Lecture Notes in Computer Science. Springer. doi:10.1007/3-540-48405-1_2.

Lamport, Leslie. 1979. "Constructing Digital Signatures from a One Way Function." CSL-98. SRI International. https://www.microsoft.com/en-us/research/publication/constructing-digital-signatures-one-way-function/.

Lyubashevsky, Vadim. 2009. "Fiat-Shamir with Aborts: Applications to Lattice and Factoring-Based Signatures." In *Advances in Cryptology – Asiacrypt 2009*, edited by Mitsuru Matsui, pp. 598–616. Berlin, Heidelberg: Springer Berlin Heidelberg.

Lyubashevsky, Vadim, Chris Peikert, and Oded Regev. 2010. "On Ideal Lattices and Learning with Errors over Rings." In *Advances in Cryptology – Eurocrypt 2010*, edited by Henri Gilbert, pp. 1–23. Berlin, Heidelberg: Springer Berlin Heidelberg.

Merkle, Ralph C. 1990. "A Certified Digital Signature." In *Advances in Cryptology — Crypto' 89 Proceedings*, edited by Gilles Brassard, pp. 218–38. New York: Springer New York.

Nguyen, Phong. 1999. "Cryptanalysis of the Goldreich-Goldwasser-Halevi Cryptosystem from Crypto'97." In *Annual International Cryptology Conference*, pp. 288–304. Springer.

Nguyen, Phong Q., and Oded Regev. 2006. "Learning a Parallelepiped: Cryptanalysis of Ggh and Ntru Signatures." In *Advances in Cryptology - Eurocrypt 2006*, edited by Serge Vaudenay, pp. 271–88. Berlin, Heidelberg: Springer Berlin Heidelberg.

Patarin, Jacques. 1996. "Hidden Fields Equations (HFE) and Isomorphisms of Polynomials (IP): Two New Families of Asymmetric Algorithms." In *Advances in Cryptology - EUROCRYPT '96, International Conference on the Theory and Application of Cryptographic Techniques*, Saragossa, Spain, May 12–16, 1996, Proceeding, edited by Ueli M. Maurer, 1070:33–48 Lecture Notes in Computer Science. Springer. doi:10.1007/3-540-68339-9_4.

Peikert, Chris. 2015. "A Decade of Lattice Cryptography." *IACR Cryptology ePrint Archive*, 939. http://eprint.iacr.org/2015/939.

Perlner, Ray A., and David A. Cooper. 2009. "Quantum Resistant Public Key Cryptography: A Survey." In *Proceedings of the 8th Symposium on Identity and Trust on the Internet*, pp. 85–93. IDtrust '09. New York, USA: Association for Computing Machinery. doi:10.1145/1527017.1527028.

Perlner, Ray, and Daniel Smith-Tone. 2020. "Rainbow Band Separation Is Better Than We Thought." Cryptology ePrint Archive, Report 2020/702.

Pomerance, Carl. 1996. "A Tale of Two Sieves." *Notices of the American Mathematical Society* 43: 1473–85.

Pornin, Thomas. 2019. "New Efficient, Constant-Time Implementations of Falcon." *IACR Cryptology ePrint Archive*, 893. https://eprint.iacr.org/2019/893.

Pornin, Thomas, and Thomas Prest. 2019. "More Efficient Algorithms for the NTRU Key Generation Using the Field Norm." In *Public-Key Cryptography - PKC 2019–22nd IACR International Conference on Practice and Theory of Public-Key Cryptography*, Beijing, China, April 14–17, 2019, Proceedings, Part II, edited by Dongdai Lin and Kazue Sako, 11443:504–33 Lecture Notes in Computer Science. Springer. doi:10.1007/978-3-030-17259-6_17.

Pradel, Gaëtan, and Chris J. Mitchell. 2020. "Post-Quantum Certificates for Electronic Travel Documents." In *Computer Security - ESORICS 2020 International Workshops, Detips, Desecsys, Mps, and Spose*, Guildford, UK, September 17–18, 2020, Revised Selected Papers, edited by Ioana Boureanu, Constantin Catalin Dragan, Mark Manulis, Thanassis Giannetsos, Christoforos Dadoyan, Panagiotis Gouvas, Roger A. Hallman, et al., 12580:56–73 Lecture Notes in Computer Science. Springer. doi:10.1007/978-3-030-66504-3_4.

Rivest, Ronald L., Adi Shamir, and Leonard M. Adleman. 1978. "A Method for Obtaining Digital Signatures and Public-Key Cryptosystems." *Communications of the ACM* 21 (2): 120–26. doi:10.1145/359340.359342.

Sakumoto, Koichi, Taizo Shirai, and Harunaga Hiwatari. 2011. "On Provable Security of UOV and HFE Signature Schemes Against Chosen-Message Attack." In *Post-Quantum Cryptography -4th International Workshop*, Pqcrypto 2011, Taipei, Taiwan, November 29- December 2, 2011. Proceedings, edited by Bo-Yin Yang, 7071:68–82 Lecture Notes in Computer Science. Springer. doi:10.1007/978-3-642-25405-5_5.

Shor, P.W. 1994. "Algorithms for Quantum Computation: Discrete Logarithms and Factoring." In *Proceedings 35th Annual Symposium on Foundations of Computer Science*, Santa Fe, NM, pp. 124–34. doi:10.1109/SFCS.1994.365700.

Silverman, Joseph H. 2009. *The Arithmetic of Elliptic Curves*. Graduate Texts in Mathematics. New York: Springer. https://books.google.co.in/books?id=Z90CA_EUCCkC.

Stehlé, Damien, and Ron Steinfeld. 2013. "Making Ntruencrypt and Ntrusign as Secure as Standard Worst-Case Problems over Ideal Lattices." *IACR Cryptology ePrint Archive*, 4. http://eprint.iacr.org/2013/004.

Tan, Teik Guan, Pawel Szalachowski, and Jianying Zhou. 2019. "SoK: Challenges of Post-Quantum Digital Signing in Real-World Applications." Cryptology ePrint Archive, Report 2019/1374.

Thomae, Enrico, and Christopher Wolf. 2012. "Solving Underdetermined Systems of Multivariate Quadratic Equations Revisited." In *Public Key Cryptography – PKC 2012*, edited by Marc Fischlin, Johannes Buchmann, and Mark Manulis, pp. 156–71. Berlin, Heidelberg: Springer Berlin Heidelberg.

Yoo, Youngho, Reza Azarderakhsh, Amir Jalali, David Jao, and Vladimir Soukharev. 2017. "A Post-Quantum Digital Signature Scheme Based on Supersingular Isogenies." In *Financial Cryptography and Data Security*, edited by Aggelos Kiayias, pp. 163–81. Cham: Springer International Publishing.

Zhao, Raymond K., Ron Steinfeld, and Amin Sakzad. 2020. "FACCT: FAst, Compact, and Constant-Time Discrete Gaussian Sampler over Integers." *IEEE Transactions on Computers* 69 (1): 126–37. doi:10.1109/TC.2019.2940949.

3 Analysis of Quantum Computing with Food Processing Use Case

Dhaval S. Jha, Het Shah, and Jai Prakash Verma
Nirma University

CONTENTS

3.1 INTRODUCTION

Quantum computing has undoubtedly pushed the frontiers of a maximum achievable computational speed. By harnessing the laws of "quantum mechanics," a quantum computer can easily outperform a conventional computer or a supercomputer [1]. Many problems that are perceived to be unsolvable can be efficiently handled by a quantum computer. A quantum computer can provide solutions to many intractable problems, within reasonable time bounds [2]. Although quantum computing technology is quite fascinating, there are many issues and challenges that need to be addressed. There are many problems that cannot be solved by conventional computers or even by modern-day supercomputers. The computational speed required for solving such problems is quite high, and so all present-day computing machines are overwhelmed by the complexity of such problems. Quantum computing provides the solution for solving such types of problems. Quantum computing is based on the concepts of Quantum Superposition and Quantum Entanglement.

A quantum computer is capable of solving highly computationally intensive problems very quickly [3]. Conventional computers process the information stored in the form of bits, whereas a quantum computer uses the concept of quantum bits (qubits). Qubits allow the states to be not only 0 or 1, but also any combination of 0 and 1. This allows a quantum computer to process more information at any given instance of time. Problems, which traditionally take exponential time to solve on a conventional computer or a supercomputer, can be solved by a quantum computer in polynomial

DOI: 10.1201/9781003267812-3

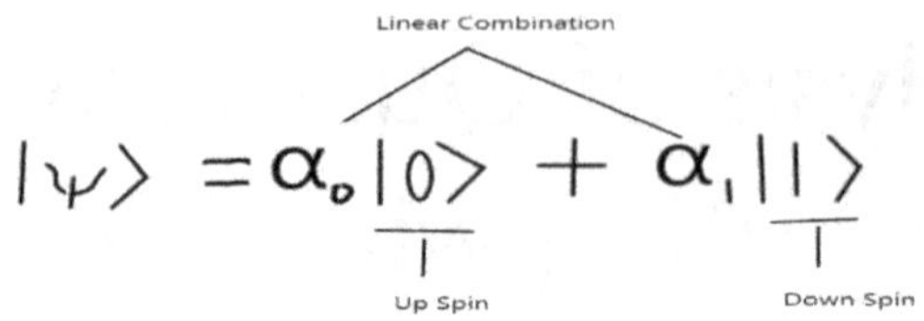

FIGURE 3.1   Linear equation for quantum computing.

time. Quantum computing has raised the frontiers of a maximum achievable computational speed. Quantum systems include the quantum circuit model, the quantum Turing machine, the adiabatic quantum computer, the one-way quantum computer, and various quantum cellular automata. The most widely used model is the quantum circuit, which is based on the quantum bit, or "qubit," which is analogous to the bit in classical computation. A qubit can exist in a quantum state that is either 1 or 0 or in a superposition of the two. However, when measured, it is always 0 or 1; the probability of either outcome is determined by the quantum state of the qubit just before measurement. A classical computer operates on bits (0 and 1), whereas a quantum computer operates on qubits (superposition of 0s and 1s). Bits are binary in nature, whereas qubits are quaternary in nature. That is why qubits can process a large amount of information in a given time frame [4]. As shown in Figure 3.1, a quantum has two states: spin-up and spin-down. Zero (0) represents spin-up and one (1) represents spin-down. Photons can also be used as qubits. The probability of each spin will be ½ for bit 0 and ½ for bit 1. As per equation 3.1, vector representations of 0 and 1 are |0 and |1, respectively.

$$|\psi\rangle = \alpha|0\rangle + \beta|1\rangle \tag{3.1}$$

3.1.1   Need for Computational Analysis of Quantum Computing

Quantum Computing technology acts as a catalyst for solving many computationally intensive problems in diverse domains [5]. The most important application of Quantum Computing can be found in the field of chemistry. In chemistry, the calculation of quantum electronic energies in molecular systems is required. This problem is extremely difficult to solve using a classical computer, whereas a quantum computer can efficiently handle this problem and provide the solution quickly. The field of chemistry requires a clear understanding of chemical reactions and their underlying mechanisms [6]. This is crucial for the production of chemicals and drugs. Quantum computing can play a vital role in speeding up this process of understanding chemical reactions, their underlying mechanisms, and thereby predicting the outcome of chemical processes. Quantum computing technology has the potential to provide solutions to many difficult problems in various domains.

3.1.2   Issues and Challenges in the Area of Quantum Computing

Quantum computing is a fascinating technology. It has an immense potential to enhance the computational speed. It provides solutions to many intractable problems within tolerable time bounds. However, the construction of such a large-scale quantum computer is a very challenging task. The issues of quantum computing are highlighted in Table 3.1. Major challenges are listed as follows:

i. **Fabrication:** A quantum computer processes the information stored in the form of quantum bits (qubits). The major challenge is to fabricate qubits, which can generate meaningful instructions for operations to be performed [7]. The complex state is to be stored in a single bit. This proves to be a major challenge in the fabrication of good-quality qubits.
ii. **Verification:** A quantum computer is not only difficult to build but also to verify. Qubits operate at very low temperatures. Due to the fragility of qubits, a complete state cannot be measured precisely. Hence, the operational verification of qubits becomes impossible.

TABLE 3.1

Quantum Computing Issues and Their Descriptions

Issues	Description
Fabrication	The complex state is to be stored in a single bit. This proves to be a major challenge in the fabrication of good-quality qubits.
Verification	Due to the fragility of qubits, a complete state cannot be measured precisely. Hence, the operational verification of qubits becomes impossible.
Qubit control	The control over qubits should have low latency. It is challenging to acquire complete control over each qubit.
Architecture	It is difficult to architect a large-scale reliable quantum computer.

iii. **Qubit Control:** For building a reliable large-scale quantum computer, it is essential to have control over the operating qubits. The control over qubits should have low latency. It is challenging to acquire complete control over each qubit.

iv. **Architecture:** It is difficult to architect a large-scale reliable quantum computer. Firstly, it requires the fabrication of good-quality qubits. Secondly, it must ensure complete control over qubits. Moreover, the errors introduced during the operation must be detected as well as corrected within a fraction of nanosecond. Therefore, an architecture that satisfies all these requirements is quite challenging to design.

3.1.3 Applications of Quantum Computing

Quantum computing technology is capable of solving intractable problems from diverse domains. Its applications range from big data, cryptography, and algorithms to chemistry, finance, and e-commerce. Quantum computing technology beats the traditional data processing techniques and algorithms. The speed-up achieved by a quantum computing algorithm is quite high compared to any conventional algorithm. Major application domains of Quantum Computing [8] are as follows: (a) Artificial Intelligence (AI) and Machine Learning. (b) Computational Chemistry. (c) Drug Design and Development. (d) Cybersecurity and Cryptography. (e) Financial Modelling. (f) Logistics Optimization. (g) Weather Forecasting.

3.1.4 Paper Organization

The rest of the paper is organized as follows. Section 3.2 shows the related and milestone work done by various researchers. It also highlights the popular researchers in this area. Section 3.3 shows the role of quantum computing in achieving high-performance computing (HPC) systems. Section 3.4 discusses the use case of requirement and the role of quantum computing for food processing systems. Section 3.5 shows the complete summary of the book chapter.

3.2 RELATED WORK

Many architectural designs, implementations, and applications of Quantum Computing technology have been proposed. This is the evidence of research opportunities available in the field of Quantum Computing. A typical quantum computer with a merely two-qubit controlled-NOT gate has been constructed [9]. The challenge is to realize Universal logic gates for Quantum Computing [10]. In the initial work, it was used in big data analytics and then how clustering of two quantum computers is done and in 2020 full-stack integration, which creates a computational device with the uses of different layers which are related to the quantum accelerator. Table 3.2 presents the recent work in the field of Quantum Computing.

TABLE 3.2

Related Research Work Done in Quantum Computing

Ref.	Year	Objective	Methodology
[2]	2020	To create a new computational device by full-stack integration of different layers that are related to build the Quantum Accelerator.	Experimental full-stack with realistic qubits. Simulated full-stack with perfect qubits.
[3]	2019	To analyse collusive attacks on MQKA (Multi-party Quantum Key Agreement) protocols and to propose an efficient MQKA protocol that requires only sequential communication of a single d-level quantum system.	The travelling-type quantum key agreement model.
[4]	2018	To present the quantum computing technology along with examples of two quantum computers: IBM-Q and D-Wave.	Mapping clustering to a specific problem.
[5]	2016	To review the available literature on Big Data Analytics using Quantum Computing	A framework for using Quantum Computing for Big Data Analytics.

TABLE 3.3

Contribution by Different Scientists in Quantum Computing

Ref.	Year	Scientist	Contribution
[11]	1982	Richard Feynman	First man to propose the idea of using quantum mechanics for building a quantum computer.
[12]	1985	David Deutsch	Constructed the Quantum Turing Machine (QTM).
[13]	1994	Peter Shor	Designed a quantum algorithm for factorizing very large numbers. It can be done in polynomial time.
[14]	1997	Lov Grover	Designed a quantum search algorithm for databases.

Table 3.3 depicts the significant contribution of different scientists in the field of Quantum Computing. It shows that initially statisticians and physics scientists worked in this area. Now, it is popular in computer science for achieving HPC systems.

3.3   ROLE OF QUANTUM COMPUTING FOR HPC

HPC derives benefits from the newly developed processors and especially the accelerators designed to serve specific purposes. It not only employs Arithmetic Logic Units (ALUs) and Floating-Point Units but also takes advantage of the vector processors which are capable of executing parallel algorithms [15]. Graphical Processing Unit (GPU) is a great example of accelerator-based technological development in HPC. Although HPC has witnessed great speed enhancement, still it is capable of tackling many intractable problems in different domains. These problems require a specialized technology. Quantum Computing is the gateway to solve such problems within reasonable time bounds. In order to solve many intractable problems, modern computer architectures should focus on the deployment of Quantum Processing Unit. It offers many advantages over ALU and GPU.

3.3.1   PROGRAMMING MODEL OF QUANTUM COMPUTING

Quantum Computing technology requires the fabrication of quantum circuits as well as the deployment of quantum algorithms. Figure 3.2 indicates the high-level view of processing that occurs in the Quantum Computing environment. The processing can be viewed as a layered architecture [16]. It is divided into six layers. Each layer interacts with its neighbouring layer(s). The functionality

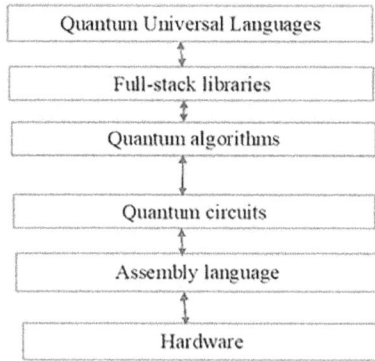

FIGURE 3.2   High-level view of the quantum computing environment.

of each layer is well-specified and unique. The Hardware lies at the bottom, whereas the Quantum Universal Languages, such as XACC and ProjectQ, occupy the topmost layer. Quantum Universal Languages make use of full-stack libraries, such as QISKit and Forest, for the deployment of quantum algorithms. These quantum algorithms are executed on the quantum circuits. The hardware is typically a quantum device. This quantum device allows the deployment of assembly language programs. The bidirectional arrows resemble the kind of producer and consumer relationship that exits between two successive layers.

3.3.2   ARCHITECTURE OF QUANTUM COMPUTING

Figure 3.3 represents the high-level architectural design of a quantum computer. A quantum computer processes the information stored in the form of qubits. We consider a general-purpose

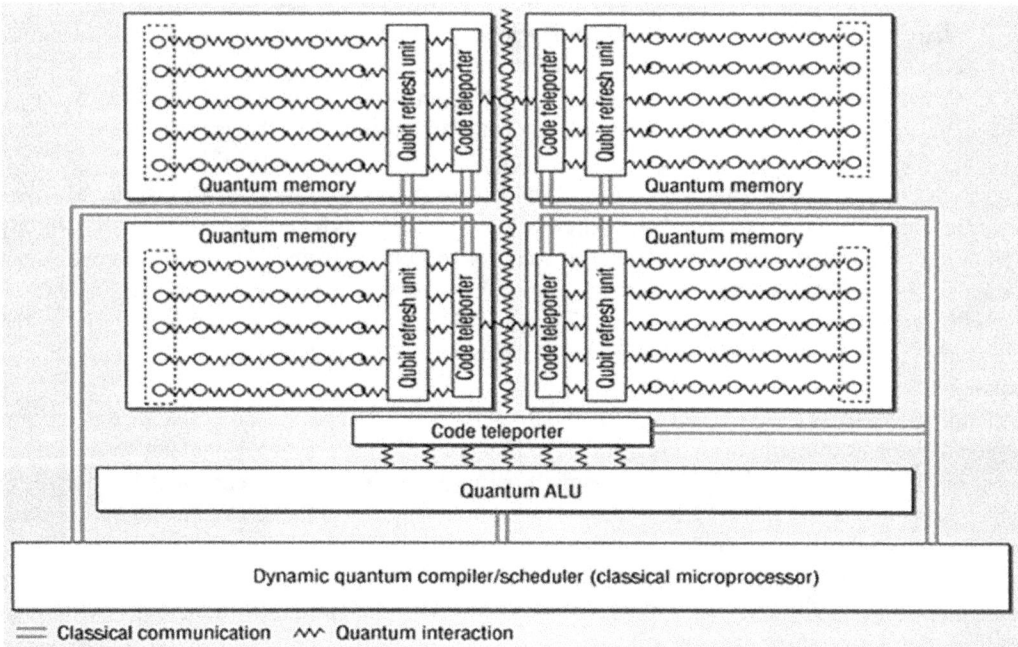

FIGURE 3.3   Architectural design of a quantum computer.

hardware. This architecture is built around three major components. These components are as follows: (a) Dynamic Quantum compiler/scheduler – It is responsible for the compilation and scheduling of a well-specified computational problem. (b) Quantum ALU – All arithmetic and logical operations are performed by this component. The results are forwarded to the quantum memory. (c) Quantum memory – This is one of the most important components of architecture. It stores all the relevant information. This information can be retrieved as and when required.

3.3.3 Methodology and Concepts for Quantum Computing

Quantum Computing is based on the laws of quantum mechanics. This leads to the interference between different states. This interference causes "Entanglement" of states [17]. All the applications of Quantum Computing must be capable of tackling the entangled states. The use of quantum tunnelling makes the quantum computer 1,000 times faster than a conventional computer. Quantum computers have the capability to accelerate the learning process in AI. However, qubits are highly error-prone. Even a single bit change can cause the system error. Quantum computers must be able to detect and correct errors. Moreover, these computers must be fault-tolerant. Controlling the interaction of qubits with the external environment and conserving their states is a difficult task. The following techniques are used to overcome these limitations:

1. **Photons:** Polarized photons can be used as a qubit as they are nearly free of de-coherence. But how can they interact with each other is still a challenge.
2. **Trapped Atoms:** These are the best option in terms of DE coherence and dephasing time.
3. **Nuclear Magnetic Resonance:** The liquid state of NMR1 allows it to handle a large number of qubits, as well as quantum algorithms and Quantum Error Correction (QEC).
4. **Quantum Dots:** The communication of quantum depends on single photon with a very low probability of being two or more. We can simulate the quantum dots to have this kind of single photon.
5. **Superconductors:** They are materials that can conduct electricity without resistance from one atom to another. Superconductors are classified into three types: charge, flux, and phase. The challenges that are being developed and investigated are a short time of DE coherence and a very low temperature of maintenance.
6. **Other Technologies:** In recent years, there were some other technologies which are being considered for improving the quantum computer architecture and their reliability.

Quantum Computing has many applications. These applications range from electric circuits, Machine Learning, analogue circuits to HPC and Wireless Networks. HPC is based on the principles of highly competent system architecture. It can solve problems of varying difficulties. Quantum Computing can be applied to Wireless Networks. This can facilitate the development of Internet of Things field.

HPC: HPC depends massively on parallel processing, which also gives high scalability to the computer. They are used for performing high computation tasks efficiently and give a non-trivial solution to the problem [18].

Analogue/Mixed Circuit: In recent computing, there are computational problems that require a greater number of qubits which is difficult to implement in the classical computer so we implement this type of problem in quantum computing. But the issue is the current quantum computing which is being made is still far from the ideal quantum computer. In the circuit, quantum processor operates analogue domain which is sensitive to noise hence error may occur and current small-scale quantum computer may not be able to tolerate such errors as error rate is low in range. For that, we need QEC codes. With that, we will need fault tolerance in the circuit. With the use of QEC and fault tolerance formalism, we can build an efficient large-scale quantum computer [19].

AI: AI is the current topic for discussion these days. One of the current research areas is Financial Trading. This research area can make use of Quantum Physics. It is the prerequisite for applying

Quantum Computing technology. Many scientists are of the opinion that, in order to develop a quantum computer, it is required that the data should follow a certain probability distribution. This is the feature of a typical quantum computer [15].

Machine Learning: Machine Learning involves the following types: **Supervised learning:** Data are labelled and the output can be predicted. **Unsupervised learning:** Data are not labelled from which we can predict. **Reinforcement Learning:** It is a type of learning where the model is left alone. It learns itself. It follows a particular algorithm and when it doesn't work, it applies another algorithm.

A large amount of data is required for training the model. By using AI and Quantum Computing, we can construct the prediction system. This indicates the diverse applications of Quantum Computing.

3.4 QUANTUM COMPUTING USE CASE FOR FOOD PROCESSING

In today's competitive world, rapid advancements are being witnessed in every walk of life. People hardly find time for leisure and relaxation. Amidst such a competitive environment, food is the most valuable entity that keeps people healthy and energetic. It is essential to assess the quality of food, apply suitable food processing techniques, and sustain the recommended quality standards of food. Many technologies of Industry 4.0 are applied in the food processing domain [20]. Approaches such as Machine Learning, Support Vector Regression, and Multi-spectral imaging are used for food processing [21]. Although these approaches are successful to some extent, the microbiological analysis of food items requires very high computing power. This can be achieved through Quantum Computing technology. Quantum Computing is capable of providing solutions to many intractable problems within reasonable time bounds. It can apply to the problem of food processing. The micro-biological analysis of food items can be easily done using Quantum Computing technology. In this paper, Quantum Computing-based solution for food processing is proposed.

3.4.1 PROPOSED SYSTEM ARCHITECTURE

As shown in Figure 3.4, the food items possess different biological values. The proposed architecture takes into account the diversity of food items. Suitable clustering algorithm(s) can be applied to these food items, based on the recommended quality parameters. Quantum computing algorithms can then be applied to each category of food items. These algorithms can also be applied to the entire set of food items. As a result, high-quality food items can be delivered to the stakeholders.

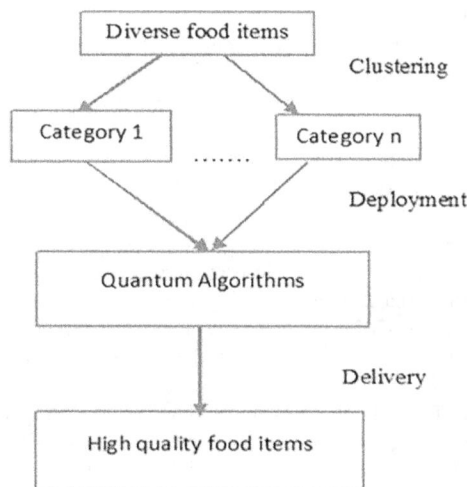

FIGURE 3.4 Proposed system architecture for food processing.

3.4.2 Applicability of Expected Outcomes

The task of food processing requires high computational power. Quantum computing technology can provide the required computational power. The proposed method can be useful to many stakeholders involved in the domain of food processing. Consumers can receive high-quality food. Suppliers can know about the quality of their food items. Analysts can provide relevant feedback, pertaining to the food processing technique being applied. This can help to improve the quality of food items. This can lead to the development of a sound food processing ecosystem.

3.5 SUMMARY

In this chapter, the basic idea and history of Quantum Computing with its features and applications have been reviewed. It is a wide field where we can implement all kinds of computational problems whether they are small scale or large scale. Many problems like food processing that are perceived to be unsolvable can be efficiently handled by a quantum computer. Although quantum computing technology is quite fascinating, there are many issues and challenges that need to be addressed. We need to focus on the error detection and fault tolerance methods in order to implement the quantum computer. Implementation of an ideal quantum computer can give us the optimal result for a particular application.

REFERENCES

1. Hey, T., "Quantum computing: An introduction." *Computing & Control Engineering Journal* 10.3(1999): 105–112.
2. Bertels, K. O. E. N., et al., "Quantum computer architecture: Towards full-stack quantum accelerators." *2020 Design, Automation & Test in Europe Conference & Exhibition (DATE)*. IEEE, 2020.
3. Sun, Z., et al., "Efficient multiparty quantum key agreement with a single d-level quantum system secure against collusive attack." *IEEE Access* 7(2019): 102377–102385.
4. Wereszczyński, K., et al., "Quantum computing for clustering big datasets." *2018 Applications of Electromagnetics in Modern Techniques and Medicine (PTZE)*. IEEE, 2018.
5. Shaikh, T. A., and Rashid, A., "Quantum computing in big data analytics: A survey." *2016 IEEE International Conference on Computer and Information Technology (CIT)*. IEEE, 2016.
6. von Burg, V., et al., "Quantum computing enhanced computational catalysis." *Physical Review Research* 3.3(2021): 033055.
7. Franklin, D., and Chong, F. T., "Challenges in reliable quantum computing". Shukla S.K., Bahar R.I. (eds) *Nano, Quantum and Molecular Computing*. Springer, Boston, MA, 2004. https://doi.org/10.1007/1-4020-8068-9_8.
8. Web Content [Available on 13-12-2021] https://analyticsindiamag.com/top-applications-of-quantum-computing-everyone-should-know-about/.
9. Cirac, J. I., and Zoller, P., "Quantum computations with cold trapped Ions." *Physical Review Letters* 74(1995): 4091.
10. Jain, S., "Quantum computer architectures: A survey." *2015 2nd International Conference on Computing for Sustainable Global Development (INDIACom)*, 2015, pp. 2165–2169.
11. Feynman, R. P., "2 – Tiny computers obeying quantum mechanical laws." Metropolis N., Kerr, D.M., Rota G.-C. (eds) *New Directions in Physics*. Academic Press, 1987, pp. 7–25, ISBN 9780124921559.
12. Deutsch, D., and Marletto, C., "Theory of information rewrites the laws of physics." *New Scientist* 222.2970(2014): 30–31, ISSN 0262–4079.
13. Web content [available on:13-12-2021], https://www.quantiki.org/wiki/shors-factoring-algorithm.
14. Grover, L. K., "Quantum search on structured problems." *Chaos, Solitons & Fractals* 10.10(1999): 1695–1705, ISSN 0960–0779.
15. Britt, K. A., and Humble, T. S., "High-performance computing with quantum processing units." *ACM Journal on Emerging Technologies in Computing Systems (JETC)* 13.3(2017): 1–13.
16. Ladd, T. D., et al., "Quantum computers." *Nature* 464(March 2010): 4.
17. IBM Research News, IBM's test-tube quantum computer makes history: First demonstration of Shor's historic factoring algorithm, http://www.research.ibm.com/resources/news/ 20011219_quantum.shtml.

18. Humble, T. S., McCaskey, A., Lyakh, D. I., Gowrishankar, M., Frisch, A., and Monz, T., "Quantum computers for high-performance computing." *IEEE Micro* 41.5(2021): 15–23, doi: 10.1109/MM.2021.3099140.
19. Bardin, J. C., "Analog/mixed-signal integrated circuits for quantum computing." *2020 IEEE BiCMOS and Compound Semiconductor Integrated Circuits and Technology Symposium (BCICTS)*, 2020, pp. 1–8, doi: 10.1109/BCICTS48439.2020.9392973.
20. Hasnan, N. Z. N., and Yusoff, Y. M. "Short review: Application areas of industry 4.0 technologies in food processing sector." *2018 IEEE Student Conference on Research and Development (SCOReD)*. IEEE, 2018.
21. Fengou, L.-C., et al., "Estimation of the microbiological quality of meat using rapid and non-invasive spectroscopic sensors." *IEEE Access* 8(2020): 106614–106628.

4 Security of Modern Networks and Its Challenges

Apurv Garg, Bhartendu Sharma,
Anmol Gupta, and Rijwan Khan
ABES Institute of Technology

CONTENTS

4.1   INTRODUCTION TO MODERN NETWORKS

Starting with the very basic, a network is defined as a collection of computers, networking devices, servers, etc., which are connected to each other and are able to share data among themselves. These devices can be connected in various arrangements, which are also called topologies. Some common network topologies are star, bus, ring, mesh, and hybrid. Modern networks are nothing but the advanced networks that are being used in today's world. There has been an explosive growth in computer networking. Since the 1970s, communication between computers has changed from a research topic to an essential part of the infrastructure [1]. Today, networking is used in every aspect of a business, which includes advertising, production, shipping, planning, billing, and accounting [1]. Most corporations are using multiple networks [1]. Computer networks nowadays have become more numerous and diverse [2]. Together, they constitute a worldwide meta network [2]. Computer networks have become an integral part of the rapidly expanding world of computing [3]. The emergence of computer networks depends on evolving communication technologies and on experiments and innovations in software tools that exploit communication [3].

Network security is used to protect our data from malware, viruses, and other unfair means. Network security involves antivirus, Virtual Private Network (VPN), encryption, policies, and firewalls. It provides the integrity and security of our data and networks. The study of network security has become an area of growing interest because of the proliferation and the paucity of security measures in most of the current networks [4]. The term enterprise network security has become more prevalent as corporations are trying to understand and manage the risks associated with business applications and practices deployed over corporate network infrastructures [5]. Network security spans over large disciplines, which range from management and policy topics to operating system kernel fundamentals [6]. One of the major goals of network security is the protection of company assets [7]. The assets are comprised of the 'information' that a company has over its network [7]. Virtually, all security policy issues are applied to the network as well as general computer security considerations [8]. If viewed from this perspective, network security is a subset of

DOI: 10.1201/9781003267812-4

computer security [8]. Whenever the data are shared from the sender to the receiver, a third-party medium is required to share the data, e.g., Internet or other communication channels. The transfer of data happens through the cooperative use of communication protocols.

The security of networks and computers has become a major concern in the past few years. This has happened because of the growing number of attacks and breaches that have occurred. New and modernized versions of spyware and viruses keep on developing. They harm the networks and systems in ways that are unimaginable. Data is the number one priority of most organizations today. Some data is invaluable as it contains confidential information about a person or an organization. This data, if breached, can cause a lot of damage to the person or the organization. To counter these security challenges, the security methods should also be regularly updated. Data should be transferred over secure channels and in an encrypted form. Various technologies like end-to-end encryption provide this type of communication security. The third party through which the data is traveling should be a completely secure and trusted one.

4.2 SECURITY OF MODERN NETWORKS

Computer network technology has developed rapidly over the past few years and people have become aware of the importance of network security [9]. For more than a decade or so, the use of technology has risen exponentially [10]. With the increase in the advancement of networks, there has been an advancement in the security techniques to secure modern networks. Modern security techniques have become a need of modern networks. This is because modern networks are more vulnerable to security attacks as the techniques to hack into modern networks have also become advanced. The systems of every organization should be regularly monitored for possible intrusions and other attacks [11]. There can be several approaches that an organization can take to implement its security model [12]. Internet of Things (IoT) is one of the most prominent technologies in today's world. Security of IoT has become a major concern and so many attacks on IoT have been invented even before its commercial implementation [13]. One of the most important aspects of an organization's security posture is user security education and training [14]. Over time, attacks have become more automated and can cause greater amounts of damage (Figure 4.1) [15].

As technology is advancing, there is a great need for some advanced security methods. Advanced security methods are used for the following reasons:

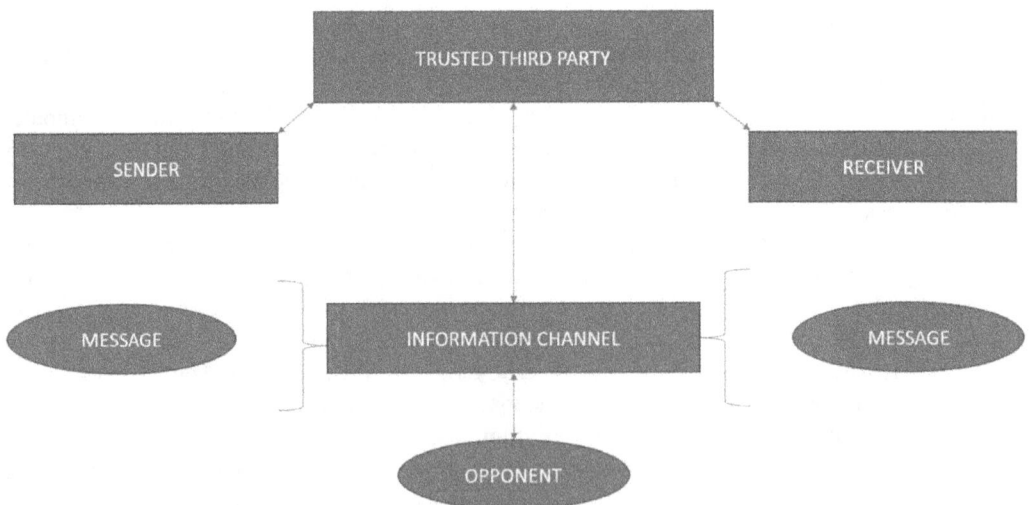

FIGURE 4.1   Basic model for network security.

- To deal with data theft
- Data tampering
- Changing user details and identity

4.2.1   How Do We Deal with Network Security?

Two of the most effective ways to deal with network security are as follows:

- **Two-Factor Authentication:** Two-factor authentication is a technique, which is being used to secure a lot of modern networks nowadays. It includes authenticating a user twice before logging in into their account. The first method of authenticating generally includes a password and the second one is generally an OTP (One Time Password)-based authentication, which involves sending a secret numeric or alpha-numeric code to the user's email address or registered mobile number. The user needs to enter this code in order to access their account.
- **Access Permissions:** When a user installs an application, it asks for various permissions to access the user's system. Most people don't pay attention to these permissions and accept them. This can lead to making the data vulnerable and can result in data leaks. So, the permission to access the system must be chosen very carefully.
- **Encryption and Decryption Techniques:** There are various methods to manipulate the actual data of the users and store it in an encrypted format on our database, this technique is called encryption. It makes the data theft to understand the actual data and use it for false purposes.
- **Making a Strong Authentication Ecosystem:** There are various effective techniques used to make user authentication more strong and secure. For example, OTP verification, 2FA (two-factor authentication), biometrics (face id and fingerprint), and security questions are used to reset the password or to change the user's personal details. 2FA works on the principle of something you know, i.e., your password and something you have on your phone. Whenever a user places a login request on any other device, the database will share the details with the login account phone and will send a verification code to your phone and on taping the right code or entering the received one-time password it will authenticate the user on the requested device.

4.3   TYPES OF SECURITY ATTACKS

Two main categories of attacks are passive attacks and active attacks.

- **Passive Attacks:**
  A passive attack is a type of attack in which an attacker tries to observe the system and scan for vulnerabilities in a system. The attacker does not modify any system data. An attack is said to be passive when an intruder tries to intercept some data traveling through a network [9].
  Types of passive attacks:
  - **The Release of Message Content:** Various methods of message transmission like telephone conversations, emails, or files transferred over the internet may contain some confidential information like passwords, for example. These conversations are sensitive and the opponent must be prevented from learning the contents of these messages.
  - **Eavesdropping Attack:** An eavesdropping attack is a type of attack in which an attacker tries to steal some information that is being transmitted between two devices over a network (Figure 4.2).

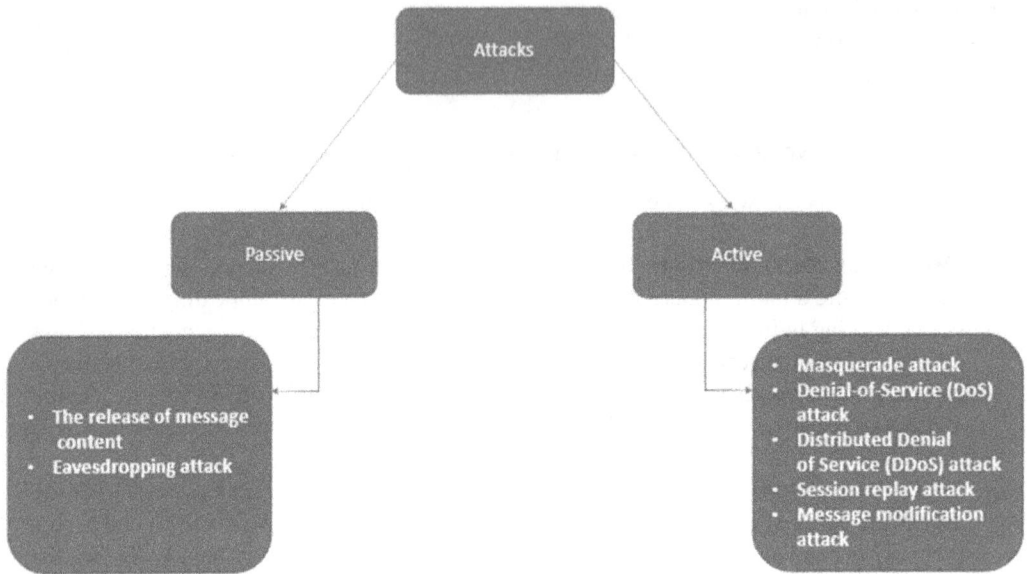

FIGURE 4.2   Types of security attacks.

- **Active Attacks:**
  Active attacks are the ones in which an attacker attempts to make an unauthorized change in the system. In this type of attack, an attacker tries to initiate a command to disturb the working of a network's normal operation [9]. Wireless Sensor Networks (WSNs) are emerging day by day as one of their applications is smart homes [16]. Active attacks can be fatal to WSNs as they can compromise sensitive information about a household.
  Types of active attacks:
  - **Masquerade Attack:** In a masquerade attack, an attacker uses a fake identity to gain unauthorized access to a computer system or any sensitive information. This attack possesses a great security problem that is a consequence of identity theft [17].
  - **Denial-of-Service (DoS) Attack:** A denial-of-service attack is an attack in which an attacker makes a network inaccessible to a user by flooding the network with requests or crashing the server. In this attack, the network resources are either not at all assigned to a user who requests the resource or the resources are delayed [18].
  - **Distributed Denial of Service (DDoS) Attack:** A distributed denial of service or DDoS attack is an attack in which an attacker uses multiple computers instead of just one to flood the target network. A network device or a computer that the intruder has is called a bot. The network of multiple computers or network devices or bots is called a botnet. In a typical DDoS attack, the hosts that are compromised send a large number of useless packets to jam a victim, or its internet connection, or both [19].
  - **Session Replay Attack:** A session replay attack is an attack in which an attacker maliciously repeats or delays a message or data in data transmission. These types of attacks are considered serious issues in the Kerberos authentication protocol [20].
  - **Message Modification Attack:** In a message modification attack, an attacker tries to modify a packet header address to direct it to an unintended destination or to modify some data on a target machine.
- **IoT Attacks:** With the increase in IoT devices, the interconnectedness between devices has increased and attackers are able to breach through an entry point and breach multiple devices through available gates.

- **Zero-Day Attacks:** A zero-day attack is an attack in which an attacker is able to find any vulnerability in software before the vendor becomes aware of it. This gives an attacker some time to exploit those vulnerabilities and take some advantage of the software before the vendor knows.

4.4   MODERN NETWORK SECURITY METHODS

Modern security methods are advanced security methods that can be used for ensuring the security of modern networks. The most effective methods are given below.

- **Cryptography**
  Cryptography is defined as the study of securing information by using the process of encryption and decryption [21]. It is a widely used tool in security engineering. It makes use of codes and ciphers for transforming information into unintelligible data [21]. Cryptography can be considered the science and study of secret writing [22].
  - Plain text is readable data that is not encrypted.
  - The ciphertext is the non-readable encrypted data.
       The algorithm converts plain text into ciphertext. This process is called encryption. The process of converting the ciphertext back to plain text is known as decryption (Figure 4.3).

There are two types of cryptographic techniques that are majorly used:

- a. **Symmetric Key Cryptography**
     In this technique, a single key is used in the process of encryption and decryption. This technique is also known as Secret Key cryptography (Figure 4.4).
- b. **Asymmetric Key Cryptography**
     In this technique, two separate keys are used for encryption and decryption. It is also called Public Key cryptography. These two keys are as follows:
     - Public Key (used to encrypt data, can be used by anyone)
     - Private Key (used to decrypt data, which can only be used by an intended user) (Figure 4.5)

- **Firewall**
  A firewall is a network security system that monitors the incoming and outgoing network traffic and determines whether to allow or block certain data packets or traffic based on a

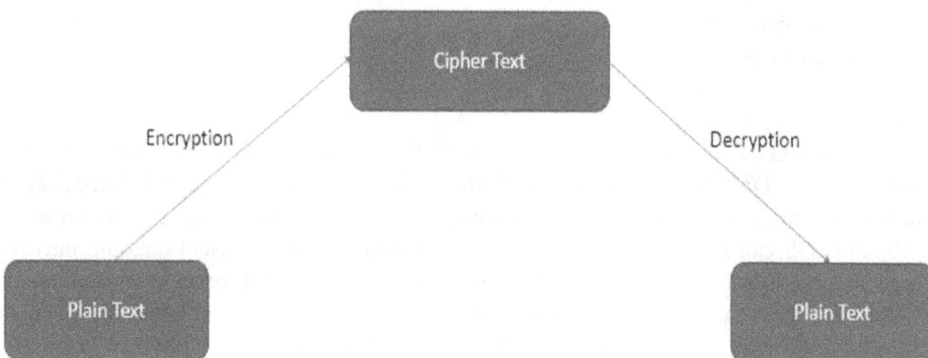

FIGURE 4.3   Basic encryption method.

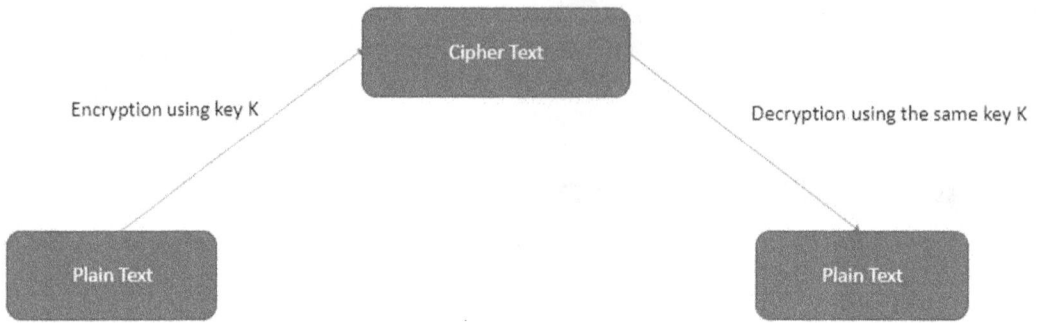

FIGURE 4.4 Symmetric key cryptography.

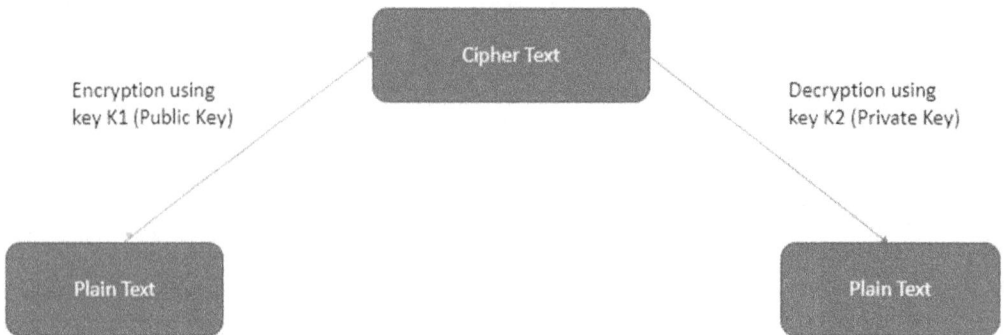

FIGURE 4.5 Asymmetric key cryptography.

set of security rules. In simple words, a firewall is a group of components that collectively form a barrier between two networks [23]. Today, firewalls have become an integrated part of the security mechanism of any organization [24]. For each network packet entering or leaving the network, a decision is made as to whether to accept or reject it [25]. Thus, a firewall's efficiency plays an important role in a network's performance [25]. Often firewalls are placed at the entrance of a private network on the Internet [26]. A firewall works on data security rules. It controls the access of all incoming and outgoing data packages over the local network and it allows or blocks the data packages based on the security rules. It protects the network from unusual traffics.

There are three types of firewalls:
- Packet filtering firewall
- Stateful Inspections
- Proxy server firewall

- **Intrusion Detection System**
   As the number of security threats and network throughput is increasing, intrusion detection systems (IDSs) have received a lot of attention in the field of computer science [27]. As technology usage is growing, intrusion detection has become an emerging area of research [28]. An IDS can be considered a security layer that is used to detect ongoing intrusive activities in information systems [29]. Intrusion detection is the process to monitor the events that occur in a computer system or network [30]. These events get analyzed for signs of security problems [30]. Intrusion detection has emerged in the past few years, so it is a relatively young technology [30]. Most of the research and development in the field of intrusion detection has occurred since 1980 (Figure 4.6) [30].

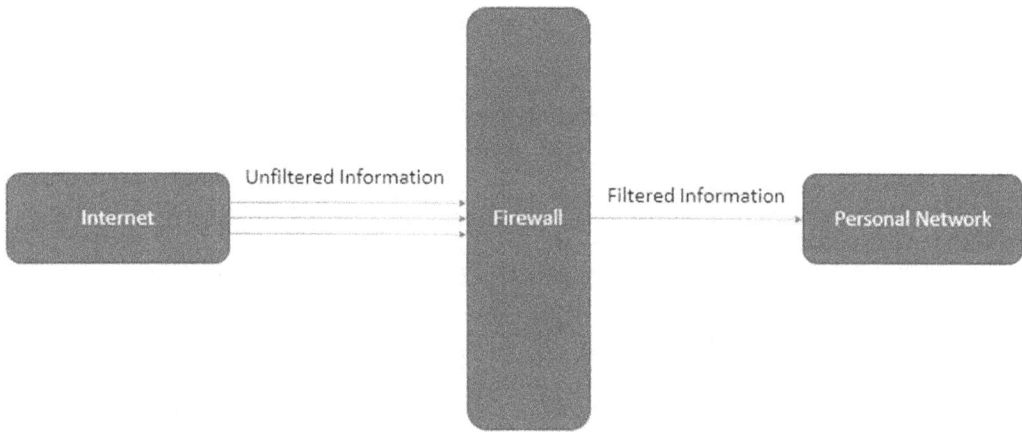

FIGURE 4.6 Basic firewall system.

The most common types of IDSs are as follows:

- Network Intrusion Detection System is a system that has the ability to analyze incoming network traffic.
- Host-based Intrusion Detection System is a system that has the ability to monitor important operating system files.

Today, every organization needs to have an IDS because of the growing number of security attacks to gain the sensitive data of clients and customers stored in an organization's database. These attacks should have a proper detection system to prevent any kind of fatality.

- **Access Control**

  Access Control constrains what a user can directly do and what the programs executing on behalf of the users are allowed to do [31]. Access control guarantees the authenticity of a person. It also guarantees that a person has proper access to whatever data he is accessing. Access control uses the process of authentication to verify a user. If a user is verified, he has to be authorized to access the data which he is trying to access; if the user is authorized only, then he can make changes to any date.

  Access control systems can use biometrics to authorize a person's identity. The authorization process can be further extended using two-factor authentication. A person may be required to enter a code or speak a phrase to gain access to the system. All this can help to reduce the risk of unauthorized access to data.

- **Data Loss Prevention**

  Data loss prevention deals with the prevention of data from getting leaked due to data breaches. Enterprises hold sensitive data that customers, business partners, regulators, and shareholders expect them to look after and protect [32]. One of the greatest threats to any enterprise is the leakage of their data through the insiders of that enterprise. Data loss prevention systems try to solve this problem by monitoring access to confidential data [33].

  Data loss prevention systems make sure that the data is not lost. In addition to this, it also provides security from unauthorized access to data. The reason a data loss prevention system is required is the growing number of data breaches in organizations that hold sensitive user data. This data has a lot of value. An organization can spend a lot of money to recover this data. This is the reason attackers are interested in this data. Therefore, a data loss prevention system can help secure the data from data breach or unauthorized access to data.

- **End-to-End Encryption**

  End-to-end encryption means that the only person with whom another person is communicating is the one who has knowledge of the contents of the message being delivered. No third party in between the communication can access the data. Large messaging providers are adopting end-to-end encryption and as a result, private communication has become readily available to increasing users [34].

  WhatsApp uses end-to-end encryption to secure its communications. It is one of the platforms, which has a large number of users. Therefore, the privacy of conversations of these users becomes the number one priority of companies like WhatsApp. End-to-end encryption is one of the most secure ways of communication today. A lot of communication services have started using this technology to secure the conversations made using them (Figure 4.7).

- **Network Segmentation**

  Network Segmentation is a security technique that tries to restrict the movement of an attacker and makes it difficult for an intruder to gain access to a network by partitioning the network into sections or segments [35]. A network is further divided into small networks or subnets by network segmentation. Network segmentation helps in decreasing the load on a server. It also helps in reducing the damage caused by any cyber-attack as the attack has a limited amount of network to spread to.

- **Email Security**

  Email security, as the name suggests, deals with the security of electronic mails, which are delivered over the internet. Email is an important means of communication today. Most organizations use email to deliver messages to their employees. These emails contain a lot of sensitive information at times. Therefore, the security of these emails should be kept on high priority. There are various protocols like Secure/Multipurpose Internet Mail Extensions (S/MIME) that can be used for email security.

- **Anti-malware/Antivirus Software**

  Antivirus software is the need for any computer system today. This is because of the advancement that has come from computer viruses. It has become an easy job for any hacker to just make a computer virus or backdoor with just a few lines of code and execute it on the target's machine with the help of some social engineering. There has also been a convergence between spam and viruses, which has led to a more insidious problem [36].

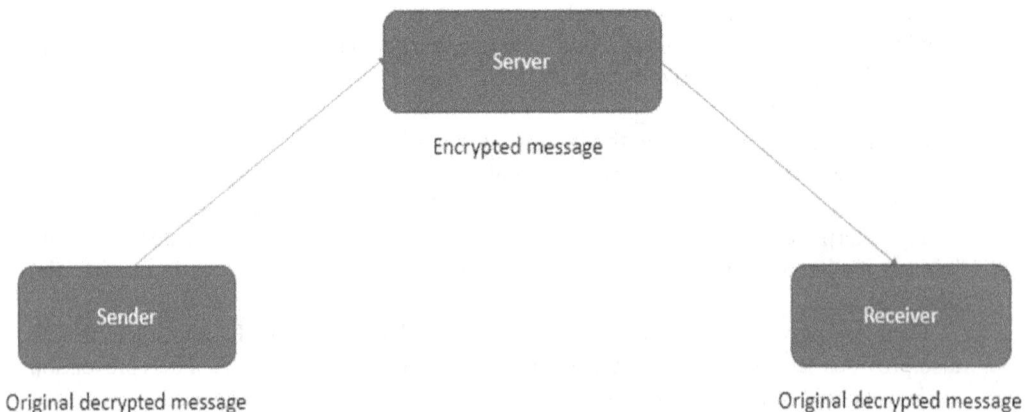

FIGURE 4.7   End-to-end encryption.

Android is one of the most widely used mobile environments today [37]. Mobile malware is one of the greatest threats today in computer security [38]. Everyone is using a smartphone nowadays. These devices contain some personal information of users, which can be very sensitive at times. Therefore, it is essential to protect every computer and mobile device with antivirus software. Antivirus software updates its systems frequently to fight new kinds of viruses that can infect a system in new ways.

Some of the antivirus software available in the market are as follows:

- Norton
- Kaspersky
- McAfee
- Malwarebytes
- Avast Antivirus
- **Virtual Private Network (VPN)**

A VPN can be considered useful in solving many issues that are associated with today's private networks [39]. A VPN allows the provision of private network services over a public network such as the Internet for an organization [40]. A VPN is a way to simulate a private network over a public network [41]. A VPN depends on the use of virtual connections [41]. Virtual connections are those temporary connections that have no real physical presence, but they consist of packets routed over various machines on the Internet on an ad hoc basis [41]. VPN provides us with online privacy. It makes our virtual address untraceable by hiding our IP address. VPN helps us to access websites that are region-restricted. VPN is capable of hiding out browsing history, IP address, location, device information, and other activities on the network. For example, Express VPN and Nord VPN.

There are four types of VPNs:

- Personal VPN
- Remote access VPN
- Mobile VPN
- Site-to-Site VPN

4.5 NETWORK SECURITY TOOLS

Some of the most prominent security tools used today are as follows:

- **Wireshark**

Wireshark is a network packet analyzer that tries to capture network packets and tries to display the packet's data as detailed as possible [42]. It is an award-winning multiplatform software. Wireshark supports more than 750 protocols [43]. Wireshark is free and open-source software. Many corporations are using Wireshark for troubleshooting. It is a packet sniffer that can capture network traffic. Wireshark can capture network packets that a user can analyze. It displays the packet information that is available about a packet. Wireshark is generally used by security engineers to analyze security problems and troubleshoot issues related to networks (Figure 4.8).

- **Metasploit**

Metasploit is an entire framework providing the infrastructure needed to automate mundane, routine, and complex tasks [45]. This framework came about primarily for providing a framework for penetration testers to develop exploits [46]. When penetration testing is being talked about, one of the first tools that come to mind is Metasploit [47].

Metasploit is a Ruby-based framework that is open-source, which is widely used for penetration testing. It has the ability to develop exploits. A user can test the developed exploits and execute them using Metasploit. Metasploit has a number of payloads and exploits to achieve its objectives. These can be used by security researchers to do their work with just

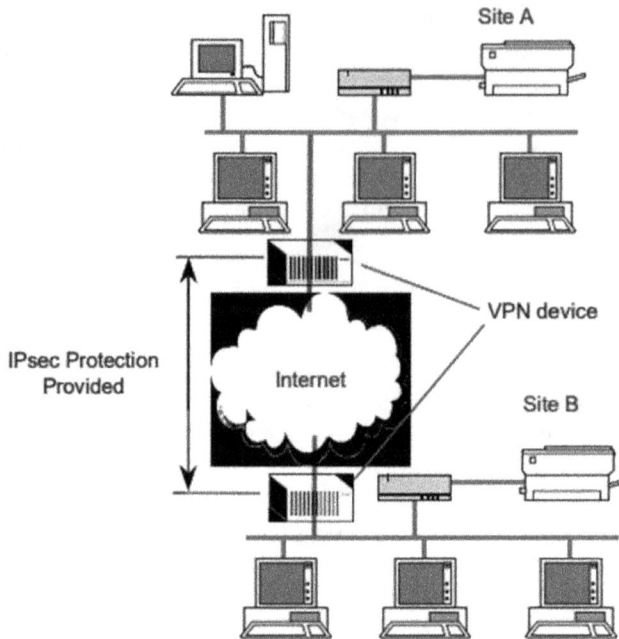

FIGURE 4.8   Site-to-site secure internet communication by the typical use of VPN [44].

a few lines of code. Metasploit is one of the most powerful tools available for penetration testing because of the ease of use and customizability it provides.

- **Nmap**

  Nmap is short for Network Mapper. It is a free and open-source tool. It is mainly used for vulnerability scanning and scanning of networks. Network scanning is the discovering of active hosts on a network and discovering information about the active hosts [48]. Nmap can be used for penetration testing of networks. It provides different scans like Transmission Control Protocol (TCP) and User Datagram Protocol (UDP) scans. It also provides the ability to scan ports operating systems.

- **Burp Suite**

  Burp Suite is a collection of tightly integrated tools that can be used in the effective security testing of modern-day web applications [49]. Web applications consist of a major of web technologies due to which they have become a target for many cyber-attacks [50]. Burp Suite is a very powerful tool for penetration testing of web applications. Burp Suite also has the ability to intercept HTTP requests [51]. These requests can be modified before they are forwarded to the web server.

- **Kali Linux**

  Kali Linux is the world's most powerful penetration testing platform, used by security professionals for a wide range of tasks such as penetration testing [52]. Penetration testing and ethical hacking are proactive ways of testing web applications [53]. Kali Linux is a Linux distribution that is derived from Debian. It is specially designed for computer forensics and advanced penetration testing. Kali Linux can be very useful if a person wants to explore the following domains:
  - Vulnerability Analysis
  - Exploitation Tools
  - Penetration Testing
  - Sniffing

- Spoofing
- Reverse Engineering
- Wireless Attacks
- Reporting Tools

Kali Linux provides over 600 utilities, which can help explore the above-mentioned domains.

Some other security tools are as follows:
- Cain and Abel
- Nikto
- Tcpdump
- Splunk
- Nessus Professional
- John the Ripper
- Forcepoint
- Paros Proxy

4.6 NETWORK SECURITY CHALLENGES

Network security challenges are the challenges that are faced during the process of making a network secure. These challenges can create a large issue if they are not paid attention to. The top security challenges are listed below.

- **Computer Virus**

  A computer 'virus' can be thought of as a program that can 'infect' other programs by modifying them to include a possibly evolved copy of themselves [54]. In the past few years, the detection of computer viruses has turned out to be a common place [55]. Viruses have the ability to replicate themselves just like in real life. They can spread from network to network and can infect a large number of machines in seconds. Some viruses can damage a system by doing things like slowing it down or formatting the hard disk of the system.

  Computer viruses can cause fatal effects on a single computer system or a network system. This is because the virus has the ability to replicate itself and spread really fast. Viruses can spread from a single system to networks and servers. Viruses have been a constant threat and challenge to the security of computers and networks for ages now. Common signs that show that a computer has been affected by a computer virus are frequent application crashes, lagging or slowing down of the system, automatic opening of applications, etc. These viruses have the ability to steal stored passwords, log keystrokes, reduce a system's performance, etc. Viruses can be specially made for a particular system to steam some specific information or perform some specific tasks which the attacker wants.

  Protection against Viruses:
  - Install trusted antivirus software and keep it updated.
  - Do not download software from untrusted sources.
  - Avoid inserting unknown USB devices and disks.
  - Scan the files before opening them.

- **Ransomware**

  Ransomware is also a theft just like other thefts or cyber-attacks like spyware or malware [56]. It is a category of malicious software which, when run, has the ability to disable a computer's functionality in some way [57]. Ransomware infections have grown exponentially over the past few years [58]. Ransomware uses encryption to hold the victim's data locked. Ransomware uses asymmetric encryption to encrypt data. The attacker demands a ransom to unlock the files. Usually, a period of 24–48 hours is given to the victim to pay the ransom. If paid, the victim pays the ransom, the data is unlocked.

A famous ransomware attack came about in 2017. This attack was called the WannaCry ransomware attack. In this attack, the attackers locked a victim's file and asked for a ransom in Bitcoin. It spread through computers using the Windows operating system. This attack was a crypto-ransomware attack.

Protection against Ransomware:
- Keep a backup of important data.
- Do not click on suspicious links.
- Avoid downloading from untrusted sources.
- Keep your antivirus software updated.

- ### IoT Security
The IoT allows devices and software to communicate over the internet [59]. The shared data contains a large amount of private information, and the security of this data cannot be neglected [59]. IoT can be considered the next era of communication [60]. As many devices are connected to IoT, there are a lot of personal devices that contain the personal data of users. The security of this data becomes the number one priority when the security of IoT is being talked about.

IoT is one of the fastest-evolving technologies today. Smart devices are everywhere, from smart phones to smart homes, IoT has applications in almost every industrial sector. This is the very fact that makes it so vulnerable. A lot of sensitive information is collected and stored by IoT devices. This data needs security as it can have a lot of damage to a person if this data gets hacked.

Protection of IoT Devices:
- Proper authentication of devices present over a network.
- Use of techniques like encryption in IoT.

- ### Phishing
Phishing attacks are most commonly initiated by sending links to spoofed websites that can harvest information [61]. These attacks become successful because they exploit a user's inability to distinguish between spoofed and legitimate sites [62]. In phishing attacks, the attacker makes fake web pages look like the official ones. The attacker sends fraudulent messages that appear to be coming from a reputable source. When a user responds to these messages and opens a web page, he is directed to a fake web page and when the user enters his details like email and password, the attacker gets access to this sensitive information.

Phishing is one of the most common types of social engineering attacks. The main aim of this attack generally is to steal credentials like passwords and emails. One of the most common sources of phishing links is email. An attacker sends a really professional-looking email generally referring to offering a discount or a job opportunity. The victim is inclined to open this link, which is the main goal of the attacker.

Protection against Phishing:
- Do not click on suspicious links.
- Always verify the source of emails.
- Check the URL of a website for authenticity before entering any credentials.

- ### Security Tools
Today, there are a lot of security tools that are available. Operating systems like Kali Linux, for example, come built-in with mostly all types of security tools. It has become easier for hackers to use these kinds of tools to solve their purpose even if they do not have enough knowledge about security systems. These tools provide a wide variety of functionality that is available with just a single click. The advancement in security tools has become a major security challenge.

- **Quality of Security Professionals:**
  Although cyber security has been gaining a lot of popularity over the past few years, there is still a lack of skilled professionals in this field. People need to be educated about the importance of the security of computers and networks. Education is needed to deal with highly skilled hackers and the advancement that has come from their hacking techniques. The same amount of growth is required for network security techniques and security professionals.

4.7 CONCLUSION

Network security is one of the most important aspects of modern networks. Increasingly attacks are discovered daily and there is a need for network security methods to grow at the same rate. The advancement in technology has led to the modernization of network attacks. Today, if an attacker just learns the basics of network security, he can take on a decent network just by using some tools that are readily available in operating systems like Kali Linux. Data can be considered one of the most powerful things in the world today. It is being produced at an alarming rate, so much so that there is even a new term for it, known as Big Data. The problem with data is that it contains a lot of personal and sensitive information about a person or an organization. If someone has access to this data, it can be very dangerous for the person or organization whose data it is. Therefore, data security becomes the number one priority of any organization holding such data. There are various challenges faced by security researchers and ethical hackers all over the world. One of the biggest challenges is the security of data. Security of data can be achieved using proper authorization and verification techniques. Using the various tools that are mentioned in this chapter, a security researcher can scan for network vulnerabilities and perform penetration testing to help secure an organization's network. Organizations should spend a good amount of money to secure their networks to protect their customer's data. There should not be any vulnerability left in an organization's network to the best of their knowledge.

REFERENCES

1. Comer, D. (2009). *Computer Networks and Internets*. Upper Saddle River, NJ: Pearson/Prentice Hall.
2. Quarterman, J. S., & Hoskins, J. C. (1986). Notable computer networks. *Communications of the ACM*, 29(10), 932–971.
3. Newell, A., & Sproull, R. F. (1982). Computer networks: Prospects for scientists. *Science*, 215(4534), 843–852.
4. Heberlein, L. T., Dias, G. V., Levitt, K. N., Mukherjee, B., Wood, J., & Wolber, D. (1989). *A Network Security Monitor (No. UCRL-CR-105095)*. Lawrence Livermore National Lab., CA (USA); California Univ., Davis, CA (USA). Dept. of Electrical Engineering and Computer Science.
5. Kaeo, M. (2004). *Designing Network Security*. Cisco Press.
6. Cole, E. (2011). *Network Security Bible* (Vol. 768). John Wiley & Sons.
7. Canavan, J. E. (2001). *Fundamentals of Network Security*. Artech House.
8. Marin, G. A. (2005). Network security basics. *IEEE Security & Privacy*, 3(6), 68–72.
9. Pawar, M. V., & Anuradha, J. (2015). Network security and types of attacks in network. *Procedia Computer Science*, 48, 503–506.
10. Rahalkar, S. (2017). *Metasploit for Beginners*. Packt Publishing Ltd.
11. Kizza, J. M., Kizza, & Wheeler. (2013). *Guide to Computer Network Security* (Vol. 8). Heidelberg, Germany: Springer.
12. Kahate, A. (2013). *Cryptography and Network Security*. Tata McGraw-Hill Education.
13. Deogirikar, J., & Vidhate, A. (2017, February). Security attacks in IoT: A survey. In *2017 International Conference on I-SMAC (IoT in Social, Mobile, Analytics and Cloud) (I-SMAC)* (pp. 32–37). IEEE.
14. Dodge, R. C., & Ferguson, A. J. (2006, May). Using phishing for user email security awareness. In *IFIP International Information Security Conference* (pp. 454–459). Boston, MA: Springer.
15. Stallings, W. (2006). *Cryptography and Network Security, 4/E*. Pearson Education India.

16. Shahzad, F., Pasha, M., & Ahmad, A. (2017). A survey of active attacks on wireless sensor networks and their countermeasures. *arXiv preprint arXiv:1702.07136*.

17. Salem, M. B., & Stolfo, S. J. (2011, July). Decoy document deployment for effective masquerade attack detection. In *International Conference on Detection of Intrusions and Malware, and Vulnerability Assessment* (pp. 35–54). Berlin, Heidelberg: Springer.

18. Jamal, T., Haider, Z., Butt, S. A., & Chohan, A. (2018). Denial of service attack in cooperative networks. *arXiv preprint arXiv:1810.11070*.

19. Chang, R. K. (2002). Defending against flooding-based distributed denial-of-service attacks: A tutorial. *IEEE Communications Magazine, 40*(10), 42–51.

20. Dua, G., Gautam, N., Sharma, D., & Arora, A. (2013). Replay attack prevention in Kerberos authentication protocol using triple password. *arXiv preprint arXiv:1304.3550*.

21. Daya, B. (2013). *Network Security: History, Importance, and Future* (Vol. 4). University of Florida Department of Electrical and Computer Engineering.

22. Denning, D. E. R. (1982). *Cryptography and Data Security* (Vol. 112). Reading: Addison-Wesley.

23. Curtin, M. (1997). *Introduction to Network Security*. Kent Information Inc.

24. Sharma, R. K., Kalita, H. K., & Issac, B. (2014, July). Different firewall techniques: A survey. In *Fifth International Conference on Computing, Communications and Networking Technologies (ICCCNT)* (pp. 1–6). IEEE.

25. Katic, T., & Pale, P. (2007, June). Optimization of firewall rules. In *2007 29th International Conference on Information Technology Interfaces* (pp. 685–690). IEEE.

26. Gouda, M. G., & Liu, X. Y. (2004, March). Firewall design: Consistency, completeness, and compactness. In *24th International Conference on Distributed Computing Systems, 2004. Proceedings.* (pp. 320–327). IEEE.

27. Liao, H. J., Lin, C. H. R., Lin, Y. C., & Tung, K. Y. (2013). Intrusion detection system: A comprehensive review. *Journal of Network and Computer Applications, 36*(1), 16–24.

28. Farnaaz, N., & Jabbar, M. A. (2016). Random forest modeling for network intrusion detection system. *Procedia Computer Science, 89*, 213–217.

29. Yu, Z., Tsai, J. J., & Weigert, T. (2007). An automatically tuning intrusion detection system. *IEEE Transactions on Systems, Man, and Cybernetics, Part B (Cybernetics), 37*(2), 373–384.

30. Bace, R. G. (2000). *Intrusion Detection*. Sams Publishing.

31. Sandhu, R. S., & Samarati, P. (1994). Access control: Principle and practice. *IEEE Communications Magazine, 32*(9), 40–48.

32. Liu, S., & Kuhn, R. (2010). Data loss prevention. *IT Professional, 12*(2), 10–13.

33. Wüchner, T., & Pretschner, A. (2012, November). Data loss prevention based on data-driven usage control. In *2012 IEEE 23rd International Symposium on Software Reliability Engineering* (pp. 151–160). IEEE.

34. Akgul, O., Bai, W., Das, S., & Mazurek, M. L. (2021). Evaluating in-workflow messages for improving mental models of end-to-end encryption. In *30th {USENIX} Security Symposium ({USENIX} Security 21)*.

35. Wagner, N., Şahin, C. Ş., Winterrose, M., Riordan, J., Pena, J., Hanson, D., & Streilein, W. W. (2016, December). Towards automated cyber decision support: A case study on network segmentation for security. In *2016 IEEE Symposium Series on Computational Intelligence (SSCI)* (pp. 1–10). IEEE.

36. Sunner, M. (2005). Email security best practice. *Network Security, 2005*(12), 4–7.

37. Mercaldo, F., Nardone, V., & Santone, A. (2016, August). Ransomware inside out. In *2016 11th International Conference on Availability, Reliability and Security (ARES)* (pp. 628–637). IEEE.

38. Sen, S., Aydogan, E., & Aysan, A. I. (2018). Coevolution of mobile malware and anti-malware. *IEEE Transactions on Information Forensics and Security, 13*(10), 2563–2574.

39. Venkateswaran, R. (2001). Virtual private networks. *IEEE Potentials, 20*(1), 11–15.

40. Sobh, T. S., & Aly, Y. (2011). Effective and extensive virtual private network. *Journal of Information Security, 2*(01), 39.

41. Scott, C., Wolfe, P., & Erwin, M. (1999). *Virtual Private Networks*. O'Reilly Media, Inc.

42. Lamping, U., & Warnicke, E. (2004). Wireshark user's guide. *Interface, 4*(6), 1.

43. Orebaugh, A., Ramirez, G., & Beale, J. (2006). *Wireshark & Ethereal Network Protocol Analyzer Toolkit*. Elsevier.

44. Karygiannis, T., & Owens, L. (2002). *Wireless Network Security*. US Department of Commerce, Technology Administration, National Institute of Standards and Technology.

45. Maynor, D. (2011). *Metasploit Toolkit for Penetration Testing, Exploit Development, and Vulnerability Research*. Elsevier.

46. Orebaugh, A., & Pinkard, B. (2011). *Nmap in the Enterprise: Your Guide to Network Scanning.* Elsevier.
47. Jaswal, N. (2014). *Mastering Metasploit.* Packt Publishing Ltd.
48. Lyon, G. F. (2008). *Nmap Network Scanning: The Official Nmap Project Guide to Network Discovery and Security Scanning.* Insecure. Com LLC (US).
49. Mahajan, A. (2014). *Burp Suite Essentials.* Packt Publishing Ltd.
50. Kim, J. (2020). Burp suite: Automating web vulnerability scanning (Doctoral dissertation, Utica College).
51. Dalziel, H. (2014). *How to Attack and Defend Your Website.* Syngress.
52. Hertzog, R., O'Gorman, J., & Aharoni, M. (2017). *Kali Linux Revealed.* Cornelius, NC: Offensive Security.
53. Najera-Gutierrez, G., & Ansari, J. A. (2018). *Web Penetration Testing with Kali Linux: Explore the Methods and Tools of Ethical Hacking with Kali Linux.* Packt Publishing Ltd.
54. Cohen, F. (1986). Computer viruses (Doctoral dissertation, University of Southern California).
55. Adleman, L. M. (1988, August). An abstract theory of computer viruses. In *Conference on the Theory and Application of Cryptography* (pp. 354–374). New York, NY: Springer.
56. Mohurle, S., & Patil, M. (2017). A brief study of wannacry threat: Ransomware attack 2017. *International Journal of Advanced Research in Computer Science, 8*(5), 1938–1940.
57. O'Gorman, G., & McDonald, G. (2012). *Ransomware: A Growing Menace.* Arizona, AZ: Symantec Corporation.
58. Hampton, N., Baig, Z., & Zeadally, S. (2018). Ransomware behavioural analysis on windows platforms. *Journal of Information Security and Applications, 40*, 44–51.
59. Zhang, Z. K., Cho, M. C. Y., Wang, C. W., Hsu, C. W., Chen, C. K., & Shieh, S. (2014, November). IoT security: Ongoing challenges and research opportunities. In *2014 IEEE 7th International Conference on Service-Oriented Computing and Applications* (pp. 230–234). IEEE.
60. Hassija, V., Chamola, V., Saxena, V., Jain, D., Goyal, P., & Sikdar, B. (2019). A survey on IoT security: Application areas, security threats, and solution architectures. *IEEE Access, 7*, 82721–82743.
61. Fette, I., Sadeh, N., & Tomasic, A. (2007, May). Learning to detect phishing emails. In *Proceedings of the 16th International Conference on World Wide Web* (pp. 649–656).
62. Parno, B., Kuo, C., & Perrig, A. (2006, February). Phoolproof phishing prevention. In *International Conference on Financial Cryptography and Data Security* (pp. 1–19). Berlin, Heidelberg: Springer.

5 Security and Performance Analysis of Advanced Metering Infrastructure in Smart Grid and Use of Blockchain in Security Perspective

Khushboo Gupta and Vinod Kumar
University of Allahabad

CONTENTS

5.1 INTRODUCTION

Traditional power grids have some limitations in efficiency and reliability. In such a case, a next-generation, efficient, scalable, manageable, cost-efficient, interoperable, secure, and reliable system is derived from the traditional power grid, which is known as the smart grid (SG) [1,2]. It uses information technology to achieve efficiency and reliability. SGs consist of the following components that are able to work in the real-time environment: (a) Smart meters (SMs) execute the following operations: (i) Collection of energy information: SMs are used to collect the energy consumption of appliances. (ii) Broadcasting of information: SMs are used to broadcast the power loss information. (iii) Reporting of information: SMs are used to report the real-time pricing information to customers [3]. (b) Sensing devices (Sensors) perform the following tasks: (i) Checking of System Performance: Check whether the system works properly or not. (ii) Detection of Malfunction: Detect any disruption in normal functioning of SGs. (iii) Transmission of Control messages: Transmit the control messages to the control center upon detection of any failure [3]. (c) **Gateways** are the intermediate devices to route the SMs data to the utility and also send control information to the SMs.

Advanced metering infrastructure (AMI) is an integral component of SG. SMs, concentrators, meter data management system (MDMS), neighborhood area networks (NANs), home area networks (HANs), and wide area networks (WANs) are used in the design of AMI of SGs. U.S. Department of Energy has published a report on SG system and in this report, it is stated that the AMI system has grabbed more attention in planning and investment [4].

DOI: 10.1201/9781003267812-5

An AMI consists of systems and networks and it performs the following tasks:

- It collects data from SMs. Further, it performs analysis on the collected data.
- It provides smart management of various power/energy-related applications and services.
- It facilitates two-way communication and smooth functioning of SG.

Due to more use of automation, remote monitoring, and controlling, SG systems become more vulnerable [5]. AMI is a combination of networks and heterogeneous systems that are highly vulnerable to passive and active attacks. In passive attacks, unauthorized entities (attacker) might access the energy consumption pattern of customers that breach the confidentiality of customers. In active attacks, the attacker might modify the information or inject the incorrect information that results in the degradation of system performance [1,3]. Various types of threats such as message replay, man-in-the-middle (MITM), and denial of service (DoS) are identified in the SG communication [6]. Security attacks can cause a significant harm on national security and society. Some adverse conditions caused by security attacks are as follows: (a) It can cause a blackout. (b) It can modify the pricing information. (c) It can change the billing information.

AMI is a heterogeneous system in which different entities/nodes have different storage, communication, and computation capability [7]. Confidentiality, integrity, authentication, and availability must be required for secure communication in AMI of SGs. Encryption and authentication protocols are used to handle the confidentiality and integrity issues [8]. AMI-Sec Task Force, which is used for developing the security requirements in AMI, was formed by the security experts and standards organizations. In AMI of smart girds, AMI Sec Task Force provides guidelines and security controls to those organizations involved in developing or implementing AMI security solutions. An efficient, scalable, and secure key management is required in AMI to generate and update the key for authentication and secure exchange of messages between nodes in AMI [7]. Due to the limitation in storage and computation capacity of SMs, simple cryptographic techniques are used in key management system for key generation and refreshment [8]. Specific refreshment policies are developed to handle the continuous change in demand response project.

A. Global Scenario

At the global level, it is clearly visible that several countries have imposed mandatory legislation for adoption of smart metering networks for achieving the goal of clean energy initiatives [9]. By 2020, under the third energy package resolution program, member countries of the European Union have decided to roll-out nearly 45 million SMs for gas and 200 million for electricity. Research analyzes that almost 72% of consumers in Europe will have SMs for electricity by 2020 [10]. In the United States (U.S.), the Smart Grid Investment Grant Program under the American Recovery and Reinvestment Act of 2009 promotes the smart metering infrastructure installations. In 2016, it is observed that electric utilities are having nearly 70 million smart metering infrastructure installations and 88% are residential users in the U.S. In 2014, it is found that nearly 6 million SMs have been deployed/operated in Canada. It is expected that a significant increase in the use of these devices will be visible in upcoming years. Under the 12th Five-Year Plan on Energy Development and the Strategic Action Plan on Energy Development (2014–2020), nearly 377 million SM base has been installed and started functioning in China by 2020 [11].

B. Characteristics of SGs

Some important characteristics of SGs are as follows [12,13]:

- **Two-Way Communications:** It facilitates two-way communication between utility systems and SMs. It should provide the facility for handling various communication modes such as unicast, broadcast, and multicast.
- **Efficient Energy Management:** It helps in the reduction of power outages and blackouts.

- **Protection against Attacks:** Since in passive attacks, attackers try to access the sensitive information such as pricing information and customer consumption patterns, confidentiality of the system is compromised in passive attacks. In active attacks, integrity of the system is compromised. So, it is required that SG provides protection against various security attacks by using various security mechanisms.
- **Automatic/Self-healing:** It must be capable to detect security attacks and perform self-healing in the case of certain failures.
- **Efficient Integration of Distributed Resources:** It performs the efficient integration of various system resources. It must be scalable to facilitate the addition of a new node in the communication system.
- **Demand Response:** During the peak time of energy demand, utility systems and customers in SGs cut their power usage.
- **Interoperability:** It is a system of interoperable systems. In SGs, transmission of secure, efficient, meaningful, and actionable messages will take place between different systems.
- **Network Integration:** SG uses a variety of communication/transmission networks such as building area networks, HANs, industrial area networks, NANs, and WANs [2].

A high-level architectural model for SGs has been developed by the National Institute of Standards and Technology (NIST) [14,15]. This model divides the SG into seven domains as follows: (a) customers, (b) markets, (c) service providers, (d) bulk generation, (e) transmission/communication, (f) distribution, and (g) operations. This model also acts as a tool for the identification of transmission paths in SGs. Routing protocol will be needed to determine the communication paths for data flows [14]. In Japan, ECHONET Lite is used as a HAN protocol for providing a standard interface for home energy management systems [16].

C. Characteristics of Communication Networks in SGs

Communication networks in SG must ensure the following characteristics for efficient communication [17]:

- **Security in Message Transmission:** In communication networks, security of sensitive information such as energy consumption information and billing information must be ensured. For secure communication, security requirements such as integrity, availability, and confidentiality must be satisfied by the communication system.
- **High Reliability and Availability of System Resources:** It is required that in communication networks, communicating nodes must be available to authorized entities in any situation. Normally, availability of system resources in wired communication is not a problematic situation but in case of wireless or powerline infrastructure it is a challenging issue because communication channels are not fixed during system operations.
- **Support for a Large Number of Communication Nodes/Meters:** Communication network provides support for various meters, utility systems, etc. Communication network must be scalable, that is, new nodes can be included in the system without causing communication overheads.
- **Automatic Management of Critical Situations/Exceptional Situations:** Communication network must be able to perform self-healing and automatic management in case of any critical situation.
- **Large Distance Coverage:** Nodes that require connection to the communication network are distributed in a wide area. To fulfill such requirements, powerline infrastructure and telecommunication systems are used in SGs.
- **Ease of Deployment:** Various mechanisms that facilitate the easy installation of system components are used in communication networks.
- **Maintainability:** After the initial installation, timely maintenance of system components is also required in communication networks to handle new situations and errors.

D. Contributions and Organization of Chapter

The contributions of this chapter are as follows:

- We provide the detail of SG and its advantage over traditional power grids. Further, we describe the role of AMI in SGs.
- We provide a brief introduction of AMI components and their working and also describe the issues and security challenges.
- We illustrate the need for key management system and also discuss the various key management protocols/schemes in AMI.
- We show a comparative analysis of various protocols/schemes on the basis of security, communication, and computation complexity.
- We identify the potential research directions in various areas such as scalability, privacy, and threats.

The organization of this chapter is as follows: In Section 5.2, AMI components, security requirements, security challenges, security attacks, and need for KMS are discussed. In Section 5.3, various key management protocols are discussed. In Section 5.4, the use of blockchain for security in AMI of SG is discussed. In Section 5.5, the comparative analysis of various KMS protocols is presented on the basis of security, computation, and communication cost. In Section 5.6, future research directions are identified. Finally, in Section 5.7, we conclude the chapter.

5.2   BACKGROUND

In this section, we explain the AMI components and their working. Further, we illustrate the security requirements and the various security challenges in AMI. Next, we present a brief introduction of various security attacks, which may cause damage in AMI systems and we also explain the need for key management scheme (KMS) in AMI. Various message transmission modes in AMI are discussed and we illustrate the various issues in the design of KMS.

A. AMI System Components

To perform the various tasks such as data collection, data analysis, resource management, and two-way communications in AMI, the following components are used:

- **SMs:** SMs are solid-state, programmable electrical devices that execute various operations, such as (a) recording of power consumption of users, (b) facilitating bidirectional communication, (c) measurement of voltage, frequency, electricity load, maximum demand, and power factor of the system in real time, (d) monitoring of power quality, (e) reporting of dynamic pricing, (f) remote switch on and off operations, (g) due to demand response (DR) mechanism, enable the users to cut down their power usage at peak time [7,8].

  SMs generally have limited computation and storage capacity [7]. SM uses Zigbee or Wi-Fi-based communication in HAN and in WAN uses the GPRS-based communication. SM can contain a crypto processor, which is able to perform various cryptographic operations for secure communication in AMI systems [18].
- **Gateways:** Gateways perform the following operations: (a) implementation of protocol conversion, (b) establishment of communication between heterogeneous networks, like communication between HAN and NAN or communication between NAN and WAN.
- **HAN:** HAN is a type of local area network that interfaces the appliances, SMs, gateways, control devices, and distributed energy resources [8].
- **NAN:** This network is a combination of multiple HANs.
- **Wide Area Transmission Infrastructure:** It enables the two-way communication/transmission between the customer domain and the utility. For implementation of wide

area communication/transmission infrastructure, various architectures are used such as power line communication systems, IP-based networks, or cellular networks [7]. In architectures, communication/transmission media like radio frequency, optical fiber, and so on [8] is used.

• **MDMS:**

Due to the important role of data transmission, data management is a key aspect in AMI of smart girds [19]. It is a database that is used for (a) durable data storage, (b) management and analysis of metering data for time-based pricing, (c) providing enhanced DR and customer services, and (d) interaction with various systems like outage management system, distribution management system, consumer information system, and geographic information system. Figure 5.1 illustrates the role of all the components.

• **Distributed Energy Resources [15]:**

These are the small-scale electricity generation systems for home use and power storage.

B. Security Requirements in AMI

According to the report published by the AMI-Sec Task Force, the major security requirements for reliable SG communications are as follows:

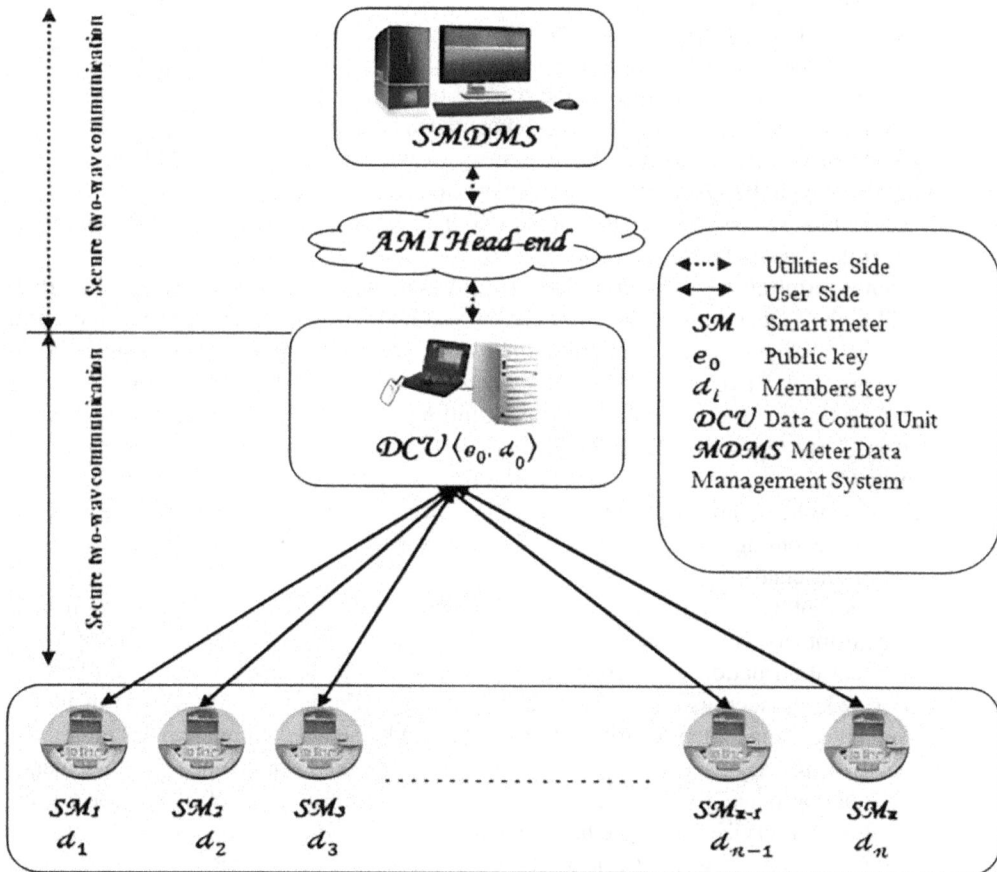

FIGURE 5.1 AMI system components.

- **Confidentiality:** AMI should provide protection of customer consumption data, business information, and transmitted messages against unauthorized access [20]. In AMI, consumption data is created in metering devices and it contains detailed information that can be used by unauthorized entities to gain insights into a customer's behavior [21].
- **Integrity:** The AMI system should ensure the integrity of control messages, business information, and customers' data. Integrity is achieved by providing protection against message alteration via false message injection, message delay, and message replay on communication network [21].
- **Availability:** The AMI should ensure that systems cannot deny access to legitimate users, i.e., system must be available to authorized users on demand. The AMI system compromises the availability aspect of the system if the system resources and data are not available at the time of need [3]. Attacks that are targeting the availability of the system resources are termed as the DoS attacks. It tries to (a) delay the information transmission process, (b) block the communication, and (c) corrupt the information, so that network resources are not available to system devices that need the exchange of information in AMI systems.

   Some other security requirements are as follows:

- **Authentication:** It is the process of determining the true identity of the system users [21]. The AMI system should provide the authentication of sender or receiver in transmission and also authenticates the source of data.
- **Authorization:** Authorization is the process of differentiating between legitimate and illegitimate users based on authentication [21]. It is also known as access control. Access control provides protection against illegitimate access to system resources. Access control refers to restrict the reading or modification of transmitted messages, issuing command to power system by unauthorized entities [21].
- **Auditability:** Auditability is the process which is carried out in AMI of SGs to rebuild the history of the system behavior from previous relevant records of actions performed on it. This security requirement helps to discover the history and to find causes of malfunctions in the system [21].
- **Nonrepudability (Accountability):** In AMI network, accountability ensures that either the sender entity or receiver entity must not deny [3]. With the proof of origin, the receiver entity can later prove the identity of the sender entity if denied. With the proof of delivery, the sender entity can later prove that data were delivered to the receiver entity. It ensures the integrity of billing information and a timely response to control messages in AMI of SG [3].

C. Security Challenges in AMI

   With the significant improvement in use of SMs, security issues/problems relating to AMI of SG are becoming more prevalent in the system [22]. Various security issues are discussed in this chapter:

- **Privacy of Customers:** Energy consumption information of customers can reveal the customers' lifestyle or behavior that results in critical safety issues in the system. Critical information obtained from customer profiling can disclose various information [23,24], such as (a) the number of persons in the house, (b) the type of home appliances, and (c) the security system in the house [3]. SG communications must preserve the privacy of customers' information anywhere and at any time [21]. Customers' satisfaction plays a very crucial role in the expansion of AMI of SGs. Poor power quality and services are some hazards that resist the implementation of AMI systems. Attackers may perform attack on the price signal and control messages obtained at the users end to damage the infrastructure. Data stored in various places for analysis and retransmission and long-distance transmission are some potential areas of data manipulation and theft [3].

- **Security against Cyber Attacks:** Due to the rising number of security attacks (active or passive attacks) against the SG, cyber security has become a crucial aspect in AMI systems that needs to be addressed carefully. Cyber security must address all the intentional attacks that can cause potential damage in the AMI infrastructure and breach the customer privacy and integrity of control signals. Due to the vulnerability in AMI systems, attackers/unauthorized entities obtain access to the control software, communication network, and modify the load condition to make the power grid unstable. In AMI, it is expected that every SM working in the network has its own digital credential. If the attacker attacks a particular SM in the AMI system, the attacker should be unable to obtain access to information of other smart devices by using the compromised meter and penetrate into the AMI communication network [3].
- **Prevention of Power Theft:** In traditional power grids, electromechanical meters (metering devices) were very less secure and easy to manipulate. In electromechanical meters, power theft is identified by the following techniques: (a) establishment of direct connection to distribution system, (b) neutral wire grounding, and (c) embedding a magnet to electromechanical meter, etc. To overcome such problems of electromechanical meters, SMs are used in AMI of SGs. In AMI of SGs, SMs are capable of performing the following operations: (a) recording of zero reading and (b) transmission of information to utility systems through AMI network. Some power theft techniques work in both electromechanical meters and SMs of AMI. Modification/tempering of data can occur: (a) at the time of data collection, (b) at the time of transmission of data, and (c) at the time of storage of data in meters. Various techniques were developed to detect and estimate power theft.

D. Security Attacks in AMI of SG

Due to higher connectivity between the AMI resources/components, various security vulnerabilities are introduced in the AMI of SGs [12]. Some major types of security attacks that are generally tried to exploit the AMI system are as follows:

- **Passive Attack:** In passive attack, attacker's objective is only to get the transmitted information. This attack does not perform any tampering in the transmitted data in AMI of SGs. Traffic analysis and snooping are the passive attacks. Illegitimate access to sensitive information or interception of transmitted messages is known as snooping. Due to the encryption of messages, messages become non-intelligible for the attackers but other sensitive information can access by the attackers by monitoring the traffic.
- **False Data Attack/Data Injection:** It is a type of active attack. In false data attack [1], an intruder tries to insert incorrect data into the AMI network to disrupt the normal functionality of the AMI system. An intruder may present itself as a SM in AMI for gaining the access of AMI network. It can be prevented by using efficient authentication schemes; intruder cannot able to obtain a security key by compromising a SM in AMI. An intruder may also intercept the network traffic and modifies the messages and retransmits them. However, intruder is unable to derive a valid digital signature in the absence of a valid security key that is embedded in SMs.
- **Replay Attack:** In replay attack [1], an attacker first performs the eavesdropping of communication process and grabs some message/information packets. Then at a later time, the attacker retransmits the grabbed message/information packets as it comes from a valid SM. In replay attack, attacker does not require any information of security keys. To prevent such type of attack, timestamp will be added as an additional input to generate a digital signature, so that any outdated data packet has been easily identified and rejected.

E. Need of Key Management in AMI of SGs

In general, the security requirements in AMI of SGs include confidentiality, integrity, availability, authentication, authorization, auditability, and accountability. Protection of privacy of customers' information, authentication of message for meter reading, and DR are some major security aspects that need to be addressed before the deployment of the AMI system. Integrity and privacy issues are solved by using various authentication protocols/schemes and cryptographic methods/techniques. As we know, a large number of devices work in the AMI, to ensure the security of these devices' key management plays a very important role in the AMI of SGs. An efficient, robust, scalable, low overhead key management system provides support for the large scale of smart devices/meters in AMI of SGs. Key management system generates the secure key and distributes it to the utility systems, SMs, and gateways by using transmission links [12]. The key management system in AMI consists of (a) key framework, (b) key generation, (c) key refreshing, (d) key distribution, and (e) storage policies [3].

F. Message Transmission Modes in AMI

Messages in AMI are classified into three different categories based on the transmission modes.

- **Unicast Communication**

  Unicast communication facilitates the transmission of messages from one end to the other end in AMI of SGs. In unicast communication, SMs sends various types of messages such as energy consumption messages, remote load control messages, and future power demand messages to the utility system. These messages can transmit in bidirectional way: from user side to management side or management side to user side [3].

- **Broadcast Communication**

  Broadcast communication facilitates the transmission of messages from one end to all other ends. The real-time pricing information of electricity sent by the utility system to SMs in AMI is an example of broadcast communication [3].

- **Multicast Communication**

  Multicast communication facilitates the transmission of messages from one point to subset of other points in the AMI system. The remote-control messages send by the utility system to SMs that subscribed the same DR project is an example of multicast communication [3]. Group members in multicast communication should be updated in a certain time depending on the real-time situation as participants in a DR project are not fixed.

G. Issues in the Design of KMS

The following are the challenges in the design of KMS based on the messages in AMI:

- The KMS should provide support for the different communication modes: unicast, broadcast, multicast, and hybrid communications. To handle all types of transmission modes, policies for key generation, refreshing, and distribution will be created, respectively [8].

- The devices that work at the user level based on the embedded system have limited storage and computation ability. The limitation in computation ability of devices at the user level affects the key generation and refreshing techniques. The number of times of key distribution should be reduced as there is time limitation in message transmission. As there is limited storage available at user-level devices, key and other data required to be stored at user-level devices in AMI should also be reduced [8].

- Users' participation in DR project is changed continuously in real time. The users can choose different DR projects according to their needs. A group can be formed by all the users participating in the same DR project. Participation of members in a group changes dynamically, so we have to pay focus on forward and backward security in multicast transmission. New participants who join in the DR project of AMI must be included in a group, and for transmission/communication, new secure keys and related data should be shared with every participant in a group. Participants in a group who leave the DR project in AMI cannot receive the messages [8].

5.3 KEY MANAGEMENT PROTOCOLS

Key management systems play a very important role in AMI of SGs. It performs the following operations: (a) key generation, (b) key distribution, (c) key refreshing or renewal, etc. For secure and efficient key management, various techniques have been developed. Some important approaches/schemes proposed by various authors are as follows:

- **Multi-Key Graph-Based Approach**

  Due to the efficient and simple implementation, key graph-based approach is the most widely known key management technique. A secure, efficient, and scalable key management approach is proposed in Refs. [25–27] for different transmission modes. The approach based on multi-key graph provides support for concurrent management of multiple DR projects for each user. Some specific methods for secure exchange of key are used in the development of individual keys between the SMs and MDMS in AMI. Periodic refreshment process is performed on the individual keys. Individual keys are used for (a) securing unicast transmission between SMs and MDMS and (b) generating the multi-group key graph for secure multicast transmission.

  A group key is refreshed periodically for DR project in AMI. Secure channels are used for the traversal of generated group key for each SM. SKM approach/scheme count on multi-key graph works on the following assumptions:

  1. Management side is denoted by the MDMS. It is accountable for the key generation, rekeying, and is secured from various potential security attacks.
  2. A default DR project is essential for all the users in AMI of SGs. MDMS uses the default DR project to broadcast the control messages or other information to all the users of AMI system.
  3. Any user can subscribe or unsubscribe the any DR project, except the specific default DR project.

     In this method, one-time function tree is used which is an advancement of the logical key hierarchy (LKH) protocol. LKH is proposed to ensure the scalability for SGs with DR projects. For each DR project in AMI, a key tree is used in LKH. In this, each node retains a copy of secure keys of relevant nodes. The MDMS and every user can perform computation of keys in internal nodes.

- **Two-Level Security Method:** In Ref. [28], a two-level security method based on two partially trusted simple servers has been proposed that implement the security without increasing the packet overhead. In this method, two security levels are as follows:

  1. **Data Encryption:** two operations have been proposed at this level—encryption by asymmetric key (public key cryptography) and randomization of data packets.
  2. Node-to-node authentication

     Data encryption between the SM and control center is handled by one server and the randomization of data transmission is handled by another server. In this method, received signal strength is used for localization of a new meter with respect to other meters whose locations are already known and one class support vector machine (OCSVM) is used for node-to-node authentication. OCSVM is a machine-learning technique for node-to-node authentication. Node-to-node authentication by OCVSM utilizes four variables: node/meter id, frequency of data reception from a particular node/SM, size of packet, and position/location of SM.

- **Enhanced Identity-Based Cryptography:** In Refs. [6,29], an efficient key management-based protocol has been proposed for secure transmission in AMI of SGs using public key infrastructure (PKI). The proposed method is capable of providing protection against various threats, such as replay, MITM, DoS, and impersonation. In the United States of America, NIST develops various SG transmission/communication-related guidelines and standards. NIST recommends using PKI for secure and efficient SG transmission/communication. This method facilitates efficient and secure authentication key management. The primary focus of this technique/method is to provide key management and authentication over the AMI of SG.

An efficient and secure smart grid key management protocol/scheme and smart grid mutual authentication (SGMA) scheme/protocol are proposed in this method. By using passwords, SGMA facilitates the efficient mutual authentication between the security and authentication server (SAS) and SMs in the SG. The network overhead caused by the control packets for key management is reduced by the proposed method. In this method, efficiency is achieved from the key refreshment protocol. In the key refreshment protocol, SAS provides periodic broadcasting of a new key generation to refresh the private/public key pairs of all the meters and any needed multicast security keys.

- **Physically Unclonable Function (PUF)-Based Approach:** In Ref. [30], a method based on physically unclonable function has been proposed for furnishing strong hardware-based authentication of SMs and secures key management in AMI of SGs. Multiple modes are used in the communication path for the exchange of messages between the SMs and utility servers. Different communication protocols are used by the nodes in AMI. Link-level security is provided by most of the protocols. Link-level security is not sufficient to provide protection of message transmission against security attacks. SMs are vulnerable to security attacks such as MITM in the absence of a proper authentication mechanism. In this method, cost-efficient PUF devices are used to provide hardware-based strong authentication. Hardware-based strong authentication provides resistance against spoofing attack (passive attack). Generation or regeneration of symmetric keys and access level passwords is performed by the PUF devices' hardware-based one-way function for SMs in AMI-systems. Strong protection against leakage of secure key is facilitated by the physically unclonable function-based secret generation function as the master key is not resided in memory.

  In AMI, many nodes are existed and different transmission modes such as unicast (one to one) and multicast (one to many) are used by the nodes for communication, it is required that KMS should be efficient to support different transmission modes.

  A Broadcast Group Key Management (BGKM) approach/method [31,32] along with a strong authentication scheme is also proposed in this work. The BGKM is a specific/distinct type of group key management approach. BGKM provides an efficient communication scheme for a subset of nodes. As only private key cryptography (symmetric key cryptography) is used in BGKM, which makes BGKM computationally cost-efficient. In communication networks, every SM has a distinctive secret that permits SM to obtain the group key. The addition of a new SM or removal of an existing SM without affecting the other SMs is efficiently handled by the BGKM approach.

- **Hybrid Encryption Technique:**

  In Refs. [7,12], a hybrid encryption-based technique is proposed that incorporates both public key (asymmetric key) and private key (symmetric key) encryptions for secure SG transmission. Hybrid encryption system utilizes the strength of asymmetric key cryptography and efficiency of symmetric key cryptography. Hybrid encryption-based technique is used to provide a light way KMS that reduces the overhead of key generation, distribution, and refreshment or renewal in AMI networks. In the proposed system, the main building blocks are as follows:

  1. **Elliptic Curve Integrated Encryption Scheme (ECIES):** It is used for the key encapsulation cryptosystem.
  2. **Advanced Encryption Scheme:** It is used for the data encapsulation system. Advance Encryption standard AES uses secure block size and different key sizes such as 128 bits, 196 bits, and 256 bits.

The proposed method ensures the data confidentiality, integrity, and authenticity and provides protection against various attacks such as false data attack and replay attack.

The proposed system has the following modules:

A. **Symmetric Encryption Module:** In this module, AES-128 with an arbitrary key (K) is used to encrypt the messages in AMI of SGs. An arbitrary key is generated by the random generation unit. This module works as a data encapsulation system. Data encapsulation system is a symmetric key encipherment system. In symmetric key cryptosystem, plain text is converted into cipher text by using a symmetric key algorithm with an arbitrary key. The same key is used for encryption of AMI messages and decryption of AMI messages in symmetric key cryptography. In this cryptography, the same key is shared between the legitimate entities. In symmetric key cryptosystem, plain text is converted into cipher text by using a symmetric key algorithm with an arbitrary key.

B. **Asymmetric Encryption Module:** Encryption of an arbitrary key which is used by symmetric key module is done by this module. It works as a key encapsulation system. Key encapsulation system is an asymmetric key encipherment system. It is used to perform the encryption of arbitrary keys by using a suitable asymmetric key algorithm (public key algorithm). Results of symmetric encryption module and asymmetric encryption module are merged to form secure messages for transmission in AMI networks.

C. **Message Integrity Module:** It produces the integrity code, which is used for the detection of any tampering in the messages.

5.4 BLOCKCHAIN IN AMI OF SG FOR SECURITY

Blockchain is a distributed, immutable ledger technology, which is capable of transferring and storing data across various devices within the system without depending on third party [33]. Blockchain technology serves as a base for the widely known virtual currency "Bitcoin" [34]. By using blockchain technology, peer-to-peer transaction of local data can be facilitated within the SG and each peer has its copy of the ledger. Unlike the traditional way to store transaction data in a single data control center, in blockchain technology these transaction data are stored in a distributed manner in all or selected nodes which take part in the network operations. Integrity of transaction records, decentralized authentication, trust, and transparency between communicating nodes are provided by the blockchain technology [35]. Figure 5.2 illustrates the peer-to-peer transaction in blockchain.

In blockchain, the consensus algorithm works on each node for verification of each transaction and if consensus is validated by more than half of the participating nodes, transaction records are stored into the block. The addition of a new block in blockchain is called mining and nodes that take part in this process are called mining nodes. Each block is linked with only those blocks that are placed before and after it and cryptographic one-way hash function (irreversible function) are used for linking of adjacent blocks. Each block in blockchain mainly consists of block header and block body. The block header mainly contains hash of previous block, timestamp, version, and Merkle root hash. Hash functions are used for computation of timestamp, Merkle tree, and encryption in blockchain. Timestamp shows the time when block is created and version indicates the format and type of data which are contained in block [33]. Merkel root hash represents the combined hash value for all the transactions. Some well-known consensus mechanisms are proof of work, proof of service, proof of stake, and proof of importance [36]. The replication of data and peer-to-peer data transfer between the devices in distributed manner provides the protection to SG from suffering single point of failure and also ensures high availability. Due to blockchain technology, data is nearly immutable in SG, which ensures the data confidentiality, integrity, and availability. Adversaries can never be able to manipulate the data in blockchain unless he/she acquires more than half of the devices in the whole system. This property provides high auditability in SG and it also makes non-repudiation

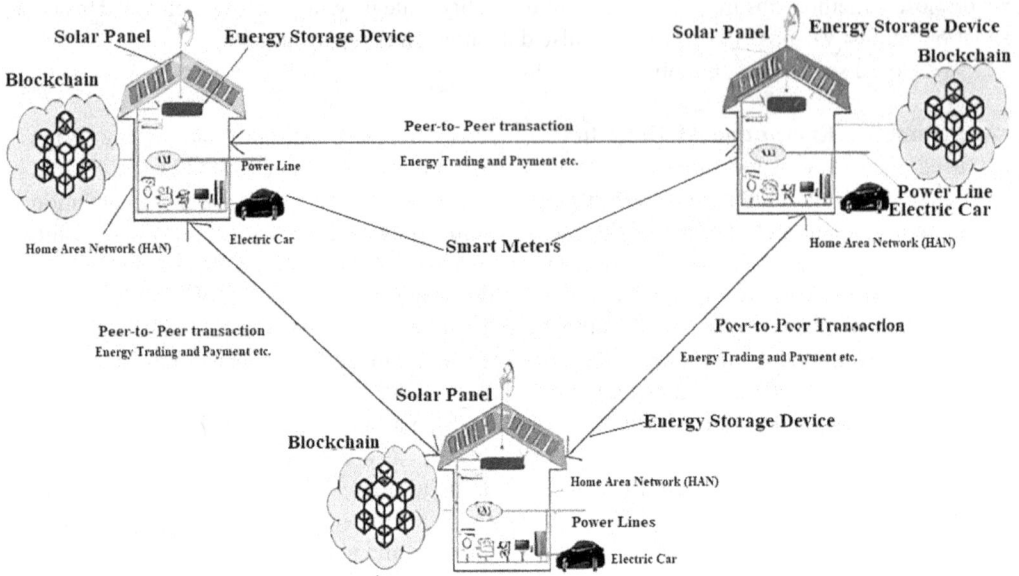

FIGURE 5.2   Peer-to-peer transaction in blockchain.

attacks nearly impossible. Authorization and authentication levels are enhanced by using public key cryptography in blockchain-based smart gird. These enhanced levels are used to ensure the privacy of customers and integrity of data. Blockchain technology provides cyber secured environment for fast low-cost computation using distributed application services and smart contract [37].

5.5   COMPARATIVE ANALYSIS

This section provides a comparative study of the works that are carried out for an effective key management system in AMI of SGs on the basis of security, computation, and communication times.

Various symbols used in key management protocols are shown in Table 5.1. Next, in Table 5.2, we show a comprehensive study of various security protocols on the basis of certain security attacks. In security attacks, we consider forward and backward secrecy, replay attack, collusion attack, man-in middle attack, etc. Further, Table 5.3 illustrates the computation complexity/cost/time on key assignment. Next, in Table 5.4, the comparative study of computation time on device join is shown and then in Table 5.5 we show the comparative study of computation time on device left. In Table 5.6, we provide the insight of communication time of various protocols.

A. Security Analysis

Table 5.2 presents the security analysis of various security protocols/schemes. In this section, we present the comparison of security protocols by taking into consideration various security aspects. These aspects are backward and forward secrecy, MITM attack, replay attack, impersonation attack, etc.

The forward secrecy is a specific feature that ensures that new users participating in DR project must be unable to gain previously used secure keys and messages. In group key management, forward secrecy ensures that evicted users must not gain any information of the new group key. The backward secrecy is also an important feature that ensures that evicted users from DR project must not be able to access future keys and messages. In group key management, backward secrecy ensures that new users must not have any

TABLE 5.1
Notations Used for Computation Time Analysis

Notations	Descriptions
n	No. of SMs
T_m	Time taken in multiplication $\left(T_m \approx 50.3 \text{ ms}\right)$
T_{bp}	Time taken in bilinear pairing $\left(T_{bp} \approx 5.8 \text{ ms}\right)$
T_h	Time taken in operations like HASH/XOR/one-way function (OFT) MOD $\left(T_h \approx 0.5 \text{ ms}\right)$
T_{Key}	Time taken in generating a key $\left(T_{Key} \approx 3.1 \text{ ms}\right)$
T_{mm}	Time taken in modular multiplication $\left(T_{mm} \approx 4.7 \text{ ms}\right)$
T_{en}	Encryption function computation time $\left(T_{en} \approx 8.7 \text{ ms}\right)$
$N_S\, u_i$	No. of demand response projects to which u_i take subscription
N_{DR}	No. of demand response projects
Q	$N_{DR} - N_S\, u_i$
h_i	Height of the new demand response project
m_j	No. of jth demand response project members
$\log_2 m_j$	Height of the one-way function tree (OFT)
$\mathcal{L}$	Size of keys and other secret parameters in bits

TABLE 5.2
Security Analysis

Security Attacks Approach/ Protocol ⇩	Authentication without Trusted Authority	Forward and Backward Secrecy	Key Confirmation	Protection against Man-in Middle Attack	Protection against Replay Attack	Protection against Impersonation Attack	Protection against Collusion Attack
Liu et al. [8]	✓	✓	✗	✓	✓	✓	✓
Wan et al. [7]	✓	✓	✓	✓	✓	✓	✗
Wan et al. [7]	✓	✓	✓	✓	✓	✓	✗
Xia et al. [18]	✗	✗	✗	✓	✓	✗	✓
Wu et al. [38]	✗	✗	✓	✗	✓	✓	✓
Kumar et al. [39]	✓	✓	✓	✓	✓	✓	✓

TABLE 5.3
Computation Time/Complexity Analysis on Key Assignment

Approach/Protocol	Computation Complexity for Key Assignment			
	SM		Smart Device	
Liu et al. [8]	nT_{en}	$\approx 8.7n$ ms	T_{en}	≈ 8.7 ms
Wan et al. [7]	$n\left(T_{bp} + T_m\right)$	$\approx 56.1n$ ms	$T_{bp} + T_m$	≈ 56.1 ms
SKM+ [7]	$n\left(2T_{bp} + T_m\right)$	$\approx 61.9n$ ms	$3T_m$	≈ 15.09 ms
Benmalek et al. [25]	$n\left(T_{bp} + T_m\right) + nT_{key}$	$\approx 59.2n$ ms	$T_{bp} + T_m$	≈ 56.1 ms
Benmalek et al. [26]	$n\left(T_{en} + T_h\right) + nT_{key}$	$\approx 12.3n$ ms	$T_{bp} + T_m$	≈ 56.1 ms
Benmalek et al. [27]	$n\left(T_{en} + h\right) + nT_{key}$	$\approx 12.3n$ ms	$T_{bp} + T_m$	≈ 56.1 ms
Kumar et al. [39]	nT_{key}	$\approx 3.1n$ ms	0	≈ 0.0 ms

TABLE 5.4

Computation Time/Complexity Analysis on Smart Device Joins the Group

Protocols	Computation Complexity at Addition of Smart Device			
	SM		Smart Device	
Liu et al. [8]	$(4n+5)T_h+(n+2)T_{en}$	$\approx 10.7n+19.9$ ms	$4nT_h+T_{en}$	$\approx 2n+8.7$ ms
Wan et al. [7]	$T_{key}+n(T_{en}+2T_h)$	$\approx 9.7\log_2 n+3.1$ ms	$n(T_{en}+T_h)$	$\approx 9.2\log_2 n$ ms
SKM+[7]	$T_{Key}+(n-1)(2T_h)$	$\approx \log_2 n+2.1$ ms	$(n-1)(2T_h)$	$\approx \log_2 n-1.0$ ms
Benmalek et al. [25]	$\log_2(T_{en}+T_h)+nT_{key}$	$\approx \log_2 9.2+3.1n$ ms	$\log_2(T_{en}+T_h)$	$\approx \log_2 9.2$ ms $= 3.2016$ ms
Benmalek et al. [26]	$\log_2(T_{en}+T_h)+nT_{key}$	$\approx \log_2 9.2+3.1n$ ms	$\log_2(T_{en}+T_h)$	$\approx \log_2 9.2=3.2016$ ms
Benmalek et al. [27]	$\log_2(T_{en}+T_h)+nT_{key}$	$\approx \log_2 9.2+3.1n$ ms	$log_2(T_{en}+T_h)$	$\approx \log_2 9.2=3.2016$ ms
Kumar et al. [39]	T_{key}	≈ 3.1 ms	0	≈ 0.0 ms

TABLE 5.5

Computation Time Analysis on Smart Device Evicts from the Group

Protocols	Computation Complexity/time at eviction of smart device			
	SM		Smart Device	
Liu et al. [8]	$(4n+5)T_h+(n+2)T_{en}$	$\approx 10.7n+19.9$ ms	$4nT_h+T_{en}$	$\approx 2n+8.7$ ms
Wan et al. [7]	$T_{key}+n(T_{en}+2T_h)$	$\approx 9.7\log_2 n+3.1$ ms	$n(T_{en}+T_h)$	$\approx 9.2\log_2 n$ ms
SKM+ [7]	$(n-1)(T_{en}+2T_h)$	$\approx 9.7\log_2 n-9.7$ ms	$(n-1)(T_{en}+T_h)$	$\approx 9.2\log_2 n-9.2$ ms
Benmalek et al. [25]	$\log_2(T_{en}+T_h)+nT_{key}$	$\approx \log_2 9.2+3.1n$ ms $= 3.1n+3.20$ ms	$\log_2(T_{en}+T_f)$	$\approx \log_2 9.2=3.2016$ ms
Benmalek et al. [26]	$\log_2(T_{en}+T_h)+nT_{key}$	$\approx \log_2 9.2+3.1n$ ms $= 3.1n+3.20$ ms	$\log_2(T_{en}+T_f)$	$\approx \log_2 9.2=3.2016$ ms
Benmalek et al. [27]	$\log_2(T_{en}+T_h)+nT_{key}$	$\approx \log_2 9.2+3.1n$ ms $= 3.1n+3.20$ ms	$\log_2(T_{en}+T_f)$	$\approx \log_2 9.2=3.2016$ ms
Kumar et al. [39]	0	≈ 0.0 ms	0	≈ 0.0 ms

information about previous group keys. An efficient security protocol must provide protection against replay, man-in middle, and other security attacks.

B. Computation Time/Complexity Analysis

The study of computation time of various security protocols is divided into three parts: (a) key assignment time complexity, (b) time complexity of device joins, and (c) time complexity of device left. These computations are performed by both SM and smart device. The result of Tables 5.3–5.5 shows that the scheme proposed in Ref. [39] is much more efficient in comparison to other protocols/schemes proposed in Refs. [7,8,25–27].

C. Communication Cost Analysis

The communication overhead of various security protocols is presented in Table 5.6. This analysis is done using unicast and multicast communication costs of device addition and device left. Scalable multi-group key management protocols proposed in Refs. [25–27] provide support for various transmission modes such as multicast, broadcast, and unicast transmissions. The proposed protocol/scheme in Ref. [8] also handles multicast, broadcast, and unicast communications/transmissions. Protocol discussed in Ref. [8] is not efficient in terms of scalability and also there is a possibility of packet loss during transmission.

TABLE 5.6
Communication Cost Analysis

Protocols	Communication Cost for Addition of a Smart Device			Communication Cost for Eviction of a Smart Device		
	Unicast Transmission	Broadcast Transmission	Overall Cost (in bits)	Unicast Transmission	Broadcast Transmission	Overall Cost (in bits)
Liu et al. [8]	$2n\mathcal{L}$	0	$\approx 1{,}024n$ bits	$2n\mathcal{L}$	0	$\approx 1{,}024n$ bits
Wan et al. [7]	$n\mathcal{L}$	$n\mathcal{L}+n$	$\approx 1{,}025\log_2 n$ bits	0	$n\mathcal{L}+n$	$\approx 513\log_2 n$ bits
SKM+ [7]	$\mathcal{L}$	n	$\approx 512+\log_2 n$ bits	0	$n\mathcal{L}+n$	$\approx 513\log_2 n$ bits
Benmalek et al. [25]	$h_j\mathcal{L}+\log_2 N_{DR}$	$\mathcal{L}+\log_2 N_{DR}$	$\approx 512\,h_j+512+2\log_2 N_{DR}$ bits	$\left(h_j+N_{DR}-1\right)\mathcal{L}$	$\left(h_j+2N_S\,u_i+Q+h_k\right)\mathcal{L}+\log_2 N_{DR}$	$\approx \left(2h_j+N_{DR}+2N_S\,u_i+Q+h_k-1\right)512+\log_2 N_{DR}$ bits
Benmalek et al. [26]	$2h_j\mathcal{L}+\log_2 N_{DR}$	$\mathcal{L}+\log_2 N_{DR}$	$\approx 1{,}024\,h_j+512+2\log_2 N_{DR}$ bits	$\left(2h_j+N_{DR}-1\right)\mathcal{L}$	$\left(h_j+2N_S\,u_i+Q+2h_k\right)\mathcal{L}+\log_2 N_{DR}$	$\approx \left(3h_j+N_{DR}+2N_S\,u_i+Q+h_k-1\right)512+\log_2 N_{DR}$ bits
Benmalek et al. [27]	$h_j\mathcal{L}+\log_2 N_{DR}$	$\mathcal{L}+\log_2 N_{DR}$	$512\,h_j+512+2\log_2 N_{DR}$ bits	$\left(h_j+N_{DR}-1\right)\mathcal{L}$	$\left(h_j+2N_S\,u_i+Q+h_k\right)\mathcal{L}+\log_2 N_{DR}$	$\approx \left(2h_j+N_{DR}+2N_S\,u_i+Q+h_k-1\right)512+\log_2 N_{DR}$ bits
Kumar et al. [39]	$\mathcal{L}$	0	≈ 512 bits	0		

5.6 FUTURE RESEARCH DIRECTIONS

Despite various efforts that are carried out for ensuring security in AMI, there remain various challenges. These challenges present the future directions for research. Figure 5.3 illustrates the future directions.

Since millions of users involved in the SG, scalability becomes an important aspect that needs to be addressed. A solution feasible for small scale may not be feasible when applying into large scale. So, one major challenge which is found in SG is how to create a scalable system to handle a large amount of data which are generated by sensors in AMI. These transmission systems provide ease in adding and evicting the smart devices. Communication in SG is very complex due to the integration of interdisciplinary areas, dynamic, non-deterministic, and heterogeneous systems. So, for coexistence of heterogeneous communication/transmission technologies and standards in SG, interoperability and standardization become the challenges.

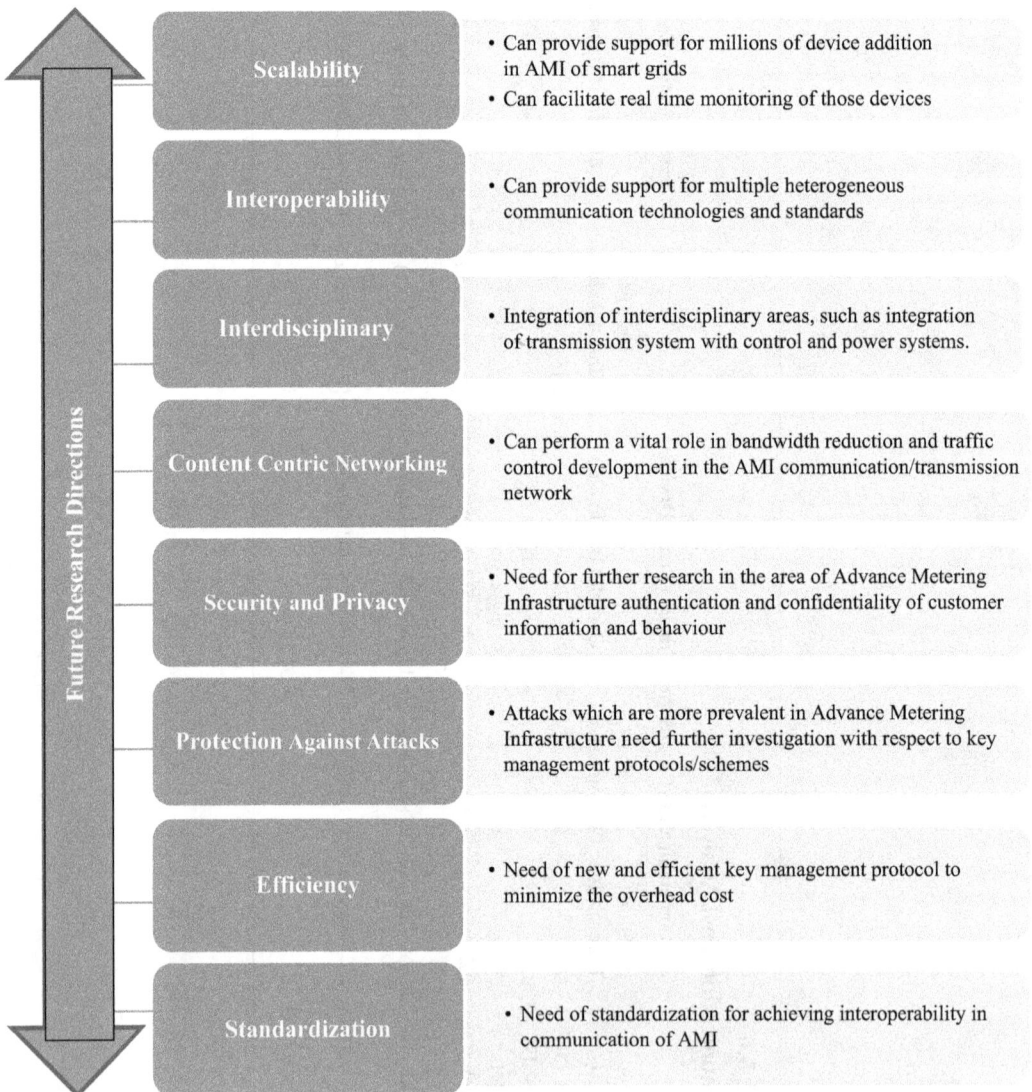

FIGURE 5.3 Future research directions.

A potential research area is also found in the use of Content Centric Networking (CCN), which is an emerging paradigm for future internet in AMI for bandwidth reduction and traffic control development. CCN provides support for multicast communication/transmission and implementation of in-network caching. Due to excessive use of remote monitoring and controlling, automation, interconnection, and integration of systems, smart gird system becomes more vulnerable. AMI is a heterogeneous system that is highly vulnerable to security attacks. DoS attack in AMI can disrupt the communication link and reception of message by flooding. Data integrity attacks alter the data timings and data, which disrupts the normal functioning of AMI systems. Thus, further exploration is required for prevalent attacks

5.7 CONCLUSION

In this chapter, we provide a comprehensive study of various key management approaches, which are used in AMI of SG. Initially, we provide a basic introduction of SGs and its characteristic, AMI components, and AMI structure. Further, we provide a brief introduction of security requirements in AMI of SGs and what challenges are found in security of AMI. Next, we explain the role of key management schemes/protocols in advance metering infrastructure of SGs. Then, we present a comparative analysis of various key management approaches on the basis of security attacks and computation complexity. This analysis of key management approaches brings new research areas for further exploration in AMI of SG.

REFERENCES

1. Deng, P., and Yang, L. "A secure and privacy-preserving communication scheme for advanced metering infrastructure." *2012 IEEE PES Innovative Smart Grid Technologies (ISGT)*. IEEE, 2012.
2. Gao, J., et al. "A survey of communication/networking in smart grids." *Future Generation Computer Systems* 28.2(2012): 391–404.
3. Ghosal, A., and Conti, M. "Key management systems for smart grid advanced metering infrastructure: A survey." *IEEE Communications Surveys & Tutorials* 21.3(2019): 2831–2848.
4. Kamto, J., et al. "Light-weight key distribution and management for advanced metering infrastructure." *2011 IEEE GLOBECOM Workshops (GC Wkshps)*. IEEE, 2011.
5. Yan, Y., et al. "A survey on smart grid communication infrastructures: Motivations, requirements and challenges." *IEEE Communications Surveys & Tutorials* 15.1(2012): 5–20.
6. Nicanfar, H., et al. "Efficient authentication and key management mechanisms for smart grid communications." *IEEE Systems Journal* 8.2(2013): 629–640.
7. Wan, Z., et al. "SKM: Scalable key management for advanced metering infrastructure in smart grids." *IEEE Transactions on Industrial Electronics* 61.12(2014): 7055–7066.
8. Liu, N., et al. "A key management scheme for secure communications of advanced metering infrastructure in smart grid." *IEEE Transactions on Industrial Electronics* 60.10(2012): 4746–4756.
9. Barbiroli, M., et al. "Smart metering wireless networks at 169 MHz." *IEEE Access* 5(2017): 8357–8368.
10. Union, E. "Directive 2009/72/EC of the European parliament and of the council of 13 July 2009 concerning common rules for the internal market in electricity and repealing directive 2003/54/ec." *Official Journal of the European Union* 211(2009): 55–93.
11. Sun, Q., et al. "A comprehensive review of smart energy meters in intelligent energy networks." *IEEE Internet of Things Journal* 3.4(2015): 464–479.
12. Khasawneh, S., and Kadoch, M. "Hybrid cryptography algorithm with precomputation for advanced metering infrastructure networks." *Mobile Networks and Applications* 23.4(2018): 982–993.
13. George, N., Nithin, S., and Kottayil, S. K. "Hybrid key management scheme for secure AMI communications." *Procedia Computer Science* 93(2016): 862–869.
14. Sabbah, A. I., El-Mougy, A., and Ibnkahla, M. "A survey of networking challenges and routing protocols in smart grids." *IEEE Transactions on Industrial Informatics* 10.1(2013): 210–221.
15. Arnold, G. W., et al. "NIST framework and roadmap for smart grid interoperability standards, release 1.0." (2010).

16. Tanaka, Y., et al. "A security architecture for communication between smart meters and HAN devices." *2012 IEEE Third International Conference on Smart Grid Communications (SmartGridComm)*. IEEE, 2012.

17. Sauter, T., and Lobashov, M. "End-to-end communication architecture for smart grids." *IEEE Transactions on Industrial Electronics* 58.4(2010): 1218–1228.

18. Xia, Jinyue, and Wang, Y. "Secure key distribution for the smart grid." *IEEE Transactions on Smart Grid* 3.3(2012): 1437–1443.

19. Zhou, J., Hu, R. Q., and Qian, Y. "Scalable distributed communication architectures to support advanced metering infrastructure in smart grid." *IEEE Transactions on Parallel and Distributed Systems* 23.9(2012): 1632–1642.

20. Cleveland, F. M. "Cyber security issues for advanced metering infrastructure (AMI)." *2008 IEEE Power and Energy Society General Meeting-Conversion and Delivery of Electrical Energy in the 21st Century*. IEEE, 2008.

21. Yan, Q., Yan, Y., Qian, Y., Sharif, H., & Tipper, D. "A survey on cyber security for smart grid communications." *Communications Surveys and Tutorials, IEEE* 14(2012): 998–1010.

22. Otuoze, A. O., Mustafa, M. W., and Larik, R. M. "Smart grids security challenges: Classification by sources of threats." *Journal of Electrical Systems and Information Technology* 5.3(2018): 468–483.

23. Murrill, B. J., Liu, E. C., and Thompson, R. M. *"Smart Meter Data: Privacy and Cybersecurity."* Congressional Research Service, Library of Congress, 2012.

24. Molina-Markham, A., et al. "Private memoirs of a smart meter." *Proceedings of the 2nd ACM Workshop on Embedded Sensing Systems for Energy-Efficiency in Building*, 2010.

25. Benmalek, M., Challal, Y., and Bouabdallah, A. "Scalable multi-group key management for advanced metering infrastructure." *2015 IEEE International Conference on Computer and Information Technology; Ubiquitous Computing and Communications; Dependable, Autonomic and Secure Computing; Pervasive Intelligence and Computing*. IEEE, 2015.

26. Benmalek, M., and Challal, Y. "eSKAMI: Efficient and scalable multi-group key management for advanced metering infrastructure in smart grid." *2015 IEEE Trustcom/BigDataSE/ISPA*. Vol. 1. IEEE, 2015.

27. Benmalek, M., and Challal, Y. "MK-AMI: Efficient multi-group key management scheme for secure communications in AMI systems." *2016 IEEE Wireless Communications and Networking Conference*. IEEE, 2016.

28. Parvez, I., Aghili, M., and Sarwat, A. "Key management and learning based two level data security for metering infrastructure of smart grid." arXiv preprint arXiv:1709.08505 (2017).

29. Nicanfar, H., and Leung, V. C. M. "EIBC: Enhanced identity-based cryptography, a conceptual design." *2012 IEEE International Systems Conference SysCon 2012*. IEEE, 2012.

30. Nabeel, M., et al. "Scalable end-to-end security for advanced metering infrastructures." *Information Systems* 53(2015): 213–223.

31. Zou, X., Dai, Y-S., and Bertino, E. "A practical and flexible key management mechanism for trusted collaborative computing." *IEEE INFOCOM 2008-The 27th Conference on Computer Communications*. IEEE, 2008.

32. Shang, N., et al. "A privacy-preserving approach to policy-based content dissemination." *2010 IEEE 26th International Conference on Data Engineering (ICDE 2010)*. IEEE, 2010.

33. Alladi, T., Chamola, V., Rodrigues, J. J., and Kozlov, S. A. "Blockchain in smart grids: A review on different use cases." *Sensors* 19.22(2019): 4862.

34. Kim, S. K., and Huh, J. H. "A study on the improvement of smart grid security performance and blockchain smart grid perspective." *Energies* 11.8(2018): 1973.

35. Pal, O., Singh, S., and Kumar, V. "Blockchain network: Performance optimization." *Applications of Artificial Intelligence and Machine Learning* (pp. 677–686). Springer, Singapore, 2022.

36. Agung, A. A. G., and Handayani, R. "Blockchain for smart grid." *Journal of King Saud University-Computer and Information Sciences* (2020).

37. Zhuang, P., Zamir, T., and Liang, H. "Blockchain for cybersecurity in smart grid: A comprehensive survey." *IEEE Transactions on Industrial Informatics* 17.1(2020): 3–19.

38. Wu, D., Member, S., and Zhou, C. "Fault–tolerant and scalable key management for smart grid." *IEEE T Smart Grid* 2.2(2011): 375–381.

39. Kumar, V., Kumar, R., and Pandey, S. K. "LKM-AMI: A lightweight key management scheme for secure two way communications between smart meters and HAN devices of AMI system in smart grid." *Peer-to-Peer Networking and Applications* 14.1(2021): 82–100.

6 Computation and Storage Efficient Key Distribution Protocol for Secure Multicast Communication in Centralized Environments

Vinod Kumar and Khushboo Gupta
University of Allahabad

Om Pal
University of Delhi

Rajendra Kumar
Jamia Millia Islamia

Pradeep Kumar Tiwari
University of Allahabad

Narendra Kumar Updhyay and Mukesh Kumar Bhardwaj
Harlal Institute of Management & Technology

CONTENTS

DOI: 10.1201/9781003267812-6

6.1   INTRODUCTION

In today's internet age, data transmission from one source to multiple receivers is widely used in many applications [1] such as audio/video broadcasting, pay-per-view and online teaching, where common secret data like key is transmitted to a group of members. In many other services like banking, military and mobile communications, privacy is the top priority. To provide one to many secure communication (i.e., single server to multiple nodes) is a challenge. In multicast communication, a secret key is constructed by a key server (KS) using the participants' shares [1–3]. Groups in communication may be static and dynamic. In the static groups, the users of the group remain unchanged; therefore, a static GK is transmitted to the users of the group [1,5]. But, in the dynamic group, the membership is changed frequently. In order to achieve the secrecy in the dynamic groups, the GK needs to be refreshed frequently whenever the member leaves and joins the group. The key updating process is known as rekeying. In group/multicast communication, maintaining forward and backward secrecy is a challenge [8].

In this chapter, an architecture for key distribution in centralized environments is proposed. Further, an efficient key distribution protocol that achieves backward and forward secrecy in the group is also proposed. In the proposed protocol, there is no need of balancing the key tree whenever the member joins/leaves the group. Moreover, our protocol ensures security against various cryptographic attacks and enhanced the performance by minimizing the computation, communication and storage overheads at KS and member sides.

This chapter is organized as follows. Section 6.2 presents related work on key distribution along with comparative analysis on the basis of performance. We propose an architecture for key distribution in centralized environments in Section 6.3. We introduce our Centralized Session Key Distribution (CSKD) protocol in Section 6.4. Section 6.5 presents the security analysis of our protocol. The performance analysis of our CSKD and other competitive protocols is presented in Section 6.6. The implementation results are illustrated in Section 6.7. Finally, we draw conclusion of this chapter in Section 6.8.

6.2   RELATED WORK

The problem of key distribution has attracted much attention of researchers in the area of information security and cryptography. Many key distribution protocols [1–21] have been designed by researchers, and we have reviewed some related protocols. Kumar et al. [3] introduced a centralized key transmission technique based on the Rivest, Shamir, Adleman (RSA) algorithm that is secure and efficient. Chou et al. [6] proposed a method for secure transmission using a secure lock. The number of keys stored by the member is minimized. However, it is only suitable for a very small group due to the high computation and communication complexity. Wallner et al. [8] focused on two main parts of key management, which include secure initialization and rekeying of the multicast group. The author proposes a hierarchical tree approach in which the rekeying cost rises logarithmically with the size of the group. The proposed approach maintains the secrecy of the group. The authors minimize the storage requirements and the number of messages required for rekeying. However, the effectiveness of the protocol depends on whether the tree remains balanced. It also requires the balancing of key tree architecture after every change in the membership.

Harney et al. [10] introduced the group key (GK) management protocol to generate the GKs and distribute them among group members. To create a single cryptographic key, the proposed GKMP protocol uses asymmetric key technology (i.e., RSA, Diffe-Hellman and Elliptic curves) to pass information between two entities. The proposed protocol has the advantages like no central key distribution site is needed, only group members have the key and sender or receiver-oriented operation.

However, there is no solution for maintaining the forward secrecy. Kumar et al. [11] presented a lightweight cryptosystem for key management in advanced metering infrastructure that provides an efficient solution with better security for smart grid. Xu et al. [12] presented a novel dynamic multicast key handling system that greatly reduces computation costs. The authors use Maximum Distance Separable (MDS) code to dynamically distribute the key. Member's computation cost has reduced drastically. The protocol requires a low computational cost while preserving a low and balanced communication and storage cost.

Poovendran et al. [13] have studied the storage of user keys on a rooted tree using information theory. They have studied rooted tree-based Key Encryption Key distribution problem as a convex optimization problem. Additionally showed that the entropy of the member removal statistics is related to the optimal number of keys allotted to a member in the group. The approach required less memory and can be used to avoid the collusion. Kumar [18] present a key establishment technique for secure and fast communication in dynamic groups. Naranjo et al. [19] designed a set of algorithms for secure communication in centralized multicast environments. The algorithm generates only one message per rekeying process. The algorithm only generates one message per rekeying process. Each member only needs to store one key. The rekeying computation cost is high. The algorithm calculates two values δ and L that must be relatively prime; otherwise, the algorithm fails and the member cannot retrieve the key transmitted by the controller of the group.

Kumar [20] present a fast-updating key technique for Pay-tv systems. Vijay Kumar et al. [21] introduced a new key supervision technique that decreases the storage and computation cost of the controller and group users for the key refresh process. However, the communication overhead is very high. Pal [22] proposed a fast key delivery method using the modulo inverse function. Kumar [23] introduced a decentralized and highly scalable key distribution technique for dynamic groups.

6.3 PROPOSED ARCHITECTURE FOR KEY DISTRIBUTION IN CENTRALIZED ENVIRONMENTS

This section describes the proposed architecture for key distribution in a centralized environment as shown in Figure 6.1. The architecture is broadly divided into two main parts: KS area and the group members area. KS further includes some sub-modules. The GK generator sub-module ensures the

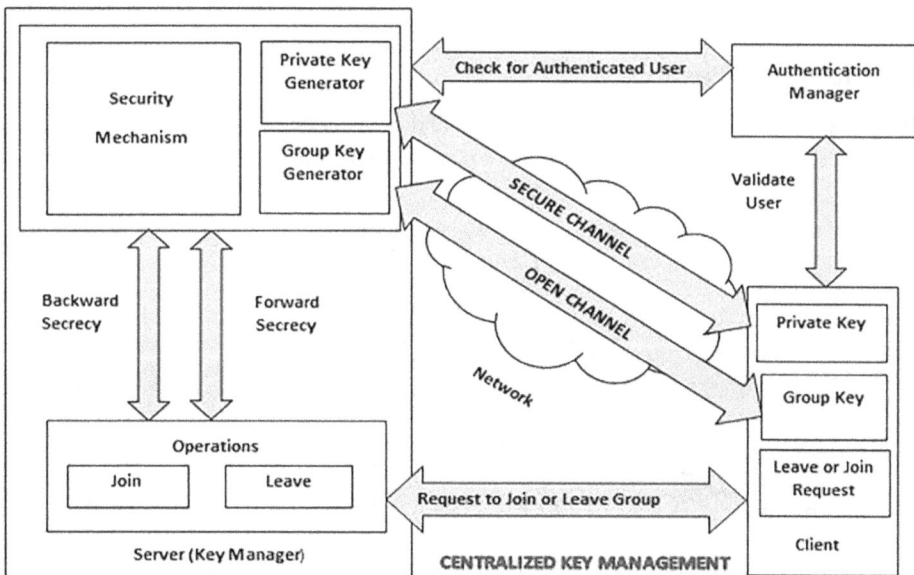

FIGURE 6.1 Proposed architecture for key distribution in centralized environments.

distribution of GK to all members through open/public channels but in encrypted form. Similarly, the private key generator distributes the private key to a specific member through a secure channel such as Secure Sockets Layer. If there are changes in the membership of the group such as joining or leaving operations are performed, the GK needs to be rekeyed to ensure confidentiality. The rekeying process is done by the operation sub-module, which guarantees the creation of new GKs as and when required while maintaining the forward secrecy at the time of leave operation and backward secrecy at the time of join operation.

To ensure that only authenticated members interact with the server, Authentication Manager verifies upcoming requests. Thus, only valid members can leave or join the group that enforces change to the GK. The member module consists of three parts: GK storage, private key storage and a module for handling leave or join requests. The GK module stores a GK at a time. It stores the most recent GK used for communication. Similarly, private key storage saves the member's private key which is used for the decryption of the GK at the client end. Leave or join requests raised by members are handled by the KS, which initiates the process of rekeying. The proposed architecture is suitable for centralized key management protocols and can also be extended to decentralized protocols by shifting the server load to further subdivisions.

6.4 PROPOSED CSKD PROTOCOL

In the proposed CSKD protocol, the KS acts as a central authority to which group members are directly connected. The KS is responsible for managing all key parameters. The KS stores the private keys of all group members and other secret parameters required for key distribution. Members are only required to store their private keys. For key distribution, the KS constructs a multicast message and multicast it to the group members. After receiving the multicast message transmitted by the KS, the group members retrieve the required parameters and generate the GK. The proposed protocol uses a one-way hash function for message authentication. Figure 6.2 describes the key distribution protocol for a centralized environment, in which the root node is the KS with the group key $\mathcal{G}_K$ and the leaf nodes are members of the group containing its private key $\mathcal{K}_n$.

There are two main computations in this protocol: one is for the formation of a multicast message for key distribution and the other is for key retrieval. The GK distribution consists of the following four main phases: Initialization phase, Initial member join phase, key update phase (adding member, removing member) and key recovery phase.

6.4.1 INITIALIZATION PHASE

Initially, KS produces two big prime numbers n and n for defining multiplicative group $\mathcal{Z}_n^*$ and computes threshold μ, as $\mu = n + n$ (the value of n is selected from $\mathcal{Z}_n^*$)

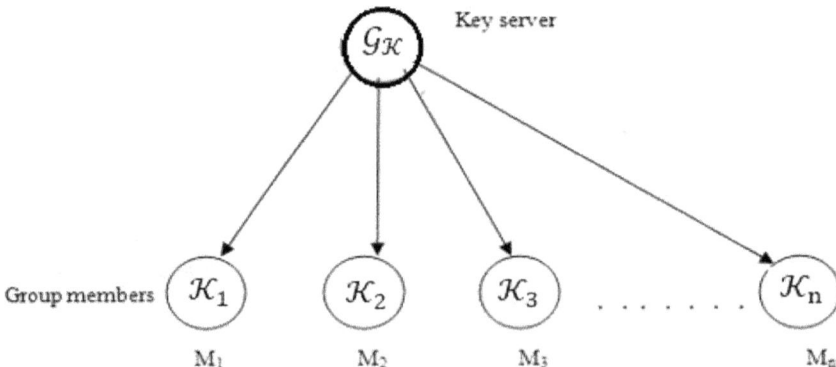

FIGURE 6.2 Key distribution protocol for centralized environments.

6.4.2   INITIAL MEMBER JOIN

If a new member initially joins the group, the following steps are executed by the KS.

1. KS selects the prime numbers from $\mathcal{Z}_n^*$ to all the members as their private keys or secret keys $\mathcal{K}_n$ such that $\mathcal{K}_n > \mu$
2. KS computes Group Key $\mathcal{G}_\mathcal{K} = n^n$ *nnn* n, where n is any element from $\mathcal{Z}_n^*$.
3. Next, the KS chooses another element $\mathcal{K}$ from $\mathcal{Z}_n^*$ and encrypt μ as

$$\lambda = \mathcal{K} \times \prod_{i=1}^{n} (\mathcal{K}_n) + \mu$$

4. After that, the KS generates the current timestamp TS_1 and generates the value of hash ($\mathcal{H}_s$) by a one-way hash function as

$$\mathcal{H}_s = h\left(TS_1 \parallel \mu \parallel \boldsymbol{n} \parallel \boldsymbol{n} \parallel \boldsymbol{n}\right)$$

The KS multicasts the parameters $\{\mathcal{H}_s, TS_1, \boldsymbol{n}, \boldsymbol{n}, \boldsymbol{n}\}$ to the members presents in the group.

6.4.3   KEY UPDATE

GK update is essential if a member leaves/joins the group. Therefore, the key update phase is divided into two sub-phases as adding member and removing member.

6.4.3.1   Adding Member

When a new member joins the group, a new secret key of a recently added member is included in the existing set of keys by the KS in its database. Therefore, the value of the product of the members' keys changes and hence its new value is calculated. The KS updates the value of μ as μ' and computes a new value of λ as λ'. After that KS update GK to maintain backward and forward secrecy. The KS computes updated GK as follows $\mathcal{G}'_\mathcal{K} = n'^{n'}$ *nnn* n, which requires updated $\boldsymbol{n}'$ and $\boldsymbol{n}'$. If the value of $\boldsymbol{n}$ changes, the value of $\boldsymbol{n}$ has to change to adjust μ for the threshold. Finally, the KS generates the updated multicast message $\{\boldsymbol{n}', \lambda', \boldsymbol{n}, \boldsymbol{n}'\}$ and multicast it to the group members.

6.4.3.2   Leaving Member

When a present group member exits the group, his or her private key is removed by the KS from the existing set of keys from its own database. The KS computes updated value of the product of the members' keys. The KS also computes the updated the values of μ and λ as μ' and λ', respectively. The KS follows the same procedure as explained in the adding member sub-phase to compute the updated GK and generate the multicast message.

6.4.4   KEY RECOVERY

After receiving multicast message sent by the KS, the key recovery procedure is executed by the members to recover the required parameters for generating the GK at their end. The key recovery procedure includes the following steps:

1. First, the group members compute $\mu = \lambda$ *nnn* $\mathcal{K}_n$
2. Next, the member generates the current timestamp TS_2 and calculates the transmission delay as $TS_2 - TS_1 = \boldsymbol{n}TS$. If the transmission delay exceeds the maximal transmission delay, the member rejects the received multicast message.
3. The member computes $\mathcal{H}'_s$ as $\mathcal{H}'_s = h\left(TS_1 \parallel \mu \parallel \boldsymbol{n} \parallel \boldsymbol{n} \parallel \boldsymbol{n}\right)$ and verifies it with the hash value received in multicast message.

4. If $\mathcal{H}'_s \neq \mathcal{H}_s$, i.e., message authentication is unsuccessful and member discards the received multicast message.
5. If $\mathcal{H}'_s = \mathcal{H}_s$, i.e., message authentication is successful and member generates GK $\mathcal{G}_{\mathcal{K}} = n^{\mu-n} \; nnn \; n$

The proposed protocol provides the fast recovery of key parameters at the member side by which group members can generate GKs much faster. For generating a GK, the members are required to perform a mod operation, a subtraction operation and a mod power function operation. Therefore, our protocol minimizes the computation cost at the server side as well as the member side. Each member is required to store only one key called the private key. On the other hand, if the group has n number of members, the KS needs to store a total of n keys, that is, one key for each member. To transmit key parameters, the KS only needs a multicast.

6.5   SECURITY ANALYSIS

In this section, we present the security analysis of our proposed CSKD protocol. The proposed protocol ensures forward, backward private key secrecy. Our protocol also provides protection against Passive Attack, Collusion Attack and Replay Attack.

6.5.1   FORWARD SECRECY

If a member is not interested in continuing communication with the group and wants to exit from the group, the KS deletes his or her secret key from the list, and it is not used to distribute updated secret parameters. Therefore, we can say that the proposed protocol ensures forward secrecy.

6.5.2   BACKWARD SECRECY

When a new member is interested to access the communication and he/she enters the group, the KS frequently updates required key parameters so that the old keys or old communications are not accessed by the new members. Therefore, we can say that our protocol ensures backward secrecy.

6.5.3   PASSIVE ATTACK

If an attacker is an outsider and has no knowledge about the GK and the member's secret key, then it is impossible to do a passive attack. Thus, we can say that our protocol ensures protection against passive attack and no outside attacker is able to access the group communication.

6.5.4   COLLISION ATTACK

Two or more members who have left the group can jointly calculate the GK by sharing the keys they have. However, this is not possible in the proposed protocol because on each leave/join in the group, the required secret parameters and GK are frequently updated by the KS. Therefore, we can say that our protocol ensures protection against collision attack.

6.5.5   REPLY ATTACK

In our protocol, an adversary cannot redistribute and delay a multicast message because each multicast message has a timestamp. Thus, our protocol ensures protection against reply attack.

6.6 PERFORMANCE ANALYSIS

The performance of our proposed protocol is analyzed on the basis of various parameters including computation, communication and storage overheads. The performance of our CSKD protocol is compared with the protocols of Vijayakumar et al. [21], Naranjo et al. [19] and Pal et al. [22]. For analysis, we have used various symbols such as those given in Table 6.1.

Table 6.2 illustrates the performance of our CSKD protocol and competing protocols. Table 6.2 shows that the computation complexity of our CSKD protocol at the server side and the member side is $(2 \times T_{as} + T_{modpow} + n \times T_m + T_{hash})$ and $(T_{mod} + T_{as} + T_{modpow} + T_{hash})$, respectively, which are less than that of protocols of Vijayakumar et al. [21], Naranjo et al. [19] and Pal et al. [22]. In our CSKD protocol, the KS and members need to store n keys and 1 key, respectively. For distrusting the key parameters, our CSKD protocol needs to send only one multicast and one unicast message.

6.7 EXPERIMENTAL RESULTS

We have implemented our proposed protocol and other competing protocols like Vijayakumar et al. [21], Naranjo et al. [19] and Pal et al. [22] using Big-Integer Class of Java Programming language.

TABLE 6.1
Symbols Used in Performance Analysis

Symbols	Description
n	Size of the group
T_{as}	Signify the computation time needs to perform addition and subtraction
T_m	Denote the computation time to perform the multiplication
T_{gcd}	Denote the computation time to calculate Greatest Common Divisor (GCD)
T_{eea}	Signify the execution time to execute extended Euclid algorithm
T_{hash}	Denote the computation time to compute hash value
T_{modinv}	Signify the computation time to compute Mod Inverse
T_{mod}	Denote the computation time to compute Mod
T_{modpow}	Signify the execution time to calculate Modulo Power

TABLE 6.2
Performance Analysis of Various Protocols Based on Different Parameters

Parameters	Naranjo et al. [19]	Vijayakumar et al. [21]	Pal et al. [22]	Proposed CSKD Protocol
Computation complexity (KS)	$2 \times T_{as} + 2 \times T_{modpow} + (n+1) + T_{eea}$	$T_{as} + T_{modpow} + n \times T_m + T_{gcd} + T_{eea}$	$T_{as} + T_{modpow} + n \times T_m + T_{gcd} + T_{eea}$	$2 \times T_{as} + T_{modpow} + n \times T_m + T_{hash}$
Computation complexity (member)	$T_{modinv} + 2 \times T_{modpow}$	$T_{mod} + T_{modinv} + T_m + T_{as} + T_{modpow}$	$T_{modinv} + T_{as} + T_{modpow}$	$T_{mod} + T_{as} + T_{modpow} + T_{hash}$
Storage complexity (KS)	$n+4$	$n+3$	$n+4$	n
Storage complexity (member)	4	3	4	1
Communication complexity	2 multicast + 1 unicast	2 multicast + 1 unicast	1 multicast + 1 unicast	1 multicast + 1 unicast

The computation time on the server side and the member side is calculated by measuring the time taken to perform key distribution and key recovery operations, respectively. We have performed experiments on the computer system with Intel Core 2 Duo 3.00 GHz CPU, 64-bit Window Operating System, 2.00 GB RAM and 200 GB Hard Disk. We have performed various experiments on different key sizes such as 64, 128, 256, 512, 1,024 and 2,048 bits. Computation time is measured in milliseconds.

Figure 6.3 shows the computation time at the KS side for key distribution. Figure 6.3 illustrate that the proposed system requires less computation time than that of protocols of Vijayakumar et al. [21], Naranjo et al. [19] and Pal et al. [22]. Figure 6.4 shows the computation time at the member side for key recovery. Figure 6.4 illustrates that the proposed method needs less computation time than that of competing protocols. Therefore, our protocol is more efficient compared with other related protocols.

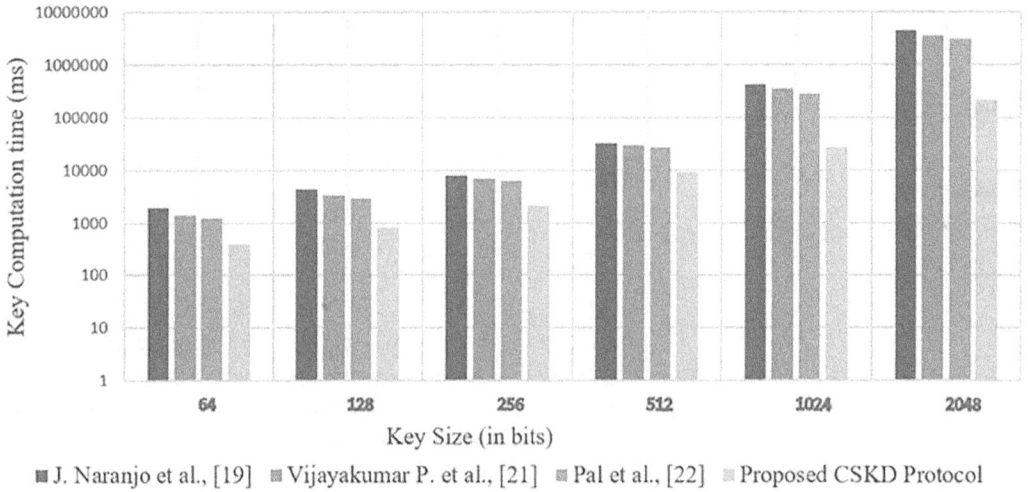

FIGURE 6.3 Computation time at the KS side for key distribution.

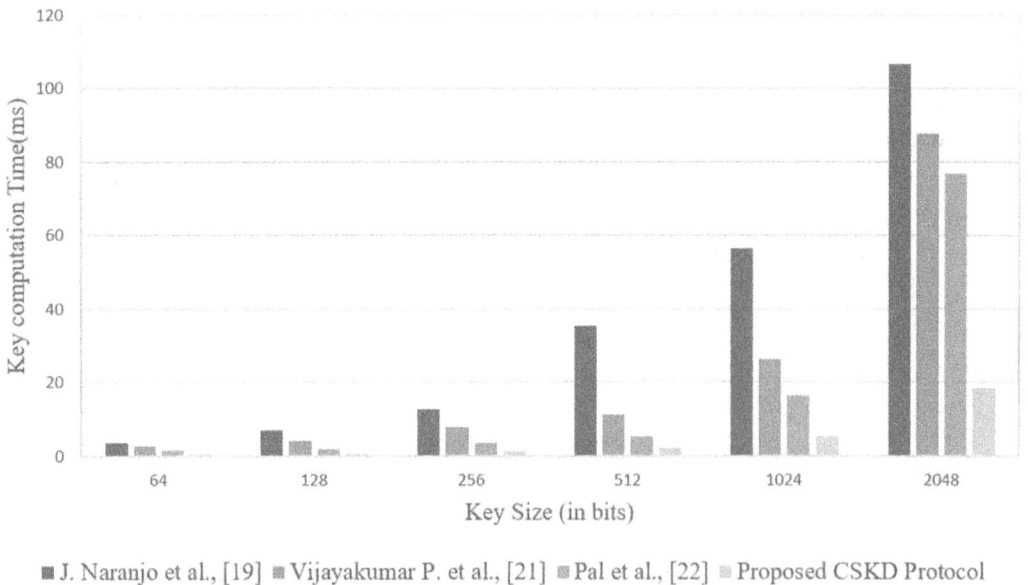

FIGURE 6.4 Computation time at the member side for key recovery.

6.8 CONCLUSION

We propose an architecture for key management in centralized environments. A key distribution protocol that is suitable for centralized architecture is also proposed. The proposed architecture is suitable for all applications where the server is solely responsible for data communication and key management alone. The proposed protocol minimizes computation, communication and storage overheads. From comparative analysis, we conclude that the presented key distribution system requires lower computation, communication and storage costs than other competing protocols.

REFERENCES

1. Rafaeli, S., & Hutchison, D. (2003). A survey of key management for secure group communication. *ACM Computing Surveys (CSUR)*, 35(3), 309–329.
2. Challal, Y., & Seba, H. (2005). Group key management protocols: A novel taxonomy. *International Journal of Information Technology*, 2(1), 105–118.
3. Kumar, V., Kumar, R., & Pandey, S. K. (2020). A computationally efficient centralized group key distribution protocol for secure multicast communications based upon RSA public key cryptosystem. *Journal of King Saud University-Computer and Information Sciences*, 32(9), 1081–1094.
4. Challal, Y., Bouabdallah, A., & Seba, H. (2005, June). A taxonomy of group key management protocols: Issues and solutions. In *Proceedings of World Academy of Science, Engineering and Technology* (Vol. 6, pp. 5–19).
5. Seetha, R., & Saravanan, R. (2015). A survey on group key management schemes. *Cybernetics and Information Technologies*, 15(3), 3–25.
6. Chiou, G. H., & Chen, W. T. (1989). Secure broadcasting using the secure lock. *IEEE Transactions on Software Engineering*, 15(8), 929–934.
7. Kumar, V., Kumar, R., & Pandey, S. K. (2020). Polynomial based non-interactive session key computation protocol for secure communication in dynamic groups. *International Journal of Information Technology*, 12(1), 283–288.
8. Harney, H. (1997). Group key management protocol (GKMP) architecture. RFC2094.
9. Wong, C. K., Gouda, M., & Lam, S. S. (2000). Secure group communications using key graphs. *IEEE/ACM Transactions on Networking*, 8(1), 16–30.
10. Wallner, D., Harder, E., & Agee, R. (1999). Key management for multicast: Issues and architectures. RFC 2627.
11. Kumar, V., Kumar, R., & Pandey, S. K. (2021). LKM-AMI: A lightweight key management scheme for secure two-way communications between smart meters and HAN devices of AMI system in smart grid. *Peer-to-Peer Networking and Applications*, 14(1), 82–100.
12. Xu, L. (2002). Dynamic group key distribution using MDS codes. In *Information, Coding and Mathematics* (pp. 27–44). Springer, Boston, MA.
13. Poovendran, R., & Baras, J. S. (2001). An information-theoretic approach for design and analysis of rooted-tree-based multicast key management schemes. *IEEE Transactions on Information Theory*, 47, 2824–2834
14. Sherman, A. T., & McGrew, D. A. (2003). Key establishment in large dynamic groups using one-way function trees. *IEEE Transactions on Software Engineering*, 29(5), 444–458.
15. Waldvogel, M., Caronni, G., Sun, D., Weiler, N., & Plattner, B. (1999). The VersaKey framework: Versatile group key management. *IEEE Journal on Selected Areas in Communications*, 17(9), 1614–1631.
16. Zheng, X., Huang, C. T., & Matthews, M. (2007, March). Chinese remainder theorem-based group key management. In *Proceedings of the 45th Annual Southeast Regional Conference* (pp. 266–271).
17. Lin, I. C., Tang, S. S., & Wang, C. M. (2010). Multicast key management without rekeying processes. *The Computer Journal*, 53(7), 939–950.
18. Kumar, V., Kumar, R., & Pandey, S. K. (2020). An efficient and scalable distributed key management scheme using ternary tree for secure communication in dynamic groups. In *Proceedings of ICETIT 2019* (pp. 152–164). Springer, Cham.
19. Naranjo, J. A. M., Antequera, N., Casado, L. G., & López-Ramos, J. A. (2012). A suite of algorithms for key distribution and authentication in centralized secure multicast environments. *Journal of Computational and Applied Mathematics*, 236(12), 3042–3051.
20. Kumar, V., Kumar, R., & Pandey, S. K. (2021). A computationally efficient and scalable key management scheme for access control of media delivery in digital pay-TV systems. *Multimedia Tools and Applications*, 80, 1–34.

21. Vijayakumar, P., Bose, S., & Kannan, A. (2013). Centralized key distribution protocol using the greatest common divisor method. *Computers & Mathematics with Applications*, 65(9), 1360–1368.

22. Pal, O., & Alam, B. (2020). Key management scheme for secure group communication. In *Advances in Data and Information Sciences* (pp. 171–177). Springer, Singapore.

23. Kumar, V., Kumar, R., & Pandey, S. K. (2020). A secure and robust group key distribution and authentication protocol with efficient rekey mechanism for dynamic access control in secure group communications. *International Journal of Communication Systems*, 33(14), e4465.

7 Effective Key Agreement Protocol for Large and Dynamic Groups Using Elliptic Curve Cryptography

Vinod Kumar and Khushboo Gupta
University of Allahabad

Om Pal
Univeristy of Delhi

Shiv Prakash and Pravin Kumar
University of Allahabad

Shiv Veer Singh
IIMT College of Engineering

Pradeep Kumar Tiwari and Animesh Tripathi
University of Allahabad

CONTENTS

7.1 INTRODUCTION

Session key agreement allows all contributors to make a deal on a common key, which is shared among group participants to make transmission secure over an insecure medium. The session key will be changed when existing participants are exiting the group or new participants entering the existing group. The protocols for key management are divided into three categories: (a) centralized, (b) decentralized and (c) distributed. Centralized group key management schemes are one of a kind in that there is only one single group controller (GC) who is accountable to generate the

DOI: 10.1201/9781003267812-7

keys and allocates them to all members of the group. However, in centralized approaches, the GC must always be available to support join and drop operations. Also, it suffers from the problem of bottleneck. In de-centralized schemes, the entire group is separated into subgroups. The subgroup controllers (SCs) control each subgroup, which makes it more robust as failure of one SC will not affect the remaining subgroups. The third approach is called contributory group key agreement or distributed schemes, in which each group participant contributes its common share to compute a group key which is shared to all participants.

In recent years, there has been a substantial growth in the popularity of group-based applications. These include audio and video conferences, pay per view, etc. The designing of key management for secure group communication has many problems such as reducing computation, communication and storage complexity, security, interactivity and non-dynamicity. These problems motivated us to conduct research in this area to solve group key management problems and accelerate the successful deployment of group applications. The major problem with traditional public key cryptography systems is that the key size must be large to meet high-level security standards. The solution to this problem is Elliptic Curve Cryptography (ECC) that provides a secure environment with superior performance in computing power and communication cost. The use of ECC also makes key management systems more efficient for many applications where storage cost is constrained. In group communication, the designing of a secure distributed group key delivery protocol that supports dynamic operations with negligible communication, storage and computation cost is the main problem. In this chapter, we have designed an effective key agreement protocol for large dynamic clusters using ECC with low communication and computation cost.

The rest of this chapter is organized as follows: Related work on key agreement protocols is presented in Section 7.2. We propose an ECC-based distributed key agreement protocol in Section 7.3. The performance analysis of our CSKD and other competitive protocols is presented in Section 7.4. The implementation results are illustrated in Section 7.5. Finally, the conclusion of this chapter is presented in Section 7.6.

7.2   RELATED WORK

The problem of key distribution has attracted much attention of researchers in the area of information security and cryptography. Many key distribution protocols [1–13] have been designed by researchers, and we have reviewed some related protocols. Kumar et al. [3] introduced a centralized key transmission technique based on Rivest, Shamir, Adleman (RSA) algorithm that is secure and efficient. Chou et al. [6] proposed a method for secure transmission using a secure lock. The number of keys stored by the member is minimized. However, it is only suitable for a very small group due to the high computation and communication complexity. Wallner et al. [8] focused on two main parts of key management, which include secure initialization and rekeying of the multicast group. The author proposes a hierarchical tree approach in which the rekeying cost rises logarithmically with the size of the group. The proposed approach maintains the secrecy of the group. The authors minimize the storage requirements and the number of messages required for rekeying. However, the effectiveness of the protocol depends on whether the tree remains balanced. It also requires the balancing of key tree architecture after every change in the membership. Harney et al. [10] introduced a group key management protocol to generate the group keys and distribute them among group members. To create a single cryptographic key, the proposed GKMP protocol uses asymmetric key technology (i.e., RSA, Diffe-Hellman and Elliptic curves) to pass information between two entities. The proposed protocol has advantages like no central key distribution site is needed; only group members have the key, sender or receiver-oriented operation, etc. However, there is no solution for maintaining the forward secrecy. Kumar et al. [11] presented a lightweight cryptosystem for key management in advanced metering infrastructure that provides an efficient solution with better security for smart grid.

7.3   PROPOSED DISTRIBUTED KEY MANAGEMENT PROTOCOL

The proposed approach provides a distributed group key agreement scheme for group communication. It uses a ternary tree which increases its scalability as well as reduces the number of rounds because a large number of group participants are positioned at the same level as the ternary tree. Furthermore, we tried to balance the ternary tree by selecting the appropriate insertion point (IP), which avoids the problem of skewed tree formation and results in low computation and communication cost. The flow graph and key tree of the proposed approach are given in Figures 7.1 and 7.2, respectively.

7.3.1   INITIALIZATION PHASE

In the initialization process, we organize the members with the formation of a subgroup of three members and the group key is calculated using the three-party Elliptic Curve Diffie Hellman (ECDH) protocol. All members are placed at the leaf nodes of a ternary tree forming a subgroup of three members each. Each member randomly chooses a secret random integer ri. Every subgroup

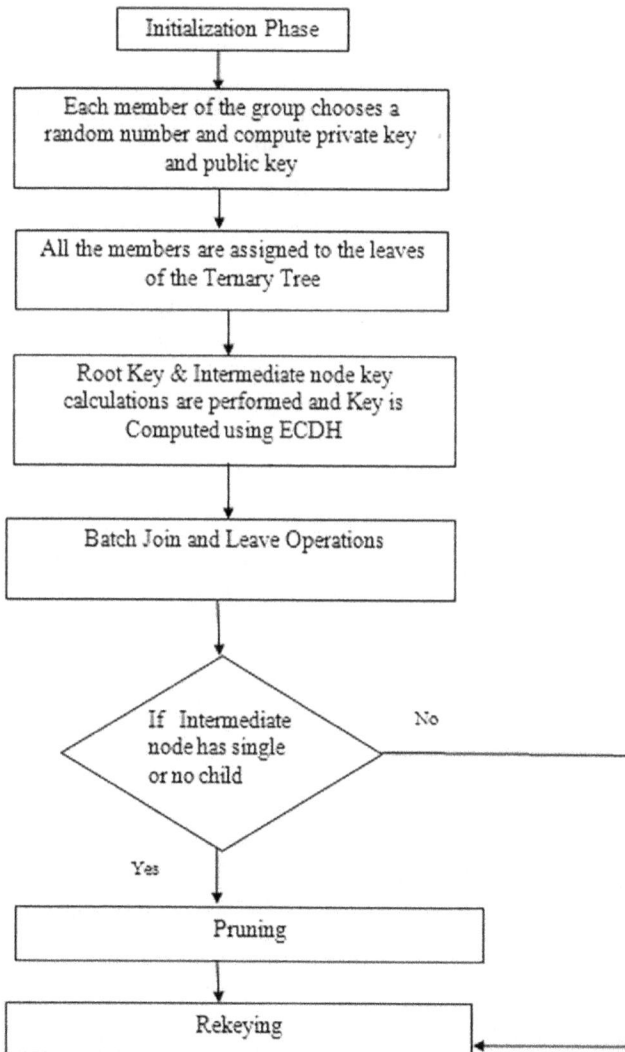

FIGURE 7.1   Flow graph of the proposed approach.

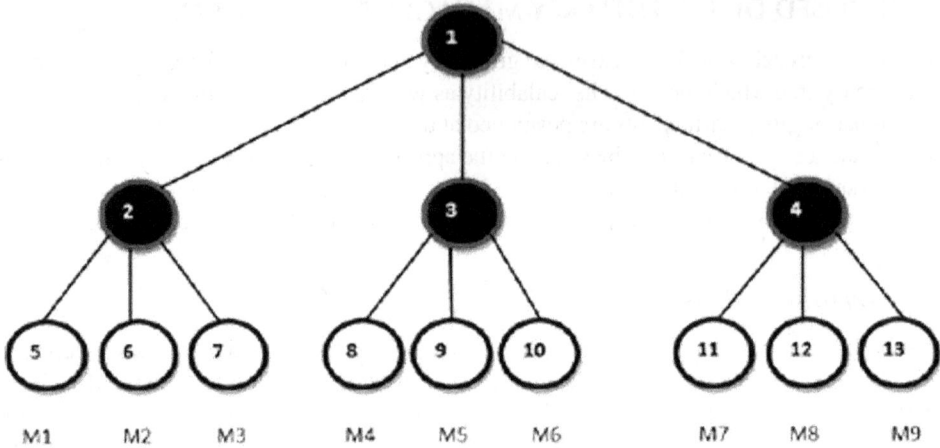

FIGURE 7.2   Key tree of the proposed approach.

forms its own subgroup key using Three Parties Diffie Hellman key exchange. With every round, the subgroup key is in the form of (rxi.ryi.rzi. G), where rxi, ryi and rzi are private keys of member and G is the generator point or base point. The same process is repeated until a single group key is formed, which is the same for all members of the group.

7.3.2   BATCH REKEYING

It is the process in which new keys are computed to maintain backward and forward secrecy after members leave and join operations. In batch rekeying, all the join and leave requests are collected for an interval and a new key is generated collectively. In the batch join/batch leave process, members can join the group or leave in batches. Five cases are designed and selected based on the number of requests to join and leave. Batch Rekeying can be classified into five cases:

Case 1: if $J=L$ then

1. Replace all the leaving members by joining members.
2. All the subgroup keys are updated with respect to the locations of joining and leaving members.

Case 2: if $J<L$ and $J=0$ then

1. Remove all 'L' leaving members from the key tree and update the subgroup keys with respect to the positions of leaving members.

Case 3: if $J>L$ and $L=0$ then

1. Create the 'TKT' with 'J' joining members
2. Find the P

Case 4: if $J<L$ and $J>0$ then

1. Out of the 'L' leaving members, 'J' joining members take the position of 'J' leaving members.
2. Remaining 'L' nodes are removed from the key tree.

Case 5: if $J > L$ and $L > 0$ then

1. Insert 'L' joining members in the positions of 'L' leaving members
2. Create Temporary Key Tree (TKT) with remaining '$J–L$' joining members
3. Find the IP

7.3.3 Procedure for Finding IP

Step 1: If root is full then perform the following steps:
- Find any Intermediate Node (IN) that has null link then IP will be that IN.
- Otherwise, find member 'M' with minimum height and at that location create the New Intermediate Node (NIN) then IP will be that New Intermediate Node (NIN) and insert member 'M' in TKT.

Step 2: If root is not full then IP will be the Root node

7.3.4 Procedure for Pruning

In pruning, some nodes in the key tree representing subgroup keys are removed if the children representing its member are eliminated during the leave operation. These removed nodes are called pruned nodes and the process is called pruning. Pruning can occur in two cases:

Case 1: If an IN has single child node in the key tree, then that IN is replaced by its child.

Case 2: If all three children of IN leave the group. The correlated IN is also removed from the key tree. This removed IN is also considered a Pruned Node.

In the proposed protocol, the communication cost is reduced as fewer number of rounds are required to compute the group key. The computation cost will be reduced as exponentiation operations used in RSA and Diffie Hellman are replaced with scalar multiplications. Batch Rekeying is performed so the cost of handling each request individually is also reduced.

7.4 PERFORMANCE ANALYSIS

In Table 7.1, the performance of our proposed scheme is compared to Ternary Tree Group Key Agreement (TTGKA) protocol. We extended the conventional TTGKA protocol by introducing mass join and mass leave operations as TTGKA handles only single join and leave operations.

The number of rounds, number of messages for communication and number of multiplications for computational cost are analysed. Here, 'n' is the number of users, 'h' is the height of ternary tree, 'm' is the number of joining members or leaving members and 'h' is the height of TKT. In our proposed approach, we solve the problem of skewed tree formation at the time of join operation though it increases our cost of join operation by logarithm but has a great impact on the calculations of subgroup keys. Our protocol outperforms the TTGKA at the time of leave operations. Table 7.1 compares the cost of each operation with respect to rounds, messages and scalar multiplication.

7.5 IMPLEMENTATION RESULTS

Table 7.2 shows the time analysis of our proposed scheme and conventional TTGKA that the proposed scheme performs better at the time of rekeying. It compares the rekeying cost with respect to key size for an initial group size of 729 members.

Figures 7.3–7.7 also show that our proposed scheme performs better with respect to the number of key sizes.

Case 1: $J = L$
Case 2: $J < L$ and $J = 0$
Case 3: $J > L$ and $L = 0$

TABLE 7.1

Performance Comparison of TTGKA and Proposed Scheme

Operation		TTGKA	Proposed Scheme
Initialization	Rounds	$\log_3 n$	$\log_3 n$
	Messages	$3\{n-1\}12$	$3\{n-1\}12$
	Scalar multiplication	$5(n-1)12+h.n$	$5(n-1)12h.n$
Single join	Rounds	1	$\log_3 n$
	Messages	2	$h+1$
	Scalar multiplication	$n+2$	$n+2$
Single leave	Rounds	$n-1$	$\log_3 n$
	Messages	$n-1$	H
	Scalar multiplication	$\boldsymbol{n-1}$	$n-1$
Mass join	Rounds		$\boldsymbol{h'+1}$
	Messages		$3\{m-1\}12+2$
	Scalar multiplication		$3\{m-1\}12+m(h'+1)+n$
Mass leave	Rounds		$\log_3 n$
	Messages		$m.h$
	Scalar multiplication		$m.n$

TABLE 7.2

Comparison of TTGKA and Proposed Scheme With Respect to Key Size and Rekeying Request

Cases	Number of Joing Members	Number of Leaving Members	Key Size in bits	Time of TIGKA (in ms)	Time of Existing Scheme (in ms)
Only joining	5	0	128	19,609	27,215
members (JM)	5	0	160	21,831	27,921
	5	0	256	25,682	29,825
	5	0	521	28,113	31,624
Only leaving	0	5	128	26,430	53,206
members (LM)	0	5	160	27,043	54,091
	0	5	256	28,520	57,128
	0	5	521	31,625	59,654
Joining members	5	5	128	36,090	71,873
(JM)=leaving	5	5	160	36,948	7,206
members (LM)	5	5	256	37,812	74,201
	5	5	521	39,157	75,311
Joining members	5	3	128	23,163	50,956
(JM)>leaving	5	3	160	24,863	51,210
members (LM)	5	3	256	26,112	53,114
	5	3	521	28,391	55,068
Joining members	3	5	128	32,075	67,782
(JM)<leaving	3	5	160	32,921	68,348
members (LM)	3	5	256	34,116	69,579
	3	5	521	35,728	71,283

Rekeying Cost		
Key Size	TTGKA	Proposed
128	27215	19609
160	27921	21831
256	29825	25682
521	31624	28113

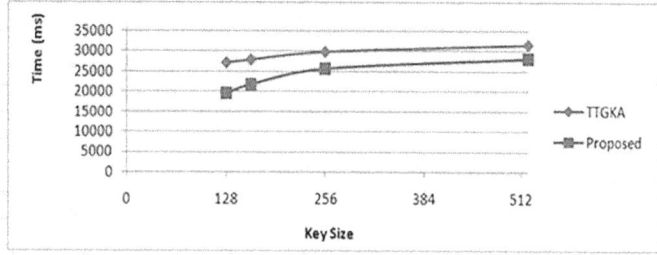

FIGURE 7.3 Performance comparison of TTGKA and proposed scheme for Case 1.

Rekeying Cost		
Key Size	TTGKA	Proposed
128	53206	26430
160	54091	27043
256	57128	28520
521	59654	31625

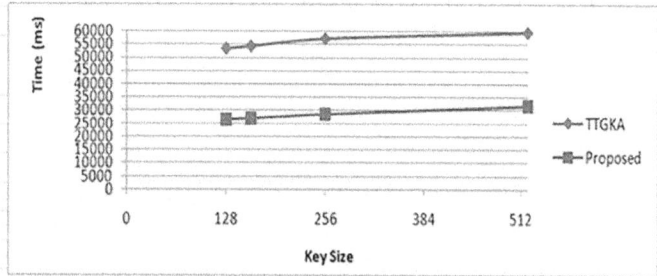

FIGURE 7.4 Performance comparison of TTGKA and proposed scheme for Case 2.

Rekeying Cost		
Key Size	TTGKA	Proposed
128	71873	36090
160	72063	36948
256	74201	37812
521	75311	39157

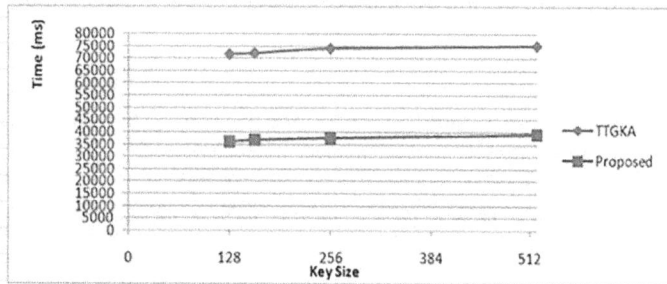

FIGURE 7.5 Performance comparison of TTGKA and proposed scheme for Case 3.

Rekeying Cost		
Key Size	TTGKA	Proposed
128	50956	23163
160	51210	24863
256	53114	26112
521	55068	28391

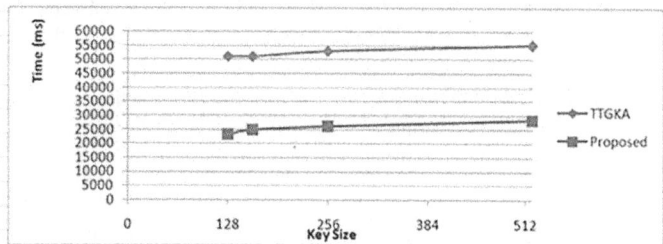

FIGURE 7.6 Performance comparison of TTGKA and proposed scheme for Case 4.

Rekeying Cost		
Key Size	TTGKA	Proposed
128	67782	32075
160	68348	32921
256	69579	34116
521	71283	35728

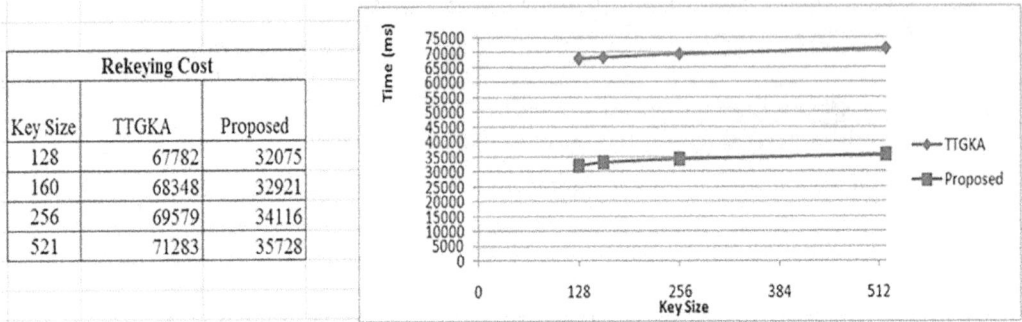

FIGURE 7.7   Performance comparison of TTGKA and proposed scheme for Case 5.

Case 4: $J < L$ **and** $J > 0$
Case 5: $J > L$ **and** $L > 0$

7.6 CONCLUSION

In this chapter, we propose an effective key agreement protocol using ECC for secure communication. To solve the scalability problem of group key management, a ternary tree is used in which each node can have a maximum of three children, but there is no restriction on the number of group members. In the proposed protocol, suitable IP is selected to keep the tree balance to solve the problem of high communication cost. Furthermore, to address the problem of high computation cost, scalar multiplications are performed using ECDH. The computation and communication cost of the proposed protocol is compared with the existing TTGKA protocol, which shows that our protocol outperforms the existing protocol.

REFERENCES

1. Rafaeli, S., & Hutchison, D. (2003). A survey of key management for secure group communication. *ACM Computing Surveys (CSUR)*, 35(3), 309–329.
2. Sharma, S., & Krishna, C. R. (2015, February). An efficient distributed group key management using hierarchical approach with elliptic curve cryptography. In *2015 IEEE International Conference on Computational Intelligence & Communication Technology* (pp. 687–693).
3. Kumar, V., Kumar, R., & Pandey, S. K. (2021). LKM-AMI: A lightweight key management scheme for secure two-way communications between smart meters and HAN devices of AMI system in smart grid. *Peer-to-Peer Networking and Applications*, 14(1), 82–100.
4. Tripathi, S., & Biswas, G. P. (2009, January). Design of efficient ternary-tree based group key agreement protocol for dynamic groups. In *First International Communication Systems and Networks and Workshops* (pp. 1–6).
5. Kumar, V., Kumar, R., & Pandey, S. K. (2020). An efficient and scalable distributed key management scheme using ternary tree for secure communication in dynamic groups. In *Proceedings of ICETIT 2019* (pp. 152–164). Springer, Cham.
6. Kim, Y., Perrig, A., & Tsudik, G. (2004, July). Group key agreement efficient in communication. *IEEE Transactions on Computers*, 53(7), 905–921.
7. Wang, Y., Ramamurthy, B., & Zou, X. K. (2006). The performance of elliptic curve based group Diffie-Hellman protocols for secure group communication. *IEEE International Conference on Communication*, 5, 2243–2248.
8. Zheng, S., Manz, D., & Alves-Foss, J. (2007). A communication–computation efficient group key algorithm for large and dynamic groups. *Computer Networks*, 51, 69–93.
9. Vijayakumar, P., Bose, S., & Kannan, A. (2011). Rotation based secure multicast key management for batch rekeying operations. *Networking Science*, 1(1–4), 39–47.

10. Lin, H.-Y., & Chiang, T.-C. (2011). Efficient key agreements in dynamic multicast height balanced tree for secure multicast communications. *EURASIP Journal on Wireless Communications and Networking*, 2011(1), 382701.
11. Kumar, A., & Tripathi, S. (2014). Ternary tree-based group key agreement protocol over Elliptic curve for dynamic group. *International Journal of Computer Applications*, 86(7), 17–25.
12. Jaiswal, P., Kumar, A., & Tripathi, S. (2015). *Design of Queue-Based Group Key Agreement Protocol Using Elliptic Curve Cryptography* (pp. 167–176). Springer Information Systems Design and Intelligent Applications, DOI: 10.1007/978–81–322–2250–7_17.
13. Renugadevi, N., & Mala, C. (2015). Pruned nodes in ternary key tree during batch rekeying in group key agreement for cognitive radio ad hoc networks. *Security and Communication Networks*, 41, 67–74.

8 Cyber Security Using Artificial Intelligence

Sapna Katiyar
Impledge Technologies

CONTENTS

8.1 INTRODUCTION

The rapid development of information and communication technologies has transformed everything into digital era. Internet has played a major role for organizations and persons, turning toward digital technologies and solutions. The development of advanced technologies like Artificial Intelligence (AI) makes digital revolution fascinating. AI is the intelligence revealed by machines, which are aware about the environment and act accordingly to accomplish something. When machines mimic the human actions for problem solving or learning, the ideally used term is AI [1,2]. The various subdomains of AI are neural network, fuzzy system, machine learning, deep learning, computer vision, natural language processing, cognitive computing, evolutionary computation and nature inspired algorithms [3]. AI is being observed as an essential element in the digital transformation of society. Future applications are anticipated to carry about massive changes. Some of the practical examples of AI are Self-driving cars, Alexa, Robo advisors, Conventional bots, OTT platform recommendations, email spam filters, etc. Basically, there are three categories of AI: Artificial Narrow Intelligence (ANI), Artificial General Intelligence (AGI) and Artificial Super Intelligence (ASI). One more way to categorize AI is on the basis of bringing value to the business. Five classes are Visual AI (Computer vision, Augmented reality), Interactive AI (Chatbots, Smart personal assistants), Text AI (Text recognition, Speech recognition), Functional AI (Robots, Internet of Things (IoT) solutions) and Analytic AI (Sentimental Analysis, Supplier risk assessment) [4,5]. Some of the benefits and risks associated with AI are depicted in Figure 8.1.

AI in cyber security is not only a smart revolution but has enormous potential. AI-enabled systems can be trained for human machine secured communication [6,7]. AI-based systems can be trained to generate threat alerts, recognize fresh malware and shield organization-sensitive information. Implementation of some ethical procedures in AI must be recommended to guarantee that technology is a friend not an enemy [8–10]. AI-based Cyber security systems have many abilities for systems, data and applications. System security implies security for network, cloud, malwares, IoT, etc. Data security is associated with the safety of information on cloud, threat detection and fresh

DOI: 10.1201/9781003267812-8

Benefits of AI		Risk associated with AI	
	Automation		High Implementation Cost
	Smart Decision Making		Human Replication
	Accuracy		It is programmable
	Exploration		Lacks creativity
	Data Collection		Accelerate Hacking
	Data Analysis		Does not improve with experience
	Solving Complex Problems		Unemployment
	Managing Repetetive Task		Global Regulations
	Minimizing Errors		AI Terrorism

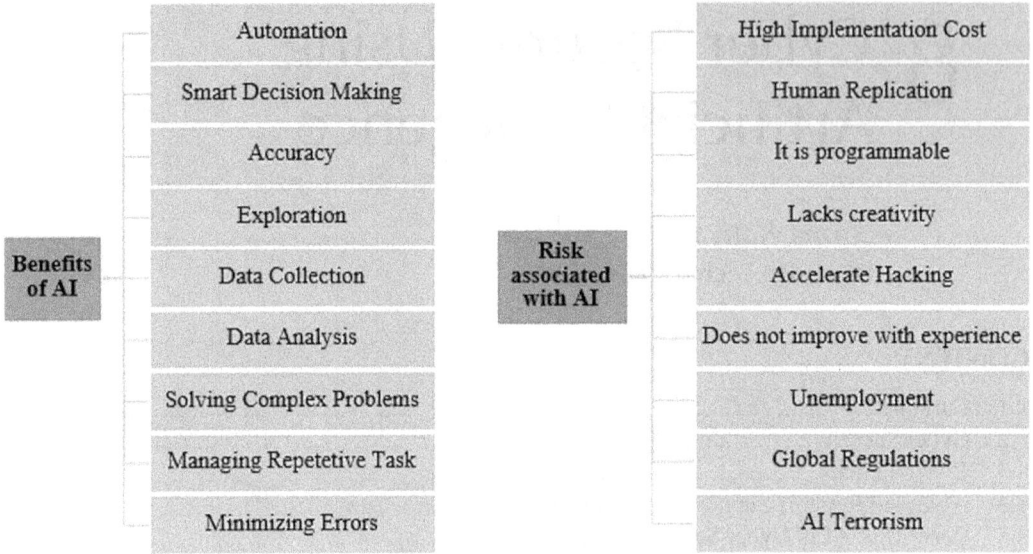

FIGURE 8.1   Benefits and risks associated with AI.

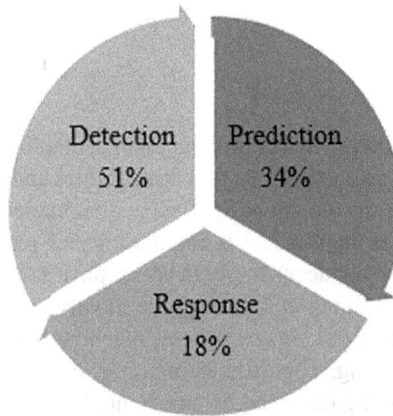

FIGURE 8.2   Key features of AI in cyber security.

attack recognition [11,12]. AI in cyber security helps organization to improve threat response time, reaction to breaches and reducing cost. AI has three key features in the domain of cyber security [13,14], which are shown in Figure 8.2.

- **Detection:** These days' organizations use AI exhaustively for precise threat detection. More than 50% of organizations have implemented AI-enabled cyber security systems. Detection via AI is the unique capability for behavior analysis to monitor and identify irregular traffic regularly. Machine learning and deep learning algorithms are mostly used.
- **Prediction:** The usage rate of the prediction function is the second highest. As per the research, 35% of the organizations use AI extensively for predicting cyber threats. AI helps in prediction among huge data on the basis of system training. Automation through AI helps organizations in vulnerability recognition and strengthens their network to protect against any Cyber-attack.

- **Response:** AI is still in the process of growing against response to threats. Approximately 18% of the organizations are using AI to counter the cyberattacks. In real-time scenario by automation, for detecting threats, a fresh defense mechanism is developed.

8.2 CYBER SECURITY

Cyber security is viewed as the technology application, processes and various means to protect systems, information, devices, programs and networks from Cyber-attack [15,16]. The objective is to reduce the threat of malicious attack and provide protection against these. One more buzzword commonly used is information security, which is a broader category and responsible to protect all data resources in digital as well as in hard copy.

In recent years, multiple high profile cyberattacks happen [17]:

- In the year 2017, Equifax breach has compromised 143 million personal user data like date of birth, address, passwords, etc.
- In the year 2018, hackers were able to access the Marriott International server and copied the data of nearly 500 million users.

From commerce to mobile computing, the word cyber security is divided into the following categories [18,19], which is shown in Figure 8.3.

- **Network Security:** Attackers may target the computer network from inside or outside the network. Various techniques are used for protection against malicious software, data breach, etc. Firewall acts as a protective barrier between client network and outside. It has the capability to block and allow network traffic depending upon security settings. Securing email is very important for securing network because phishing attacks are very common [20]. Therefore, some programs must be designed which can scan incoming as well as outgoing e-mails to supervise such phishing attacks.
- **Application Security:** The aim is to protect the device or protecting softwares. Successful security measure is taken at the design stage, i.e., before the device is deployed. In some cases of application security, a strong password from a user is expected. Sometimes, two-step authentications, security questions, etc. are required to make sure that the user is authenticated one [21].

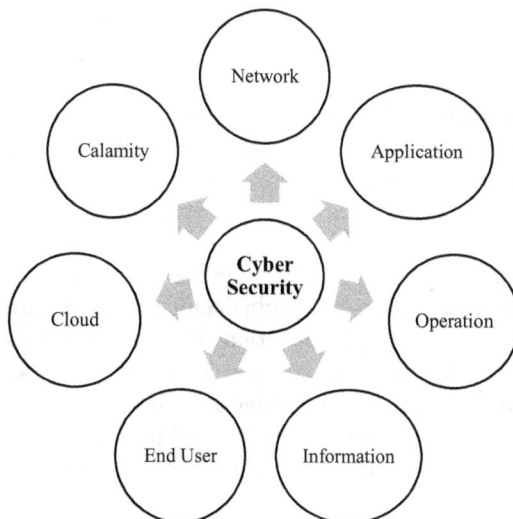

FIGURE 8.3 Categories of cyber security.

- **Operational Security:** It is related to risk management related to data resources. Authenticated users in any organization must be educated about the best practices to store and secure personal and professional data in a network. Usually, risk management professionals are employed by organization to make sure that a backup plan is in standby.
- **Information Security:** It ensures that information integrity and confidentiality are maintained during storage and in transportation [22].
- **End user Security:** Educating end users in any organization may protect the system from accidentally introduced viruses. They can erase or delete suspicious email attachments or plug in unknown USB drives etc.
- **Cloud Security:** Nowadays, everyone stores the majority of their data in cloud. Many online systems like MS one Drive, Google Drive, Apple iCloud, etc. are used for storage instead of hard drive. Therefore, these platforms must be secure and hence right cloud security actions must be taken in terms of end user interface, human error which may expose network, backup strategy, information storage safety, etc. [23].
- **Calamity Recovery:** Every organization must have some policies for disaster recovery to efficiently recover its operations and data. It is required to operate the organization on the same operating capacity as it was running earlier. Sometimes, organizations have a business continuity plan to meet the operating requirements even in the absence of certain resources [24].

8.3   CYBER THREATS

It is a kind of malicious act, in general, which may steal data, damage data, or interrupt digital life. Attackers may use malicious code for compromising others computer and, hence, cyber threats must be scanned and necessary. Cyber-attack avoidance techniques are required by every association [25–27]. Cyber security encountered mostly the following threats:

- **Cyber-Crime:** It includes the act done by group or individual for interruption or monetary gain.
- **Cyber Terrorism:** It is used to damage electronic-based system to create terror or panic situation.
- **Cyber-Attack:** Sometimes, it is related to information gathering, which is politically provoked.

Some of the significant cyber security threats are mentioned in Figure 8.4, which must be known to every cyber security professional.

- **Malware:** It is a malicious software and the most common cyber threat. Hackers used to destroy users' computers when they clicked on malicious email attachments, links, etc. Once it is active, it may block network access, install fake softwares, disrupt system, etc. A variety of malwares are virus, spyware, adware, Trojans, ransomware and botnets [28].
- **Phishing:** It is a kind of false communication in the form of mainly email to get the users sensitive information such as password, login details and credit card details.
- **Structure Query Language (SQL) Injection:** Hacker's objective is to introduce malicious code into server through SQL, to take control of database. Infected server may release sensitive information enclosed in database [29].
- **Emotet:** It is one of the most expensive and harmful malwares. CISA (Cyber security and infrastructure security agency) depicts it as developed modular baking Trojan and basically it is downloader.

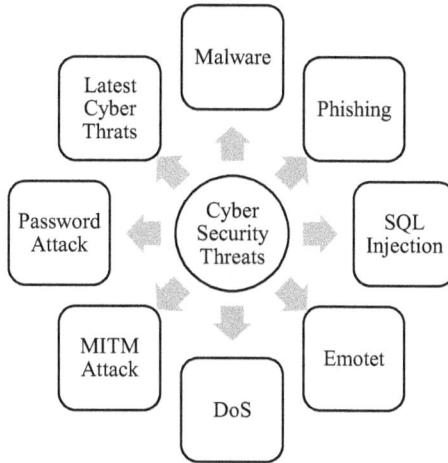

FIGURE 8.4 Cyber security threats.

- **Denial of Service:** In this Cyber-attack, hackers create the situation of flood. Computer system finds too much traffic and is not capable to fulfill the lawful request [30]. In such situations, network becomes impractical and any organization is not capable to carry out essential functions.
- **Man in the Middle Attack:** In this Cyber threat, a hacker intercepts two parties communication. Therefore, traffic can be interrupted and Cyber criminals may steal and segregate information. Mostly, this attack occurs during the use of an unsecured WiFi network, where the hacker could intercept information or use malware.
- **Password Attacks:** Having the correct password means access to the rich information. One common example of a password attack is community engineering. Two possibilities are here, either accessing a password database or blindly relying on human interaction to break standard security practices.
- **Few Latest Cyber Threats:** Dridex Malware attack was observed in the year 2019, which affected the infrastructure, public, government and industry worldwide. In this view, U.K. Cyber Security Center suggested that devices must be patched and antivirus must be updated and turned on. In the year 2020, Romance Scams happened where hackers used chat rooms, dating sites and apps. It affects a large number of populations along with a huge amount of financial loss.

8.4   AI-BASED SYSTEMS SUPPORT CYBER SECURITY

AI-enabled systems support data security, network security and endpoint security [31,32]. Three key categories are there for AI in cyber security: Detection, Prediction and Response.

Motivating factors for AI integration in cyber security are as follows [33,34]:

- **Pace of Impact:** According to reports analyzed for major attacks, average impact time on any organization is about 240 seconds. These days' attacks are for different motives also, and they can be adjusted according to the target. Such type of Cyberattacks impacts very quickly even without much human interaction.
- **Operational Complications:** Majority of organizations are using cloud for data. Cloud-based computing platforms are quick and deliver services within msec duration. Therefore, AI-enabled systems may strengthen logic-driven capabilities.

- **Skill Gap:** As per the survey of Frost and Sullivan, shortage of cyber security profession-als is there at the global level. Therefore, organizations must automate processes at a very fast pace. AI plays a major role here in three possible ways:
  - Robustness
  - Response
  - Resilience

Application of AI for Cyber-attack detection and response can be bifurcated into three parts, shown in Figure 8.5. Data collection is done via consumer environment while it is processed by a system which is managed by a security vendor [35]. Detection system gives alert for malicious activity, which is used to turn on an action in response. Organizations have reported that with time cyberat-tacks are rising because of the implementation of IoT, transmission of mobile devices and varying threat landscape [36,37]. Two security measures can be taken in this scenario:

- Speed up the defenders
- Slow down the hackers.

Generally, cyber security professionals use AI to resolve five essential questions, as depicted in Figure 8.6.

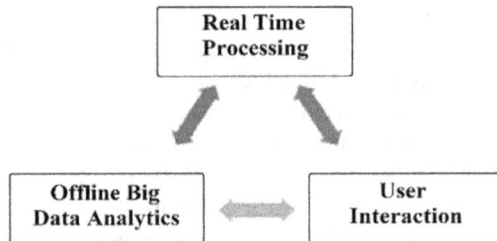

FIGURE 8.5   AI in cyber security detection and response.

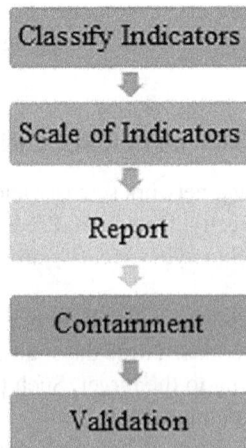

FIGURE 8.6   General process of AI in cyber security.

8.5 BENEFITS OF AI IN CYBER SECURITY

AI techniques have the capability to replicate human intelligence. It has massive prospective in the field of AI. Nowadays, cyberattacks are raising, even the level of complexity is more. Every organization has a team of security experts but sometimes they are not capable to handle that degree of threats [38,39]. If they went unnoticed, they may cause huge loss and sometimes damage the whole network. Therefore, security professionals must have strong support from intelligence machines, softwares and latest technologies for flourishing safety of concerned organizations from cyberattacks. AI-driven systems can be trained to generate threat alerts, recognize malwares, identify fresh malwares and safeguard sensitive information of an organization. Some of the significant benefits of integrating AI with cyber security are as follows [40,41]:

- **AI Identifies Threats Faster:** Almost every year, cyber criminals initiate a large number of attacks with different objectives. Unknown threat must be identified quickly; otherwise, it may collapse the entire network. AI is a verified technology, capable of examining relationships among threats within second or minute.
- **AI Accelerates Detection:** Threats recognition at the initial stage can protect any organization. AI technologies are efficient to scan the entire network quickly for any potential threat and shorten safety tasks unlike humans.
- **AI Eliminates Time-Consuming Task:** AI identifies threats quickly, provides precise risk analysis and simplifies security efforts. Decisions can be made instantly to remediate cyber threats.
- **AI Is Capable to Handle Huge Data:** Mostly, mid-scale organizations have massive traffic because a number of activities were occurring on their network. Therefore, data safety and security is must from malicious softwares and people [42]. Only the cyber security professionals cannot handle the entire traffic and hence AI-enabled algorithm do it quickly among the massive information and traffic. Hidden threats can also be screened and handled.
- **AI Provides Enhanced Vulnerability Management:** Key to protect any organization network is vulnerability management. Many threats can be viewed almost every day and hence it must be identified and handled rapidly. Researchers can contribute efficiently to vulnerability management by using AI and accessible security measures.
- **AI Provides Enhanced Security:** Organizations may face numerous kinds of threats as cyber criminals tend to change their strategies. Prioritization among safety task must be done to deal with potential threat first [43]. Hence, deploying AI on network manages various natures of threats and prioritizes them to safeguard the whole organization network.
- **AI Helps in Securing Validation:** Most of the time users are creating an account by submitting various personal details to avail any service or buy or sell products. AI helps in securing authentications every time user is doing login in account [44]. Multiple tools are used by AI for recognition of genuine login such as fingerprint, face recognition, CAPTCHA, security questions and OTP.

8.6 AI-BASED CYBER SECURITY TOOLS

Cyber criminals employed AI within deep locker malware due to which many organizations suffered from brutal downfall. This malware is capable to bypass high cyber security systems because it uses AI models for hitting networks via face recognition and speech recognition [45]. This incident has drawn the attention of integrating AI into the cyber security domain. Therefore, to protect networks from such a kind of attacks, it is said that the AI for Cyber security is a necessity now. Figure 8.7 shows some of the AI-based tools used by many organizations.

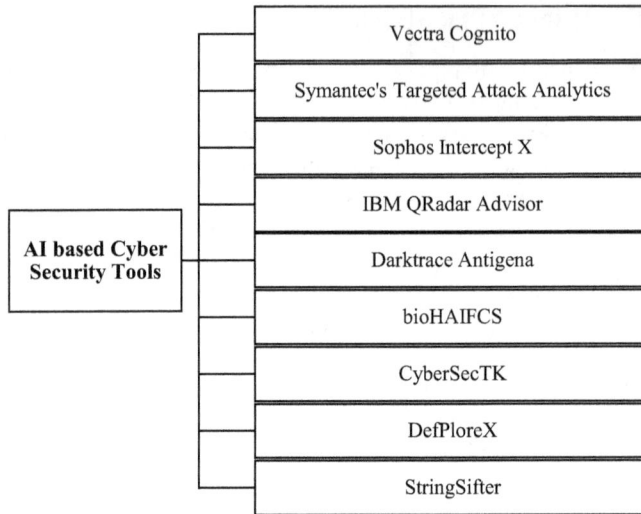

FIGURE 8.7 AI-based cyber security tools.

Nowadays, every organization even the startup is spending a lot in creating AI-enabled systems. These systems have the capability to analyze a huge volume of data to help cyber security professionals for potential threat identification. This, in turn, activates to take precautionary measures to resolve them. Various tools are available now which are using AI for improving cyber security [46]. They have been proven as a boon to achieve best security; some of the popular tools are as follows:

- **Vectra Cognito:** This platform is having the capability to detect hackers in real-time scenarios using AI. This tool is automated for the threat recognition and classification. Cognito uses logs, network metadata, cloud events and behavioral detection algorithm to analyze for detection of buried attackers among workloads and IoT devices. Cognito platform contains two portions:
  - **Cognito Detect:** It is used for determination of hidden trackers. This function is performed in real-time scenarios using data science, machine learning and deep learning.
  - **Cognito Recall:** It helps to find out exploits available in historical data. Threats investigation is speedup with actionable context.
- **Symantec's Targeted Attack Analytics (TAA):** This tool was developed by Symantec Security. The objective is to expose silent and targeted attack. AI and Machine Learning were applied for the processes, knowledge base and capabilities. This developed tool TAA was successfully used to oppose the Dragonfly 2.0 attack. This attack was for energy organizations to get entrée for operational network.
- **Sophos's Intercept X Tool:** A British hardware and software safety Company is Sophos. Intercept X signifies the use of deep learning neural network and its working is the same as the human brain. Earlier than a folder is performed, intercept X try to recover millions of characteristics from folder and then carry out exhaustive examination. Finally, decision is made whether a folder is genuine or harmful. This task is performed within 20 milliseconds.
- **Darktrace Antigena:** One of the efficient approaches for self-defense is Darktrace. Antigena is capable to extend the significant functionality of Darktrace, for detecting and duplicating digital antibodies' functions. Therefore, they can deactivate threats and viruses. Whole task is performed without the involvement of human, information about past attacks, regulations, etc.

- **IBM QRadar Advisor:** It is based on IBM Watson technology. AI is used to auto inspect indication of any susceptibility to counter cyberattacks. QRadar advisor makes use of cognitive reasoning to get significant insight and accelerates the response method. With this tool, security professionals can evaluate threats and reduces the probability of missing them. Some of the significant features of this tool are as follows:
  - Automatic incident examination
  - Detection of high priority risk
  - Provide elegant reasons
  - Key insights about users
  - Key insights for significant assets
- **bioHAIFCS:** It is a bio-inspired hybrid intelligence framework. It involves evolutionally computational intelligence algorithms for shielding critical network applications like military information system.
- **StringSifter:** It is a machine learning tool and it performs automatic string ranking based on the significance of malware investigation.

8.7 GROWTH OF AI IN CYBER SECURITY

Overall growth of AI in cyber security marketplace is backed by the troublesome digital technologies among industries upright, budding cyber threats and growing demand of advanced cyber security privacy and solutions [45]. AI in cyber security market was appreciated in the year 2019 at USD 8.6 billion and is expected to reach in the year 2026 by a growth of USD 43.6 billion with a growth of approx. 25% over the forecast period 2020–2026. Key factors of AI technology contributing to industrial growth are as follows:

- Expending digitization rate
- Modifying traditional banking to digital
- Growing investment by government
- Reducing data breach
- Advanced training and resources
- Increasing awareness among users and skilled AI professionals
- Integrating advanced technologies with AI
- Continuous supervision and adaptation of versatile security vulnerabilities

A report has been prepared to analyze the impact of COVID-19 on AI in cyber security on the basis of mentioned factors:

- Boost in data usage
- Increase in internet access due to work from home and e-healthcare
- Imposed lockdown in worldwide

Most of the organizations have implemented work from home model to provide services and hence during lockdown a sudden increase in cyberattacks has been observed. Crises have been exploited by cyber criminals by developing fresh cyber-associated attacks [47]. It has raised the demand of AI-incorporated cyber security solutions. Integration of AI in cyber security market is segmented into many categories, demonstrated in Figure 8.8.

- AI in cyber security market by offering implies hardware, software, services, etc.
- AI in cyber security market by deployment means on cloud or on premises
- AI in cyber security market by security is related to cloud security, network security, endpoint security, application security, etc.

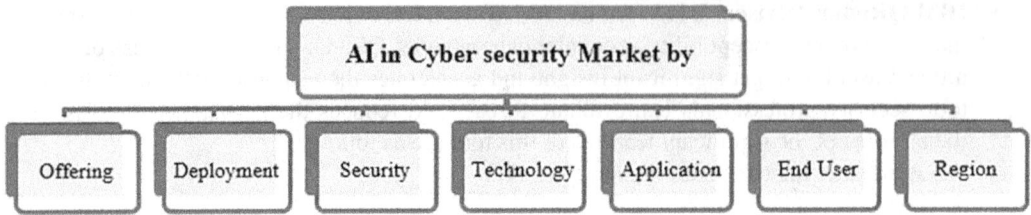

FIGURE 8.8 AI in cyber security market segment.

- AI in cyber security market by technology consists of machine leaning, deep learning, natural language processing, computational intelligence, etc. [48,49].
- AI in cyber security market by applications comprises various tasks like
  - Data loss prevention
  - Threat identification
  - Security management
  - Vulnerability management
  - Fraud detection
  - Access management
  - Threat and compliance management
  - Intrusion detection system
  - Threat intelligence
  - Antivirus
  - Web filtering
  - Patch Management, etc.
- AI in cyber security market by end user means Enterprise, Healthcare, Defense, Government, Manufacturing, Retail, Automotive, Education, Transportation, Energy, Infrastructure, etc.
- AI in cyber security market by market depicts Europe, Asia Pacific, North America, Rest of the World, etc.

Cyber security industries and professionals are working very hard and twisted to apply every tool to encounter the challenges faced by cyberattacks. It is certainly bringing out the new trend of AI-powered solutions [50,51]. As per the current report published by International Data Corporation (IDC), worldwide spending on cyber security reaches around USD 100 billion in 2019, which is growing annually. Some of the top cyber security organizations, which are using AI and Machine Learning, are shown in Figure 8.9.

8.8   CHALLENGES AND LIMITATIONS

AI provides numerous benefits to cyber security systems. But still some challenges are there due to which it is still not a key security tool [41,52]. Some of the factors are as follows:

- **Cyber Criminals Also Use AI:** AI technology is of dual use, i.e., it can be used for defense or offense. Cyber security is controlled by regulations while hackers have freedom. The cost of budding applications diminishing; therefore, it is very easy for hackers to control the technology for destructive purposes. Hackers are using AI technology for quick, accurate, efficient and vicious hacking to introduce fresh malicious attack [9,32]. Hackers investigate and advance their malware so that it can breach AI-enabled security tools, and more advanced attack can be held.

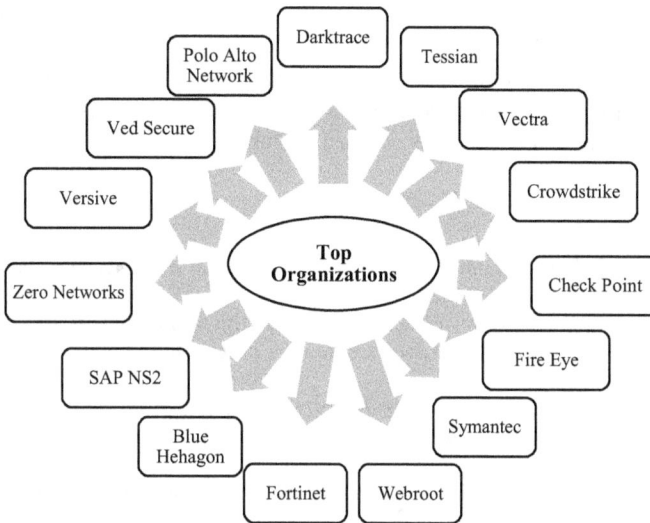

FIGURE 8.9   Top cyber security organizations using AI and ML.

- **Resources:** Organizations are required to spend lots of money, time and security professionals for resources. It may be memory required, training programs, computing power and information to develop and prolong AI systems [29].
- **Datasets:** These are required to train AI models. Cyber security professionals must have good hands-on various kinds of datasets of malware codes, anomalies and malicious codes. Sometimes, organizations have deficiencies in terms or resources and time related to precise datasets.
- **High Adoption Barriers:** Organizations must construct and maintain AI systems by investing sufficient funds. With the developing technologies, these costs decrease and hence developing quality servers became inexpensive. By that time, efficiency of automating security grows and became the prerequisite for organizations to run on cloud [43]. Some of the significant roadblocks for its acceptance and its deployment are talent acquirement, complexity in data and AI tool employments.
- **Neural Fuzzing:** This is the process of vulnerability identification of testing a huge amount of arbitrary input data. This process occurs within software. Though fuzzing is very beneficial but hackers may identify some target weakness via neural network. Microsoft developed a process for this approach to advance softwares for securing more codes which is difficult to exploit.

8.9   CONCLUSION

In today's professional environment, possession of data and network safe and secure from cyber-attacks is not an easy task. Frequency, volume and nature of Cybercrimes are getting complex; therefore, the techniques used for cybersecurity must be more robust and intelligent. It will help to strengthen defense mechanisms for real-time decision making. AI and Machine Learning have enormous potential for making a better world to enrich our lives. Emerging technologies can help cyber security professionals to improve network security, cyber threat identification and eliminate breach risk. Hackers can also use AI to make Cybercrimes in a sophisticated manner and penetrates into various domains of organizations. Many organizations are successfully using AI-enabled cyber security tools, which work efficiently for threat detection, prevention and indication. Globally, it has a huge impact in a constructive way on market growth as well. AI in cyber security market would be appreciated by about USD 43.6 billion till the year 2026 and will keep growing at a fast rate. Researchers are still skewing the domain of cyber security with the application of emerging technologies particularly around Intrusion Detection Systems. The objective is to accomplish justifiable, adjustable, flexible and authenticated cyber security solutions.

REFERENCES

1. Liu J, Kong X, Xia F, Bia X, Wang L, Qing Q, Lee I, "Artificial Intelligence in the 21st Century," *IEEE Access*, vol. 6, pp. 34403–34421 (2018).
2. Cheng S, Wang B, "An Overview of Publications on Artificial Intelligence Research: A Quantitative Analysis on Recent ss," *International Joint Conference on Computational Sciences and Optimization*, pp. 683–686 (2012).
3. Katiyar S, Towards Hybridization of Nature Inspired Metaheuristic Techniques for Collision Free Motion Planning," *International Journal of Computer Science and Network (IJCSN), Technologies*, vol. 7, pp. 192–198 (2018).
4. Chen J, See KC, "Artificial Intelligence for COVID-19: Rapid Review," *Journal of Medical Internet Research*, vol. 22 (2020).
5. Kotsiantis SB, "Supervised Machine Learning: A Review of Classification Techniques," *Informatics*, vol. 31, pp. 249–268. (2007).
6. Okutan A, Eyüpoğlu C, "A Review on Artificial Intelligence and Cyber Security," *International Conference on Computer Science and Engineering*, pp. 304–309 (2021).
7. Patil P, "Artificial Intelligence in Cybersecurity," *International Journal of Research in Computer Applications and Robotics*, vol. 4 (2016).
8. Truong TC, Diep QB, Zelinka I, "Artificial Intelligence in the Cyber Domain: Offense and Defense," *Symmetry (Basel)*, vol. 12, pp. 1–24 (2020).
9. Siddiqui Z, Husain MS, Yadav S, "Application of Artificial Intelligence in Fighting Against Cyber Crimes: A Review," *International Journal of Advanced Research in Computer Science*, vol. 9, pp. 118–122 (2018).
10. Sadiku MNO, Fagbohungbe OI, Musa SM, "Artificial Intelligence in Cyber Security," *International Journal of Engineering Research and Advanced Technology*, vol. 06, no. 05, pp. 01–07, 2020,
11. Lu Y, "Artificial Intelligence: A Survey on Evolution, Models, Applications and Future Trends," *Journal of Management Research and Analysis*, vol. 6, pp. 1–29 (2019).
12. Oche JO, "The Risk of Artificial Intelligence in Cyber Security and the Role of Humans," *Texila International Journal of Academic Research* (2019).
13. Wiafe I, Koranteng FN, Obeng EN, Assyne N, Wiafe A, Gulliver SR, "Artificial Intelligence for Cybersecurity: A Systematic Mapping of Literature," *IEEE Access*, vol. 8 (2020).
14. Dilek S, Cakár H, Aydán M, "Applications of Artificial Intelligence Techniques to Combating Cyber-Crimes: A Review," *International Journal Artificial Intelligence and Applications*, vol. 6, pp. 21–39 (2015).
15. Walt EV, Eloff JHP, Grobler J, "Cybersecurity: Identity Deception Detection on Social Media Platforms," *Computers and Security*, vol. 78, pp. 76–89 (2018).
16. Li Y, Liu Q, "A Comprehensive Review Study of Cyber-Attacks and Cyber Security; Emerging Trends and Recent Developments," *Energy Reports*, vol. 7, pp. 8176–8186 (2021).
17. Rahim NHA, Hamid SB, Laiha M, Kiah M, Shamshirband S, Furnell S, "A Systematic Review of Approaches to Assessing Cybersecurity Awareness," *Computer Science Kybernetes*, vol. 44, pp. 606–622 (2015).
18. Al-Mhiqani MN, Ahmad R Mohamed W, Hasan A, "Cyber-Security Incidents: A Review Cases in Cyber Physical Systems," *International Journal of Advanced Computer Science and Applications*, vol. 9, pp. 499–508 (2018).
19. Kruse CS, Frederick B, Jacobson T, Monticone DK, "Cybersecurity in Healthcare: A Systematic Review of Modern Threats and Trends," *Technology in Health Care*, vol. 25, pp. 1–10 (2017).
20. Bhamare D, Zolanvari M, Erbad A, Jain R, Khan K, Meskin N, "Cybersecurity for Industrial Control Systems: A Survey," *Computers and Security*, vol. 89 (2020).
21. D Wu et al., "Cybersecurity for Digital Manufacturing," *Journal of Manufacturing Systems*, vol. 48, pp. 3–12 (2018).

22. Solms RV, Niekerk JV, "From Information Security to Cyber Security," *Computers and Security*, vol. 38, pp. 97–102 (2013).
23. Thakur K, Qiu M, Gai K, Ali ML, "An Investigation on Cyber Security Threats and Security Models," *International Conference on Cyber Security and Cloud Computing*, pp. 307–311 (2015).
24. Nishant R, Kennedy M, Corbett J, "Artificial Intelligence for Sustainability: Challenges, Opportunities, and a Research Agenda," *International Journal of Information Management*, vol. 53, p. 102104 (2020).
25. Roberts MJ, "The Cyber Threat and Globalization: The Impact on US National and International Security," *Journal of Strategic Security*, vol. 11 (2019).
26. Humayun M, Niazi M, Jhanjhi N, Alshayeb M, Mahmood S, "Cyber Security Threats and Vulnerabilities: A Systematic Mapping Study," *Arabian Journal for Science and Engineering*, vol. 45, pp. 3171–3189 (2020).
27. Razzaq A, Hur A, Ahmad HF, Masood M, "Cyber security: Threats, reasons, challenges, methodologies and state of the art solutions for industrial applications," In *IEEE Eleventh International Symposium on Autonomous Decentralized Systems (ISADS)* (2013).
28. Cui Z, Xue F, Cai X, Cao Y, Wang GG, Chen J, "Detection of Malicious Code Variants Based on Deep Learning," *IEEE Transaction on Industrial Informatics*, vol. 14, pp. 3187–3196 (2018)
29. Kieyzun A, Guo PJ, Jayaraman J, Ernst MD, "Automatic creation of SQL injection and cross-site scripting attacks," In *Proceedings of the 31st International Conference on Software Engineering*. IEEE Computer Society (2009).
30. Kustarz C, "System and Method for Denial of Service Attack Mitigation Using Cloud Services," Google Patents (2016).
31. Coulter R, Han QL, Pan L, Zhang J, Xiang Y, "Data-Driven Cyber Security in Perspective – Intelligent Traffic Analysis," *IEEE Transactions on Cybernetics*, vol. 50, pp. 3081–3093 (2020).
32. Raban Y, Hauptman A, "Foresight of Cyber Security Threat Drivers and Affecting Technologies," *Foresight*, vol. 20, pp. 353–363 (2018).
33. Radanliev, P, De Roure, D, Page, K, Nurse JRC, Santosh O, Burnap P "Cyber Risk at the Edge: Current and Future Trends on Cyber Risk Analytics and Artificial Intelligence in the Industrial Internet of Things and Industry 4.0 Supply Chains," *Cybersecurity*, vol. 3 (2020).
34. Trifonov R, Nakov O, Mladenov V, "Artificial Intelligence in Cyber Threats Intelligence," *International Conference on Intelligent and Innovative Computing Applications*, pp. 1–4 (2018).
35. Abomhara M, Køien GM, "Cyber Security and the Internet of Things: Vulnerabilities, Threats, Intruders and Attacks," *Journal of Cyber Security*, vol. 4, pp. 65–88 (2015).
36. Wilner AS, "Cybersecurity and Its Discontents: Artificial Intelligence, the Internet of Things, and Digital Misinformation," *International Journal: Canadas Journal of Global Policy Analysis*, vol. 73, pp. 308–316 (2018).
37. Kuzlu M, Fair C, Guler O, "Role of Artificial Intelligence in the Internet of Things (IoT) cybersecurity," *Discover Internet of Things*, vol. 7 (2021)
38. Sedjelmaci H, Guenab F, Senouci SM, Moustafa H, Liu J, Han S, "Cyber Security Based on Artificial Intelligence for Cyber-Physical Systems," *IEEE Network*, vol. 34, no. 3, pp. 6–7 (2020).
39. Collins C, Dennehy D, Conboy K, Mikalef P, "Artificial Intelligence in Information Systems Research: A Systematic Literature Review and Research Agenda," *International Journal of Information Management*, vol. 60 (2021).
40. Sarker IH, Furhad MH, Nowrozy R, "AI-Driven Cybersecurity: An Overview, Security Intelligence Modeling and Research Directions," *SN Computer Science*, vol. 173 (2021).
41. Li JH, "Cyber Security Meets Artificial Intelligence: A Survey," *Frontiers of Information Technology and Electronic Engineering*, vol. 19, pp. 1462–1474 (2018).
42. Farivar F, Haghighi MS, Jolfaei A, Alazab M, "Artificial Intelligence for Detection, Estimation, and Compensation of Malicious Attacks in Nonlinear Cyber Physical Systems and Industrial IoT," *IEEE Transactions on Industrial Informatics*, vol. 16, pp. 2716–2725 (2020).
43. Yu S, Carroll F, "Implications of AI in National Security: Understanding the Security Issues and Ethical Challenges," *Artificial Intelligence in Cyber Security: Impact and Implications*, pp. 157–175 (2021)
44. Tao F, Akhtar SH, Jiayuan Z, "The Future of Artificial Intelligence in Cybersecurity: A Comprehensive Survey," *EAI Endorsed Transactions on Creative Technologies*, vol. 8 (2021).
45. Soni N, Sharma EK, Singh N, Kapoor A, "Artificial Intelligence in Business: From Research and Innovation to Market Deployment," *Procedia Computer Science*, vol. 167, pp. 2200–2210 (2020).
46. Naik B, Mehta A, Yagnik H, Shah M, "The Impacts of Artificial Intelligence Techniques in Augmentation of Cybersecurity: A Comprehensive Review," *Complex and Intelligent Systems* (2021)

47. Morel B, "Artificial Intelligence and Key to the Future of Cybersecurity," *Proceedings of ACM Conference in Computer Communication and Security*, pp. 93–97 (2011).

48. Zhang Y, "Research on Artificial Intelligence Machine Learning Character Recognition Based on Online Machine Learning Method," *IEEE International Conference of Safe Production and Informatization*, pp. 649–652 (2020).

49. Das R, Morris TH, "Machine Learning and Cyber Security," *International Conference on Computer, Electrical & Communication Engineering*, pp. 1–7 (2017).

50. Shafique K, Khawaja BA, Sabir F, Qazi S, Mustaqim M, "Internet of Things (IoT) for Next-Generation Smart Systems: A Review of Current Challenges, Future Trends and Prospects for Emerging 5G-IoT Scenarios," *IEEE Access*, vol. 8, pp. 23022–23040 (2020).

51. Branitskiy A, Kotenko I, "Applying Artificial Intelligence Methods to Network Attack Detection," *AI in Cybersecurity. Intelligent Systems, Reference Library*, vol. 151 (2019)

52. Chen X, Yu L, Wang T, Liu A, Wu X, Zhang B, Lv Z, Sun Z, "Artificial Intelligence-Empowered Path Selection: A Survey of Ant Colony Optimization for Static and Mobile Sensor Networks," *IEEE Access*, vol. 8, pp. 71497–71511 (2020).

9 Cloud Computing
An Overview of Security Risk Assessment Models and Frameworks

Subarna Ghosh
Maharishi International University

Nafees Mansoor
University of Liberal Arts Bangladesh (ULAB)

Mohammad Shahriar Rahman
United International University (UIU)

CONTENTS

9.1 INTRODUCTION

Cloud computing has become a dominant model in the realm of information technology (IT), since it has the potential to transform a large portion of the IT industry [1]. On the other hand, the term 'Cloud' is used as a metaphor for the Internet, where resources, information, software, and hardware are shared [2]. Primarily, there exist three major service models for cloud computing, namely, software as a service (SaaS), platform as a service (PaaS), and infrastructure as a service (IaaS). In SaaS, the consumer controls only the application configurations, whereas, in PaaS, consumers can

DOI: 10.1201/9781003267812-9

control only the hosting environment. On the other hand, the consumer controls everything except the data center infrastructure in IaaS.

Furthermore, cloud computing has four main deployment models, which are public, community, private, and hybrid clouds [3]. These models specify the cloud infrastructure look and define the relationships between the infrastructure and the users. It is worthwhile to mention that the cloud providers mainly build large data centers at low cost, organize and provide computing resources, and achieve efficient resource utilization [4,5]. Hence, the main advantages of using cloud computing are scalability, ubiquitous availability, and maintenance costs [2].

Considering the usages and advantages, a good number of research works have been conducted lately on cloud computing. However, there remain very few survey analyses on the cloud pre-migration models and risk management frameworks. Hence, here, we have studied existing models and risk management frameworks for cloud computing and analyzed the performances to identify their strengths and limitations. For this purpose, cloud clients (CCs) are categorized into two types. One is the client who is yet to be migrated into the cloud, while the other type is the existing CC. For the first type of users, we have analyzed two cloud security assessment models, namely, Cloud Adoption Risk Assessment Model (CARAM) and Consultative, Objective, and Bi-functional Risk Analysis (COBRA). The model abbreviated as CARAM in Ref. [6] is a qualitative risk assessment model, which helps clients to select the appropriate cloud service provider (CSP) based on the risk factors regarding security, privacy, and service. On the other hand, COBRA in Ref. [7] consists of a range of risk analysis, consultative, and security review tools, which have been developed in full cooperation with one of the world's major financial institutions and followed by many years of research. Comparing the performance of both the models, it has been observed that the CARAM offers a more granular level of risk assessment and selects a CSP based on security, privacy, and service. Hence, in CARAM, the decision is accepted to be more precise in terms of the risk values on cloud customers' acceptance level. On the other hand, decision accuracy in COBRA may exhibit less precision and result in a vague outcome.

On the other hand, for existing CCs, we have considered three different cloud risk management frameworks for performance analysis purposes. The cloud security risk management framework abbreviated as CSRMF in Ref. [8] offers the clients to identify the impacts of cloud-specific security risks on the business objectives of the organization. Hence, this framework facilitates both organizations and CSPs to identify, analyze, and evaluate security risks in cloud computing platforms and establish the best solution for mitigating them. It has been anticipated that the CSRMF provides an adequate level of confidence in CC for organizations and cost-effective productivity for CSP [8]. The work presented in Ref. [9] proposed a cloud security risk assessment framework for CSPs that pulls cloud customers at the early stages of the risk assessment process. On the other hand, the framework presented in Ref. [10] proposes a security framework that can be used by CSPs to perform a risk analysis, risk assessment, and risk mitigation.

The rest of the chapter is organized as follows. Section 9.2 provides an overview of the existing models and frameworks. Furthermore, this section also highlights the strengths and limitations of these studied models and frameworks. Next, in Section 9.3, the performance comparisons of the mentioned models and frameworks are presented with some suggestions. The chapter is concluded in Section 9.4 highlighting a few open research recommendations.

9.2 EXISTING SECURITY RISK ASSESSMENT MODELS & FRAMEWORKS

We have gone through multiple risk assessment models and frameworks and then we have chosen two cloud assessment models (CARAM, COBRA) and three cloud security risk assessment frameworks for further review. At the end of this paper, we have drawn a comparative analysis between the risk assessment models and frameworks.

9.2.1 CLOUD RISK ASSESSMENT MODELS

9.2.1.1 Cloud Adoption Risk Assessment Model

CARAM [6] is a risk assessment model derived from the core standards of the European Network and Information Security Agency (ENISA) [11] and Cloud Security Alliance (CSA) [12]. CARAM helps Cloud Service Customers (CSCs) to choose a CSP based on assessing the risk factors they might have with the CSP. The framework for this model has the following components:

1. Questionnaire for the potential CSC. Through this questionnaire, a CSC can assess the risk level corresponding to the incident probability.
2. Analyzing and classifying discrete answers in the Consensus Assessment Initiative Questionnaire (CAIQ), provided by the CSP.
3. A model for mapping both questionnaires for the CSC and CSP to risk values.
4. Providing a multi-criteria decision approach to CSC for comparing multiple CSPs and eventually selecting the most appropriate one by assessing the relative risk through analyzing a few risk factors for Security, Privacy, and Quality of Service.

The classical questionnaire from ENISA for CSCs is based on 35 incident scenarios and 53 vulnerabilities, and the risk level is evaluated by a qualitative scale as below (Figure 9.1):

- **Low Risk:** 0–2
- **Medium Risk:** 3–5
- **High Risk:** 6–8

The problem with the generic scale is that practically these risk values do not cover many risk factors related to different CSPs and CSCs. These values also cannot be applied for a specific CSP and CSC pair. To fill up this gap, CARAM has established a significant role. At first CARAM maps, quantitative risk level value (probability and impact value pair) instead of qualitative scale is used by the ENISA as follows:

- **Very Low Risk:** 1
- **Low Risk:** 2
- **Medium Risk:** 3
- **High Risk:** 4
- **Very High Risk:** 5

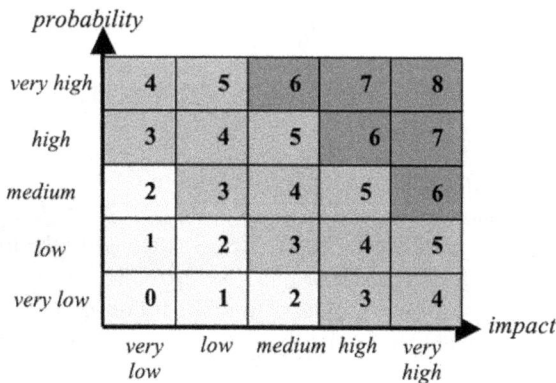

FIGURE 9.1 ENISA definitions of risk levels (ranging from 0 to 8).

R.1 LOCK-IN

Probability	HIGH	Comparative: Higher
Impact	MEDIUM	Comparative: Equal
Vulnerabilities	V13. Lack of standard technologies and solutions	
	V46. Poor provider selection	
	V47. Lack of supplier redundancy	
	V31. Lack of completeness and transparency in terms of use	
Affected assets	A1. Company reputation	
	A5. Personal sensitive data	
	A6. Personal data	
	A7. Personal data - critical	
	A9. Service delivery – real time services	
	A10. Service delivery	
Risk	HIGH	

FIGURE 9.2 Sample risk factor 'Lock-in' and its assessment parameters.

Figure 9.2 is from ENISA, where for the 'Lock-in' risk the Probability (P) is High and the Impact (I) is Medium. The values for P and I are 4 and 3, respectively, according to the CARAM quantitative risk scale. Practically, the probability and impact have a great dependency on the vulnerabilities and assets.

Thus, CARAM introduces new equations keeping the original P and I as a baseline,

$$\beta_i = P_i * \vartheta_i \tag{9.1}$$

$$\delta_i = I_i * \alpha_i \tag{9.2}$$

where β_i and δ_i are adjusted probability and adjusted impact, ϑ_i is the vulnerability index of a given CSP, and α_i is the asset index of a given CSC. We can get ϑ_i and α_i from the following two equations:

$$\vartheta_i = \frac{\sum_{k=1}^{53} v_{ki} * \epsilon_k}{\sum_{k=1}^{53} v_{ki}} \tag{9.3}$$

$$\alpha_i = \frac{\sum_{k=1}^{23} \alpha_{ki} * \gamma_k}{\sum_{k=1}^{23} \alpha_{ki}} \tag{9.4}$$

Here, v_{ki} is 1 if k is available in the total 53 vulnerabilities of the ENISA's risk scenario list and 0 otherwise. α_{ki} is 1 if k is available in the total 23 assets of the ENISA's risk scenario and 0 otherwise. The asset-related parameter γ_k is derived from the CSC's response to the following question from the questionnaire and it is 1 for the answer 'Yes' and 0 for 'No' CSC's Question to calculate γ_k: Does the service that you seek will involve any asset of yours that fall in the same category as asset k? To compute the vulnerability-related parameter ϵ_k, CSP's response to CAIQ is evaluated. CAIQ has been developed by the CSA, and CARAM uses CAIQ to derive the value for ε_k. The main objective of CAIQ is to gather data from CSP containing their compliance with the standards and their level of security in infrastructure. The expected answer from the CSP is Boolean, i.e., either 'Yes' or

'No'. However, the CSPs, very often, answer the questionnaire in STAR (Security, Trust, Assurance, and Risk) registry elaborately rather than the Boolean format, and hence it becomes difficult to fit the non-Boolean answers in the automated analysis. To overcome this issue, CARAM introduces a mechanism by which the CSP's non-Boolean answers can be mapped into the Boolean format. Besides a machine learning algorithm for data mining provided by the Waikato Environment for Knowledge Analysis (WEKA) [13], the tool is also used in CARAM to automate the categorization (Boolean Yes or No) for the Boolean answers in CAIQ questions which has an accuracy rate of 84 CSPs, the following equation is used to derive the value of ϵ_k:

$$\varepsilon_i = \frac{\sum_{m=1}^{n} \psi_{m,k} * q_m}{\sum_{m=1}^{n} \psi_{m,k} * b_m} \tag{9.5}$$

In Equation 9.5, the implementation value q_m is 0 if the answer to any questions of CAIQ is 'Yes' and the value is 1 if the answer is either 'No' or 'Not Available'. $\psi_{m,k}$ maps the CAIQ questions to vulnerabilities and n refers to the number of questions. $\psi_{m,k}$ is 1 if m is related to vulnerability k and it is 0 otherwise. Again, for the answer 'Not Applicable', the value b_m is 0 and 1 otherwise. The value of ε_k can be a minimum of 0 when all the controls related to the vulnerability are implemented and the non-zero value of ε_k means that the controls associated with the vulnerabilities are not implemented and its maximum value is 1, which implies that the CSP has no measures against the available vulnerabilities k. CARAM also shrinks 35 incident scenarios provided by ENISA to the following three categories of cloud risks to simplify selecting the best CSP according to the CSC's requirements.

- Security
- Privacy
- Service

The corresponding incident probability (β) and impact (δ) of the Security (β_s, δ_s), Privacy (β_r, δ_r), and Service (β_e, δ_e) can be derived from the following equations:

$$\beta_r = \frac{\sum_{i=1}^{35} \beta_i * r_i * \omega_{ri}}{\sum_{i=1}^{35} r_i * \omega_{ri}} \tag{9.6}$$

$$\beta_a = \frac{\sum_{i=1}^{35} \beta_i * S_i * \omega_{si}}{\sum_{i=1}^{35} S_i * \omega_{si}} \tag{9.7}$$

$$\beta_e = \frac{\sum_{i=1}^{35} \beta_i * e_i * \omega_{ei}}{\sum_{i=1}^{35} e_i * \omega_{ei}} \tag{9.8}$$

$$\delta_r = \frac{\sum_{i=1}^{35} \delta_i * r_i * \omega_{ri}}{\sum_{i=1}^{35} r_i * \omega_{ri}} \tag{9.9}$$

$$\delta_a = \frac{\sum_{i=1}^{35} \delta_i * S_i * \omega_{si}}{\sum_{i=1}^{35} S_i * \omega_{si}} \tag{9.10}$$

$$\delta_e = \frac{\sum_{i=1}^{35} \delta_i * e_i * \omega_{ei}}{\sum_{i=1}^{35} e_i * \omega_{ei}} \tag{9.11}$$

Here, ω and α are weight factors for incident and impact probabilities, respectively, and the values can be between 0 and 1. After calculating probability and impact, a higher resolution qualitative scale with ten values, as shown in Table 9.1, compared to ENISA's five values can be mapped.

Hence, the probability and the impact value are converted as risk level is fine-tuned for each incident scenario and so is the quantitative scale as shown in Figure 9.3 (between 0 and 18 instead of 0 and 8 in Figure 9.1).

TABLE 9.1

High-Resolution Qualitative Scale of Risk Level from CARAM

0, 0.5	Negligible
0.5, 1.0	Extremely low
1.0, 1.5	Very low
1.5, 2.0	Low
2.0, 2.5	Below average
2.5, 3.0	Above average
3.0, 3.5	High
3.5, 4.0	Very high
4.0, 4.5	Extremely high
4.5, 5.0	Not recommended

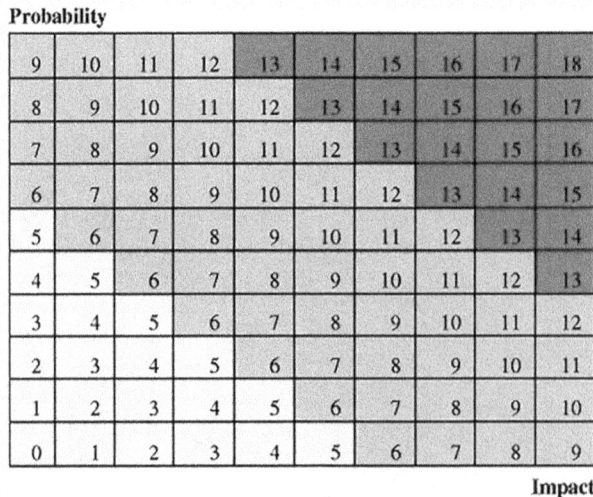

Probability

9	10	11	12	13	14	15	16	17	18
8	9	10	11	12	13	14	15	16	17
7	8	9	10	11	12	13	14	15	16
6	7	8	9	10	11	12	13	14	15
5	6	7	8	9	10	11	12	13	14
4	5	6	7	8	9	10	11	12	13
3	4	5	6	7	8	9	10	11	12
2	3	4	5	6	7	8	9	10	11
1	2	3	4	5	6	7	8	9	10
0	1	2	3	4	5	6	7	8	9

Impact

FIGURE 9.3 Revised ENISA definitions of risk levels (ranging from 0 to 18).

CSC (the CSC that needs the relative risk assessment) then provides CARAM the maximum acceptable risk values for Security (R_{smax}), Privacy (R_{rmax}), and Service (R_{emax}). CSC may also provide a list of number of CSPs that are to be excluded from the assessment. Thus, CARAM creates a set F of suitable CSPs out of total assessed CSPs, S (the CSPs that have filled up CAIQ), thus $F \subset S$ and $pi \in F$, if $(pi \in / U) \wedge (R_{rmax} > R_{ri}) \wedge (R_{smax} > R_{si}) \wedge (R_{emax} > R_{ei})$.

Here, R_{si}, R_{ri}, and R_{ei} are the risk values of Security, Privacy, and Service of the CSP, pi. The value of F can be zero, one, or more than one. Zero value means there is no feasible CSP for the CSC. One value means there is only one feasible CSP for the CSC. In case of multiple values of F, all the dominating CSPs are removed from the set F and a new set F' is created with the non-dominating CSPs based on the lower value of R_{si}, R_{ri}, R_{ei}. Then, the new feasible preference list of CSPs from the new set F is provided to the CSC.

9.2.1.2 Consultative, Objective, and Bi-Functional Risk Analysis

COBRA consists of a range of risk analysis, consultation, and security review tools that were developed largely in recognition of the changing nature of it and security and the demands placed by business upon these areas. Cobra was designed to give organizations the means to perform a self-assessment of their IT security without the need for external assistance from consultants. COBRA and its default methodology evolved very fast to tackle these issues properly; it was developed in full cooperation with one of the world's major financial institutions and followed by many years of research. It was and is becoming largely expected that security reviews should be business related to cost-justified solutions and recommendations. Another issue most of the late 90s is the search by many organizations for a better and more visible return on their security budgets.

The risk assessment process, using COBRA, is extremely flexible. A substantial number of approaches are supported. However, the default process usually consists of three stages

- Questionnaire Building
- Risk Surveying
- Report Generation

During the first stage, via module selection or generation, the base questionnaire is built to fit the environment and requirements of the user.

- Consideration of specific aspect of security risk
- Performing risk analysis in various proposed scenarios
- Analysis of all risk areas even if some are not of real significance to the organization

The second stage is the survey process – Risk Consultant questions are answered by appropriate personnel and the information is securely stored. For the third stage, risk assessments and 'scores' are produced for individual risk categories, individual recommendations are made and solutions offered, and potential business implications are explained.

- Recommended solutions and specific additional security control suggestion
- A descriptive assessment and relative risk score for each risk category in each area considered

Benefits

- COBRA provides a variety of tools for risk assessment, which means most of the processes are automated. This makes the risk assessment process very easy.
- The methodology has very simple steps and hence this is very easy from an implementation perspective.

Limitations

- Is based on various questionnaire or survey, i.e., opinion-based; the participants may or may not be well aware of the recent developments in the concerned area.
- Is a generalized one; hence, there is still a need to develop or extend the methodology for the particular requirements phase.
- What is the accuracy level of this methodology is also not mentioned?
- Risk assessment technique is not clearly mentioned.
- COBRA does not talk about the security attributes.
- Threats and vulnerabilities play an important role in the process of risk assessment, but how these are taken into consideration is not given in the methodology.

9.2.2　Cloud Risk Assessment Frameworks

9.2.2.1　Cloud Security Risk Management Framework

CSRMF represents a business-oriented framework for organization and represents the attacks and risks of cloud-specific and explains how to mitigate them and countermeasures. This framework describes overall the process to identify the business objectives for the organization and the risks and how to analyze them from the knowledge base and profiling this and identifying those risks and their countermeasures to mitigate them if they exceed the tolerance level and then monitoring them to validate the treatment and enhance the knowledge base if a new risk is found and establish an adequate level of confidence in CC and a reliable and affordable efficiency for CSP. The proposed framework goals at

- Identifying the risks
- Analyzing and evaluating identified risks
- Applying the best treatment actions to reduce the likelihood and/or impacts of risks
- Monitoring validation of treatment actions against the identified risks regularly
- Establishing a dynamic relationship between the organization and CSP during the risk management process to ensure the compliance with SLA

Identifying Business Objectives: In this first stage, it identifies the organization's objectives by Strengths, Weaknesses, Opportunities, and Threats (SWOT) analysis which are internal strengths and weaknesses and external opportunities and threats and then should follow the SMART model (Specific, Measurable, Attainable, Relevant, and Time-bound), which consists of a set of questionnaires for each step. The risk tolerance level is finalized by the organization and at last, all information stored in the risk knowledge base is used as a profile of the organization (Figure 9.4).

Risk Identification: It is the second stage and depends on the application area, nature of projects, resources, the desired outcome, and the required level of detail. There is no single scientific method that guarantees the identification of all risks. Due to that reason, this framework proposes a hybrid approach of two techniques (documented knowledge acquisition and brainstorming) for risk identification.

- Documented Knowledge Acquisition uses CC risk domain documents (most useful provided by ENISA) and then stored in a cloud security risk knowledge base.
- Brainstorming has two stages: idea generation and idea evaluation. In idea generation, generate ideas to address the problem and in idea evaluation, prioritize ideas.

Finally, a final list of possible risks is generated based on the two techniques and the previously stored risk knowledge base.

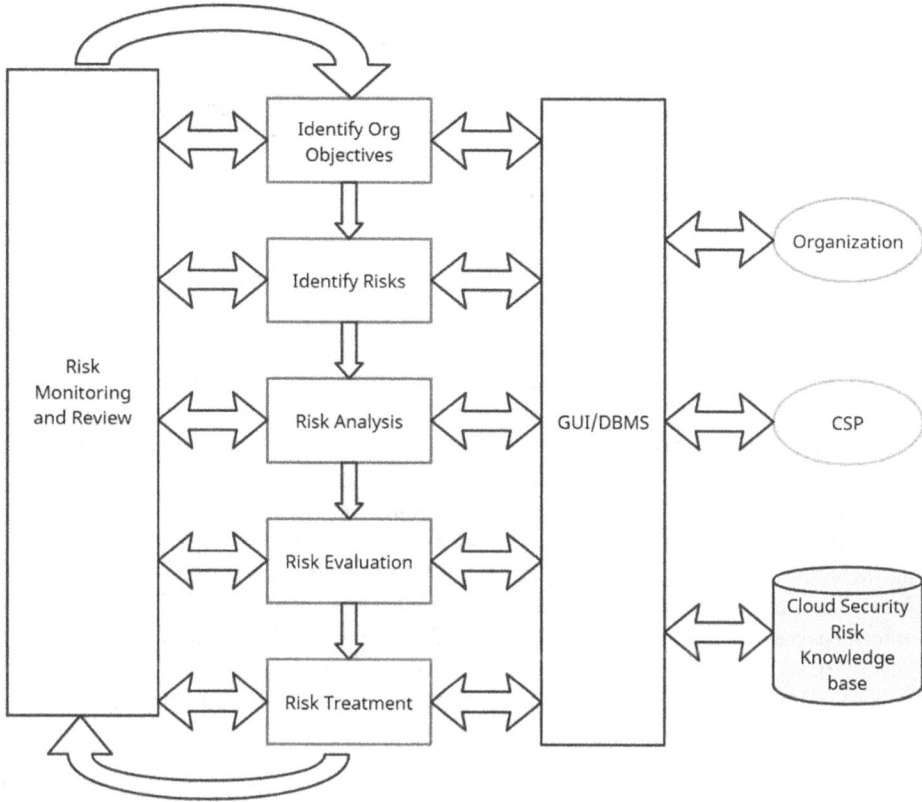

FIGURE 9.4 Block diagram of ENISA estimation of risk levels.

Risk Analysis: Risk analysis is a third stage where it requires the estimation of the likelihoods $L(r_i)$ of each risk r_i and the impacts $I(r_i, o_j)$ of risk r_i on objective o_j according to the weight w_j of each objective o_j. Objective weight defines the importance of an objective o_j, $(0 \leq w_j \leq 1, \Sigma_j w_j, j = 1, 2,..., m)$. Risk likelihood $L(r_i)$ represents the probability of occurrence of risk o_j, $(0 \leq L(r_i) \leq 1, i = 1, 2,..., n)$. Risk impact $I(r_i, o_j)$ is the effect of each risk on each objective $(0 \leq I(r_i, o_j) \leq 1)$ where 0 means no loss of satisfaction in o_j, 1 means total loss of satisfaction in o_j, m and n are numbers of objectives and risks, respectively. This framework uses a quantitative approach and consensus-based estimation technique for risk analysis which is the Delphi method. This method has three essential characteristics: (a) structured and iterative information flow, (b) anonymity of the participants to alleviate peer pressure and other performance anxieties, and (c) iterative feedback of the participants. Here, the moderator collects information from a selected group of Subject Matter Experts. When estimates reach a consensus (85% or more), the moderator reports the final estimates to be used in the next phase [14].

Risk Evaluation: Risk evaluation is a process where it estimates the risk level (risk severity) and whether the risk is tolerable by the organization or not. Risk tolerance level criteria are already approved by the experts and stakeholders in the first stage. If the risk is greater than the tolerable level, then the risk needs treatment or improved countermeasures. Using Equation 9.12, the level of risk level (r_i) for risk r_i is estimated as

$$\text{Level}(r_i) = \sum_{j=1}^{m} \omega_j I(r_i, o_j), \ 0 \leq \text{Level}(r_i) \leq 1 \qquad (9.12)$$

Risk level (r_i) ranges are between 0 and 1, where 0 means r_i has no effect (min. severity) on the organization's objectives and 1 means r_i has a significant effect (max. severity) on the organization's objectives. A risk r_i may be considered acceptable (tolerable) if a level is less than threshold α, otherwise requires treatment. This threshold ($0 \leq \alpha \leq 1$) is predetermined by the organization. By applying this condition, an organization can achieve an acceptable Global Risk Level (GRL) by Equation 9.13.

$$\text{GLR} = \sum_{i=1}^{n} \text{Level}(r_i) \tag{9.13}$$

Risk Treatment: CSRMF introduces a risk mitigation approach for risk treatment to reduce GRL by reducing the level of risk level(r_i) for each unacceptable risk. Risk likelihood $L(r_i)$ can be minimized by changes in business systems and processes, preventative maintenance, or quality assurance and management. However, risk impact $I(r_i, o_j)$ can be reduced by contingency planning. Risk mitigation actions can be determined using a combination of documented knowledge acquisition and brainstorming techniques.

Risk Monitoring: In this last stage, CSRMF monitors and evaluates the risk treatments and current control activities. So, the estimation of the risk level reduction is needed to calculate after applying the countermeasure techniques. Here, the Delphi technique is used to calculate the risk reduction matrix which is denoted as LevelRed($r_i \mid c_k$). If the risk level reduced of risk r_i after applying the countermeasure c_k is LevelRed($r_i \mid c_k$) has ranged between 0 and 1 where 0 means no reduction and 1 means risk elimination. The Combined Risk Reduction (CRR) of a risk r_i which helps to determine whether the risk treatment is successful or not can be calculated from Equation 9.14.

$$\text{CRR}(r_i) = 1 - \prod_{k=1}^{p} (1 - \text{LevelRed}(r_i \mid c_k)), \ 0 \leq \text{CRR}(r_i) \leq 1 \tag{9.14}$$

This CRR has a value range between 0 and 1 where 0 means no reduction and 1 means risk elimination. Finally, the global risk reduction (GRR) for organization is estimated from the given Equation 9.15.

$$\text{GRR} = \sum_{i=1}^{n} n\text{CRR}(r_i) \tag{9.15}$$

After applying the mitigation, the value of GRR value should be reduced. The organization should monitor the system continuously to find out if the applied treatment is valid for the occurrence of the identified risks and list down if there is any new risks may occur.

9.2.3 Information Security Risk Management Framework

The framework in Ref. [10] follows the Plan, Do, Check, Act (PDCA) cycle and contains seven processes: selecting relevant critical areas, strategy and planning, risk analysis, risk assessment, risk mitigation, assessing and monitoring program, and risk management review, as shown in Figure 9.5. The first phase, Architecting and Establishing the Risk Management Program (Position Level Actions Network [PLAN]) involves selecting a relevant critical area that highlights 12 critical areas to address both the strategic and tactical security 'pain points' within any cloud model. The CSP selects applicable critical areas for the specific cloud environment before moving to the strategy and planning process. Strategy and planning involve assigning a responsible body for the risk management, establishing directions, defining goals, requirements, and scope, and planning to achieve the defined goals and requirements. The Implement and Operate (Do) phase include risk analysis, risk

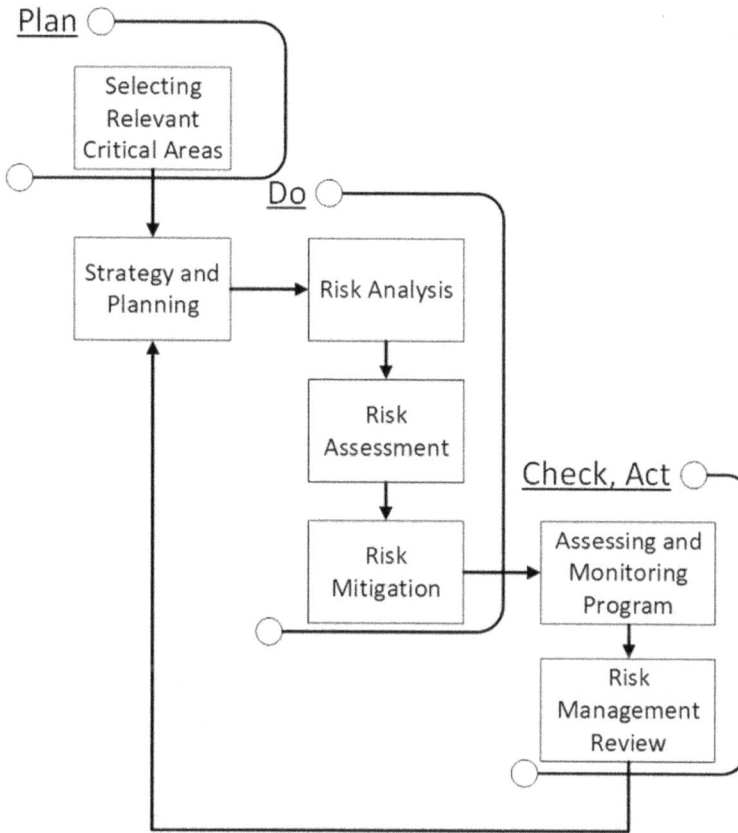

FIGURE 9.5 Cloud computing risk management framework [10].

assessment, and risk mitigation processes. The Monitoring and Review (Check, Act) stage involves 'assessing and monitoring program and risk management review' to ensure the effectiveness of the overall risk management program. The framework was applied to manage the risks of a SaaS platform, which provides a real-time logistic web-based application software service to support the logistic industry in China.

9.2.4 Security Risk Assessment Framework

A cloud security risk assessment framework (in Figure 9.6) is proposed in Ref. [9]. They relied on ISO27005 standard to define the main steps of the framework. The proposed framework considers both the cloud customer and the CSP during the risk assessment process and defines their responsibilities. It limits the involvement of the customer in the evaluation of security risk factors to avoid the complexity that can result due to the involvement of the customer in the whole risk assessment process.

The framework consists of two main parts: the CSP and the CCs. The CSP side contains four main entities: CSP risk assessment manager (CSPRAM), CSP and CC communicator (CSP3C), CSP security requirements classifier (CSPSRC), and CSP database interface. The CCs side contains the cloud client risk assessment assistance (CCRAA) only.

The CSPRAM is the CSP body responsible for managing the risk assessment processes as a whole. The CSP3C is used to handle all communication with the CCs during the process. The CSP3C transfers information received from the CCs to the CSPSRC. The CSPSRC categorizes this

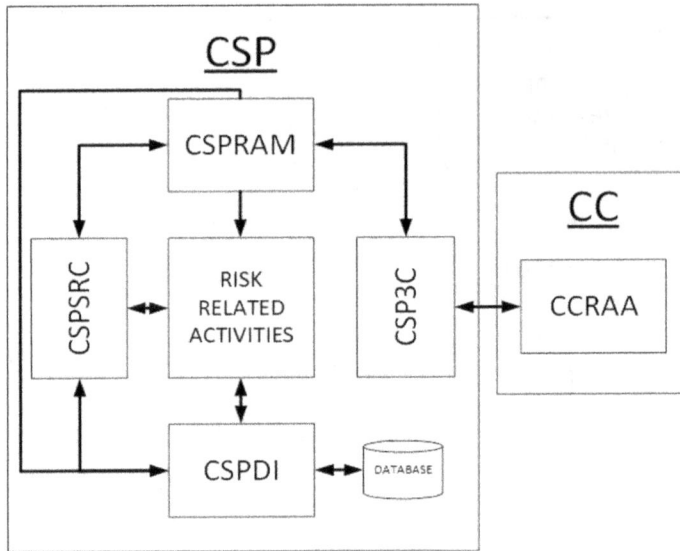

FIGURE 9.6   Cloud security risk assessment framework [9].

information and saves them in the database. It makes decisions on the security requirements and the importance of threats received from the CCs. The CCRAA is the CC entity responsible for communication with the CSP.

The context establishment process involves three sub-processes: setting the basic criteria (risk evaluation, impact, and risk acceptance criteria), setting the scope and boundaries, and organizing the information security risk management process, CCs are expected to assess their data and define their criteria for risk evaluation, risk impact, and risk acceptance. The CSP establishes the context and considers the criteria and expectations of the CCs at this phase. The next phases include risk identification, risk analysis, and risk evaluation, which constitute the risk assessment process. Each CC is expected to evaluate its assets, identify the threats and vulnerabilities, and submit the information to the CSPRAM through the CSP3C. The CSP conducts its risk identification and integrates the CCs' risks and prioritizes them. The risk treatment phase involves preparing a risk treatment plan and implementing appropriate security controls to address the risks. In the risk acceptance phase, residual risks should be identified and the CSP should ensure that the treatment plan has decreased the risk to an acceptable level based on the CCs risk acceptance criteria. The last phase requires the CSP to monitor and review the risks and their factors to consider any change in the next iteration. The framework in Ref. [9] performed an initial experiment on the framework by using it for a scenario that assesses the security risks of a SaaS provider, which provides a website package for medium and small organizations that sell products on the Web.

9.3   PERFORMANCE ANALYSIS OF THE EXISTING MODELS AND FRAMEWORKS

In the previous sections, we have tried to discuss in detail the risk assessment models and frameworks. CARAM is a risk assessment model, which is proposed exclusively to deal with CC and CSP, whereas COBRA is a more generalized one and this methodology may not portray the full scenario of CSC and CSP explicitly. CARAM is based on the well-standard questionnaire provided by ENISA and CSA and eventually, CARAM proposes a more granular risk level which helps to select a CSP based on security, privacy, and service, hence the decision based on CARAM provides more accuracy in terms of cloud customers' acceptance level of the risk values. In the case of COBRA, the accuracy level is not proposed precisely, which may result in a vague outcome. Hence, we may depict that CARAM provides more reliability and constancy than COBRA.

At the time of analyzing risk management frameworks, we have used the qualitative content analysis technique. Qualitative content analysis is a 'research method for the subjective interpretation of the content of text data through the systematic classification process of coding and identifying themes or patterns' [5]. This analysis technique will be used to review the two frameworks and analyze their strengths and limitations. The qualitative content analysis will be done based on the following thematic areas:

9.3.1 Does the Framework Effectively Address Both Phases of Risk Management (Risk Assessment and Risk Treatment)?

The framework [10] addressed both the risk assessment and risk treatment phases. However, some components of context establishment and asset identification, which are the primary steps to be performed during risk management [15], are not considered.

Context establishment involves setting criteria of risk evaluation, impact, and risk acceptance. It also includes setting the scope and boundaries of the risk assessment and establishing an appropriate body that performs the risk management. Although assigning a responsible body and defining the scope are included, the other components of context establishment are overlooked in the framework. The quality of the results produced when performing context establishment and asset identification has a crucial influence on subsequent steps such as identifying loss, vulnerabilities, possible attacks, and defining countermeasures [15]. Therefore, ignoring these components hurts the effectiveness of risk management. In addition, control recommendation, which is included in the risk assessment process, is a component of risk treatment [16]. It is better to group controls recommendation in the risk mitigation process.

The framework [9] effectively addresses both risk assessment and risk management although it is named as a risk assessment framework and gives much focus on the risk assessment phase. The framework follows a qualitative risk assessment approach. The qualitative approach helps to prioritize the risks and identify areas for immediate improvement [16]. However, it doesn't provide the specific quantifiable magnitude of the risk impacts, thus making the cost-benefit analysis calculation of recommended controls difficult [16].

According to CSRMF, it introduces both risk assessment and risk treatment which are the crucial part of risk management. In risk assessment, it includes risk identification, analysis, and evaluation concerning the organization's business objectives, which are identified in the first stage of the framework. In addition to this, it enhances overall risk management by incorporating the Risk Treatment and Risk Monitoring and Reviewing stages that help to mitigate the risks by applying the countermeasures and also monitoring the validation of all applied treatments.

9.3.2 Does the Framework Enable the CSP and the Customer to Efficiently Assess and Mitigate Cloud Security Risks?

The framework [10] is applicable to all cloud service models and cloud deployment models. However, it considers only the CSP and overlooks the role of the customer in the risk management processes. The customer is the real owner of the data assets. The only one who knows the real value of the data and the realistic impact of data security breaches is the customer [7]. Hence, ignoring the customer's involvement will result in inaccurate security risk level evaluation and inefficient risk management (Table 9.2).

Alruwaili and Gulliver [17] criticized the framework that it lacks details on important elements such as the cloud risk control matrix, threats and vulnerabilities, proactive detection and response, risk control strategies, secure service level agreement parameters, and compliance and monitoring process. The research [18] also noted that the framework is very similar to traditional

TABLE 9.2

Comparison among Frameworks Addressing Both Phases of Risk Management

Zhang et al. Framework	Albakri et al. Framework	CSRMF Framework
Yes	Yes	Yes
• Some components of context establishment and asset identification are not considered	• Doesn't provide the specific quantifiable magnitude of the risk impacts • Makes the cost-benefit analysis calculation of recommended controls difficult	• Introduce risk assessment where it uses a quantitative approach to risk analysis and evaluation • Risk treatment applied to mitigate risk which exceeds tolerance value according to the business objectives

quality management. In addition, the framework lacks a risk communication feature, which is an important component.

The framework [9] considers both the CSP and the customer and mitigates the major limitations of traditional risk assessment methods. However, it targets only SaaS as claimed by the authors. Therefore, it may not be efficient to assess and mitigate the risks of PaaS and IaaS environments. The involvement of the client in the risk assessment process is limited to a minimum in order to decrease complexity.

Although minimizing the client's involvement makes the process easier for the CSP, it may create a negative impact on the effectiveness of certain phases and may cause client dissatisfaction. Ignoring the involvement of clients, especially on risk treatment and risk acceptance, and providing them only reports on these issues can limit the effectiveness of the risk treatment. As clients are part of the problem, making them part of the solution will be important. The clients should also have a chance to comment on whether the selected controls meet their expectations and whether the residual risk is acceptable or not.

This limitation will have a higher impact in the case of PaaS and IaaS and may make the framework inefficient for these models. In addition, the framework assumes that the client can perform a risk assessment, which may not be the reality for all SaaS clients.

Mainly this framework focuses on customer's business objectives, and customers can conduct cost-value analysis and take decisions regarding the migration into the Cloud Platform. On the other hand, CSPs can improve their productivity and profitability by providing reliable cloud service and they also can gain knowledge about cloud risk identification with respect to different customers' business objectives and how to manage those cloud risks as well. By statistical analysis of previous knowledge-based mitigation data and risk management in CC, we can achieve a higher level of confidence in future Cloud technology (Table 9.3).

TABLE 9.3

Comparison among Frameworks Basis on Enabling CSP and Customer to Assess and Mitigate Risks

Zhang et al. Framework	Albakri et al. Framework	CSRMF Framework
No	Yes	Yes

• Considers only CSP
• Lacks details on important elements
  1. The cloud risk control matrix.
  2. Threats and vulnerabilities, proactive detection, etc.
	• Targets only SaaS
	• Involvement of the client in the risk assessment process is limited
	• Clients should also have a chance to comment
		• Consider business objective and tolerance level of risk from CC as input
		• Use statistical analysis using Delphi to assess the risk level
		• CSP improves their productivity and profitability by managing the risk analysis process

Value analysis and take decisions regarding the migration into the Cloud Platform. On the other hand, CSPs can improve their productivity and profitability by providing reliable cloud service and they also can gain knowledge about cloud risk identification with respect to different customers' business objectives and how to manage those cloud risks as well. By statistical analysis of previous knowledge-based mitigation data and risk management in CC, we can achieve a higher level of confidence in future Cloud technology.

Finally, the review and analysis of the risk management frameworks mentioned in this section showed that the frameworks follow different approaches and have their strengths and limitations. Despite their limitations, these frameworks can be used in different cloud environments to assess and mitigate cloud security risks. However, their applicability to managing the security risks of a given cloud environment should be evaluated before applying them. Their limitations should also be considered or mitigated.

9.4 CONCLUSION AND FUTURE DIRECTIONS

As we have studied the current practices, we would suggest the following points to the CSP and CC:

- CSP should include CC during risk assessment.
- CSP may use the mentioned Delphi method to mitigate risk.
- At the time of migration, CC may select their CSP based on the CARAM.

We also envisage the following directions relevant to future research:

- A standard classifier may be introduced for homogeneous clients based on business interest.
- The Cloud domain is an evolving technology. CSPs do update their domain with new assets and hence every now and then it unfolds new vulnerabilities and risk factors. Thus, the questionnaire for the CSP and CC may also need to be updated simultaneously and continuously to have a more impact and accurate assessment.

In conclusion, no sooner had the demand for cloud computing increased than the demand for its security also increased. If the outcome of our study is being considered, the future CSPs along with the CCs will be benefited.

ACKNOWLEDGMENT

The authors thank their M.Sc. students, Mukta Khanam, Anik Chowdhury, and Md. Adnan Rashidul Islam, at the United International University (UIU) for their contributions.

REFERENCES

1. Madhavaiah, C., & Bashir, I. (2012). Defining cloud computing in business perspective: A review of research. *Metamorphosis*, 11(2), 50–65.
2. Namboodiri, V., & Ghose, T. (2012, June). To cloud or not to cloud: A mobile device perspective on energy consumption of applications. In *2012 IEEE International Symposium on a World of Wireless, Mobile and Multimedia Networks (WoWMoM)* (pp. 1–9). IEEE.
3. Mell, P., & Grance, T. (2011). The NIST definition of cloud computing.
4. Armbrust, M., Fox, A., Griffith, R., Joseph, A. D., Katz, R. H., Konwinski, A.,... & Zaharia, M. (2009). *Above the Clouds: A Berkeley View of Cloud Computing* (Vol. 17). Technical Report UCB/EECS-2009–28, EECS Department, University of California, Berkeley.
5. Chen, Y., Paxson, V., & Katz, R. H. (2010). *What's New about Cloud Computing Security* (pp. 2010–2015). University of California, Berkeley Report No. UCB/EECS-2010–5 January 20.2010.
6. Cayirci, E., Garaga, A., Santana de Oliveira, A., & Roudier, Y. (2016). A risk assessment model for selecting cloud service providers. *Journal of Cloud Computing*, 5(1), 1–12.

7. Pandey, S. K. (2012). A comparative study of risk assessment methodologies for information systems. *Bulletin of Electrical Engineering and Informatics*, 1(2), 111–122.
8. Youssef, A. E. (2020). A framework for cloud security risk management based on the business objectives of organizations. arXiv preprint arXiv:2001.08993.
9. Albakri, S. H., Shanmugam, B., Samy, G. N., Idris, N. B., & Ahmed, A. (2014). Security risk assessment framework for cloud computing environments. *Security and Communication Networks*, 7(11), 2114–2124.
10. Zhang, X., Wuwong, N., Li, H., & Zhang, X. (2010, June). Information security risk management framework for the cloud computing environments. In *2010 10th IEEE international Conference on Computer and Information Technology* (pp. 1328–1334). IEEE.
11. Catteddu, D. (2009, December). Cloud computing: Benefits, risks and recommendations for information security. In *Iberic Web Application Security Conference* (pp. 17–17). Springer, Berlin, Heidelberg.
12. Cloud security alliance, security CSA trust and assurance registry (STAR). https://cloudsecurityalliance.org/star/, Last accessed: 03-October-2021.
13. WEKA; machine learning tool. https://www.cs.waikato.ac.nz/ml/weka/, Last accessed: 03-October-2021.
14. Youssef, A. (2020). A Delphi-based security risk assessment model for cloud computing in enterprises. *Journal of Theoretical and Applied Information Technology*, 98(1), 151–162.
15. Beckers, K., Schmidt, H., Kuster, J. C., & Faßbender, S. (2011, August). Pattern-based support for context establishment and asset identification of the ISO 27000 in the field of cloud computing. In *2011 Sixth International Conference on Availability, Reliability and Security* (pp. 327–333). IEEE.
16. Calder, A., & Watkins, S. G. (2010). *Information Security Risk Management for ISO27001/ISO27002*. It Governance Ltd.
17. Alruwaili, F. F., & Gulliver, T. A. (2014). Safeguarding the cloud: An effective risk management framework for cloud computing services. *International Journal of Computer Communications and Networks (IJCCN)*, 4(3), 6–16.
18. Xie, F., Peng, Y., Zhao, W., Chen, D., Wang, X., & Huo, X. (2012, October). A risk management framework for cloud computing. In *2012 IEEE 2nd International Conference on Cloud Computing and Intelligence Systems* (Vol. 1, pp. 476–480). IEEE.

10 Generating Cyber Threat Intelligence to Discover Potential Security Threats Using Classification and Topic Modeling

Md Imran Hossen and Ashraful Islam
University of Louisiana at Lafayette

Farzana Anowar
University of Regina

Eshtiak Ahmed
Tampere University

Mohammed Masudur Rahman
Bangladesh University of Engineering and Technology

CONTENTS

DOI: 10.1201/9781003267812-10

10.1 INTRODUCTION

10.1.1 BACKGROUND AND MOTIVATION

Cybersecurity is one of the key concerns among users [1–3]. Cybercriminals are delivering more strong and advanced security threats constantly to gain security control over the networks, i.e., the Internet [1,3,4]. During the previous decades, computer security remained in only preventing computer viruses and another most common security issue was email spamming [5,6]. In recent years, cybercriminals have been making targeted attacks so that they can obtain the whole control from the users. Over the online community, topics related to computer security are the most popular and community members discuss prior and possible security threats in social media and hacker forums [1,2,7,8]. Hacker forums are the most valuable source for computer security-related posts and blogs and these posts usually contain vital information about possible computer and cybersecurity threats or holes [1,2].

Cyber Threat Intelligence, abbreviated as CTI, can be defined as knowledge and intelligence, which assists in collecting, analyzing, understanding, and preventing possible cyber-attacks based on gathered data from various security-related sources, e.g., hacker forums and by transferring analyzed data to possible cyber threats in meaningful representations [2,3,8]. Gartner, Inc. [9] describes

> Threat intelligence is evidence-based knowledge, including context, mechanisms, indicators, implications, and action-oriented advice about an existing or emerging menace or hazard to assets. This intelligence can be used to inform decisions regarding the subject's response to that menace or hazard.

In general, CTI usually compiles data from diverse sources (usually of unstructured nature) such as the National Vulnerability Database, social media platforms, e.g., Facebook, Twitter, and computer security forums, and uses scientific methods to convert it to the meaningful representation of threats [2,3,10,11]. In this way, CTI can provide insights into emerging security threats and may help prevent possible and potential cyber-attacks [1,12].

From the literature, Le et al. [13] proposed a novelty detection model for automatically gathering CTI and it learns the features of CTI by looking at threat descriptions in publicly available sources. They were able to categorize cyber threat tweets with an F1-score of 0.643 outperforming binary classification models. Ampel et al. [14] collected and categorized hacker forum source code into eight different classes using a proactive Deep Transfer Learning for Exploit Labeling. However, Zhao et al. [15] address two challenges in identifying the unseen types of indicators of compromise (IOC) and automatic generation of CTI with domain tags. They proposed a novel deep learning-based tool called TIMiner that generates CTIs with domain tags and extracts IOC using word-embedding and syntactic dependence. The tool exceeds an accuracy of 84% and 94% on domain recognizer and IOC extraction. Koloveas et al. [16] draw an integrated threat management lifecycle framework named inTIME to collect, rank, extract, leverage, and share CTI and security artifacts using natural language understanding methods. They used machine learning algorithms that provide a complete process of handling cyber threats. Moreover, Gao et al. [17] claimed the novelty of modeling a heterogeneous graph convolutional network-based approach called HinCTI. They improved threat type identification of infrastructure nodes and the relation between them.

10.1.2 PROBLEM STATEMENT AND GOAL

Hackers are constantly posting and sharing information about diverse cybersecurity-related topics on online forums [1,8]. The posts on these forums can hold data that may help in the exploring and discovery of CTI. However, manual analysis of this highly unstructured data could be tedious and ineffective [4,11,18]. The objective of the work in this chapter, therefore, is to conceptualize and develop a set of methods that collects, processes, and analyzes hacker forums data to discover emerging threats. Once we have constructed datasets, classification tasks could be applied to the posts so that classifiers can distinguish security-relevant posts from irrelevant ones. Furthermore, we can categorize security-relevant posts in more specific cyber threat categories, e.g., credential leaks, keyloggers, and DDoS (Distributed Denial of Service) attacks.

Moreover, we utilize the topic modeling algorithms for the whole datasets as well as on each security-relevant class so that we can explore latent topics available in the datasets. For example, if we discover a category that holds a discussion about ransomware, and if our topic modeling algorithms find keywords like *WannaCry* and *Petya* for this topic, we would know that these are the current trends in ransomware-based attacks.

The contributions of this research are two-folded.

- Designing and developing a set of procedures to collect data and process and analyze them from open-source hacker forums.
- Generating CTI from processed data to discover emerging threats by utilizing the knowledge of Information Retrieval. This stage can be done in three phases as follows:
  - Identify security-relevant posts from non-relevant/irrelevant ones *(Classification problem)*.
  - Categorize security-relevant posts according to different threat categories *(Classification problem)*.
  - Explore key topics in the dataset using topic modeling algorithms *(Unsupervised learning)*.

10.2 METHODOLOGY

This research work is implemented by performing several sequential stages. The following stages are involved in developing our intended goal, i.e., mining and classifying hacker forums' data for fathering CTI. Figure 10.1 demonstrates the steps of the methodology for extracting CTI from the hacker forums.

10.2.1 DATA COLLECTION

We first collect data from the posts shared in hacker forums that are posted by computer and network security-loving persons, analysts, and experts. Collecting data from hacker forums is a challenging task for several reasons: (a) the forums usually restrict the access to their content only to registered users making it difficult to build a web crawler to collect data automatically and (b) some forums use CAPTCHAs to prevent automated programs that are used to collect data. To make things worse, forums often employ invitation-only registration; one can register by invitation only from already-registered users. Alternatively, we may use some publicly available datasets on the Internet.

However, we find some decent forums that do not have such restrictions, e.g., [19,20]. In addition, we only use the data from the second forum since the first one does not seem to contain a good number of posts. We build two separate crawlers (using Python) to collect posts from these forums, which build the corpus eventually. Moreover, we utilized a leaked dataset labeled by the experts that was used as the ground truth dataset for this work from *https://nulled.io*.

10.2.2 PREPROCESSING AND DATASET CONSTRUCTION

Generally, data collected from the web is highly unstructured. Some preprocessing is necessary before we create the final dataset. Hence, first, we clean up the HTML tags and extract the title and content of each post found in the forums; second, we remove stop words, special characters, and punctuation after transforming all the text into lowercase. Further, lemmatization is also used so that we can obtain a more clean and optimized text collection to build the datasets. Once the data is reasonably cleaned up, we represent each post as a single-line document to build a corpus.

Our target is to build two datasets for both binary classification and multi-class classification. We build separate datasets for each security forum we crawled and preprocessed. Both datasets

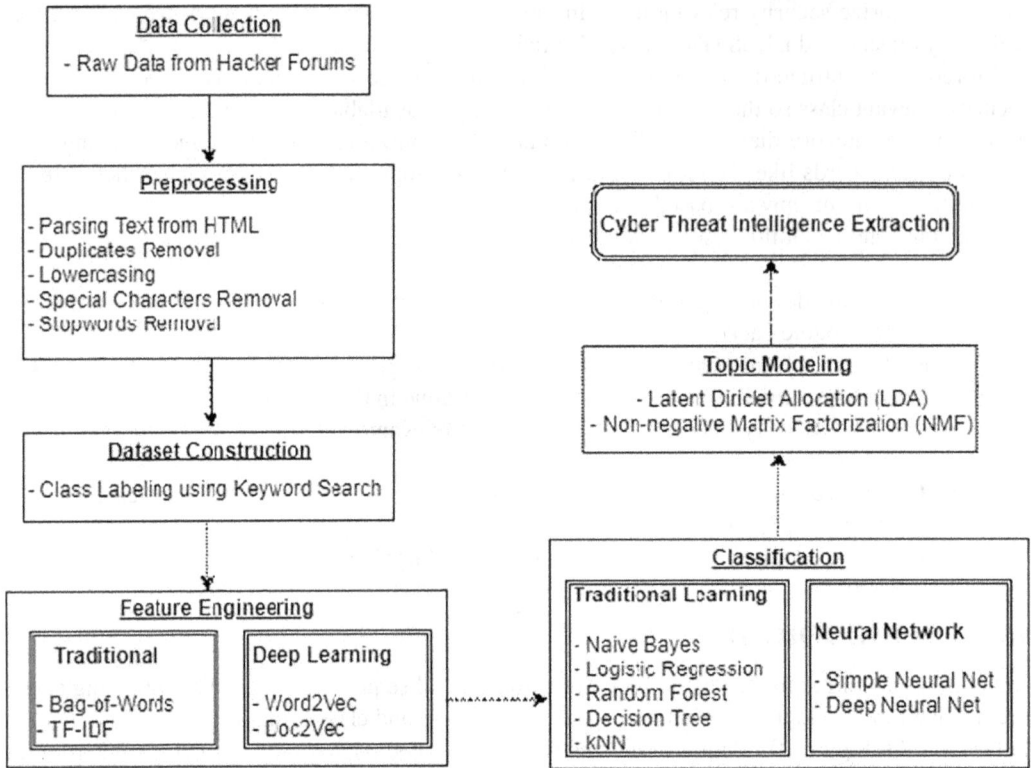

FIGURE 10.1 Steps for extracting CTI.

are constructed using the keyword search method where we label each sample based on keyword relevancy.

A list of very common cybersecurity keywords used for labeling the relevant class are as follows: *"adware", "antivirus", "botnet", "backdoor", "crack", "crimeware", "crypter", "ddos", "downloader", "dropper", "exploit", "firewall", "hijack", "infect", "keylogger", "malware", "password", "ransomware"," reverse", "shell", "rootkit", "scanner", "shell", "code", "security", "spam", "spoof", "spyware", "trojan virus", "vulnerability", "worm", "zero-day", "stealware", and etc.*

We attempt the following approaches for building both the datasets.

10.2.2.1 Binary Dataset Construction

The binary dataset has two labels or classes, i.e., relevant and irrelevant. The samples bearing the posts containing keywords relevant to computer security are labeled as relevant. Some common relevant keywords that we approach are "exploit", "keylogger", "reverse shell", "antivirus", "ransomware", etc. Other samples are labeled as irrelevant as they contain irrelevant keywords such as "song", "tv show", "browsing", and "football". Since the number of security keywords is not comprehensive, we label a post as irrelevant if it satisfies these two properties:

- None of the keywords from the relevant class is present.
- Any computer security-related keywords are absent.

Table 10.1 shows two samples from the binary dataset for a better understanding where relevant keywords are underlined. Figure 10.2 presents the distribution of sampler for each class, i.e., relevant and irrelevant. Table 10.1 contains examples of posts labeled in the binary dataset.

TABLE 10.1

Two Samples Collected from Binary Dataset

Post	Class Label
"... Here is a.php file from an <u>exploit</u>, downloaded from... Apache APR is prone to a <u>vulnerability</u> that may allow <u>attackers</u> to cause a <u>denial-of-service</u> condition..."	Relevant
"... Good work, its good to learn how to do things yourself. It's a very rewarding experience..."	Irrelevant

Source: For dataset construction, posts were gathered from a variety of sources on the Internet.

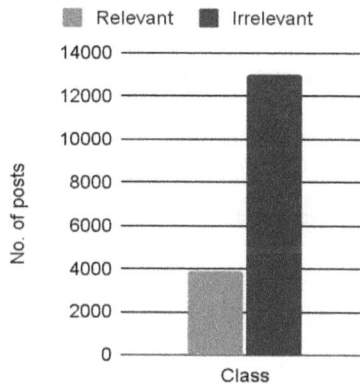

FIGURE 10.2 Samples (posts) distribution per class, i.e., relevant and irrelevant.

10.2.2.2 Multi-Class Dataset Construction

At this phase, we build another dataset from the samples labeled as relevant in the binary dataset. This dataset contains the samples labeled in six categories and the categories are as follows:

- Credential Leaks (Keywords: username, password, pass list, etc.)
- Keylogger (Keyword: keylogger)
- DDoS Attack (Keywords: ddos, denial of service, server, etc.)
- Remote Access Trojans (Keywords: rat, trojan, remote access, etc.)
- Cyrpters (Keywords: crypter, fud, etc.)
- SQL Injection (Keywords: sql, injection, id=, etc.)

Table 10.2 contains examples of posts labeled in the multi-class dataset.

10.2.3 Feature Engineering

Machine learning algorithms cannot directly work with textual data [1,12,21,22]. We need to vectorize the documents before feeding them as inputs to any learning algorithms, either supervised or unsupervised [18,23]. To this end, we explore the best of both worlds for this purpose. First, we use standard feature engineering methods, for example, *"Bag-of-Words (BOW)"* and *"Term Frequency-Inverse Document Frequency (TF-IDF)"*-based weights. Further, both binary term weights and term-frequency weight measurements are used for the BOW model. Moreover, for TF-IDF-based weighting, we use two varieties – word-level unigram and word-level bigram features.

TABLE 10.2

Few Samples Collected from the Multi-Class Dataset Constructed from the Relevant Class

Post	Class Label
"Leaked data for *** USERNAME / PASSWORD hope0507 : *** jimsam100234:***"	Credential Leaks
"The Best Keylogger logs all keystrokes, mouse clicks, applications... sent to you via mail so you can monitor your computer without being there"	Keylogger
"If you're not interested in creating a botnet, with a RAT for example... Which use servers to DDoS targets ... Written on mobile-phone, sorry for the formatting"	DDoS attack
"Initial checker $newHost=$host.'/interface/ipsconnect.php'; $sql='SELECT COUNT(*) FROM members'; ... $response=SendPost($newHost, $data)";	SQL Injection
"Plasma HTTP BotNet allows you to remotely control computers, known as Clients... Persistent (Miner is re-injected if terminated) (...truncated)"	Crypter
"Application name: Imminent Monitor Application description: Advanced RAT (The others cracked/ leaked here... Vendor's URL: http://www.hackforums.net/showthread.php?/"	Remote Access Trojans

Source: For dataset construction, posts were gathered from a variety of sources on the Internet.

Second, we conduct the experiment with two contemporary and more advanced feature engineering techniques based on deep learning, namely, *Word2Vec* and *Doc2Vec* models. *Word2Vec* model is based on word-embedding and it identifies and learns the semantics between words in a given corpus [24]. On the Internet, there exist various pre-trained *Word2Vec* models. When incorporating sentiment analysis data, the *Doc2Vec* model works quite well [25–27]. None of them, however, have been trained on the cybersecurity-related corpus. For this reason, we do not use any trained models and train a *Word2Vec* model (with *feature_size=100*) on our own dataset. Since the *Word2Vec* model by default does not provide document-level vectors, we obtain a 100-dimensional vector for each word in a document, sum them up, and calculate the final averaged document vector. We do the same for all the documents in the dataset. We also train a *Doc2Vec* model on our dataset, which directly learns and provides the document-level vectors [28].

10.2.4 SUPERVISED METHOD: CLASSIFICATION

We first carry out classification tasks to see how accurately cybersecurity-relevant posts can be separated and identified from the non-security posts. This is an important step since the majority of the posts in the hacker forum have nothing to do with security and serve as noise. Filtering out the noises provides us the chance to focus more on what we actually care about and that is security-relevant posts. It also allows us to further inspect and investigate the security-relevant posts for potential cyber threats. We also evaluate the performance of classification for our multi-class dataset.

In our experiment, we use five different classification algorithms along with two neural network learning classifiers. The classifiers that we use are Logistic regression, Decision Tree, Random Forest, Naive Bayes, and k-nearest neighbors (kNN). We also use one shallow neural network-based and one deep neural network-based classifiers.

10.2.5 UNSUPERVISED METHOD: TOPIC MODELING

Topic modeling is the process of identifying and determining topics in a set of documents. It is an unsupervised learning approach and can be a powerful tool to find possible latent topics (in the form of keywords) in a large unlabeled dataset [29,30]. For instance, say we have a dataset that contains newspaper articles from some major online news agencies. We provide the number of topics to the algorithm; it will provide the top k (number of topics) per topic. By looking at the keywords for each topic, we may be able to tag topics like politics, technology, and culture. It can also be useful for a labeled dataset to further explore key topics in each class. We use two popular commonly used topic modeling algorithms in this work. The algorithms are (a) Latent Dirichlet allocation (LDA) and (b) Non-negative Matrix Factorization (NMF). We consider "frequency weights" as a feature to LDA as we find TF-IDF weights do not seem to make any difference for this algorithm. In contrast, TF-IDF weight-based features are used for the NMF algorithm.

10.3 EXPERIMENTAL SETUP

Data that is used for this work is collected on a server with 12 Intel R Xeon R E5-2667 CPUs, and 64 GB of RAM running Debian Linux operating system with Linux kernel version 4.18. For classification purposes, the programming codes were developed and tested on a server with 6 Intel R Xeon R E5-2667 CPUs, an NVIDIA GeForce RTX 2070 GPU, and 96GB of RAM running the Arch Linux operating system.

The collected data were processed using standard regular expressions. However, for lemmatization, we use the *spaCy* library. We also utilize several open-source Python libraries throughout this research work. We use the *scikit-learn* libraries to develop the classifiers. The neural network-based classifiers are coded in *Keras*. We use both *gensim* and *scikit-learn* libraries for topic modeling. The *gensim* library is also used for training *Word2Vec* and *Doc2Vec* models.

10.4 EXPERIMENTAL RESULTS

In the first phase of our experiment, we utilize five standard and two neural network-based classifiers. In all of our experiments, we split the whole data for organizing datasets for both training (67%) and testing (33%). We validate our code on two datasets: the IMDB movie review dataset (binary) [31] and 20 newsgroups dataset (multi-class) [32]. We find that our results are comparable with the given benchmark results. Specifically, the accuracies vary from 76% to 88% for different classifiers with different feature sets on IMDB movie review datasets. For the 20 newsgroups dataset (only 8 classes have been considered), the accuracies vary from 79% to 89%.

Table 10.3 lists the performance metrics based on accuracy for different classifiers for *forums. hak5.org* datasets. Table 10.4 presents the performance metrics in accuracy for different classifiers for the *nulled.io* datasets or our ground truth datasets.

For topic modeling, we utilize two popular algorithms: (a) LDA and (b) NMF. These algorithms require some k (number of topics) to be passed as an input. Finding the optimal value of k requires some experimentation. However, our goal is about determining the best value of k, rather we want to explore the datasets and get some more insights into the datasets. We use k as 10 for binary datasets and apply the algorithm as a dataset as a whole. For multinomial datasets, we run LDA and NMF separately on individual categories. In any case, we show only the top 05 keywords per category. The topic modeling results are listed in Tables 10.5–10.8 (showing only 05 topics in each table and "Credentials" category for the simplicity of the presentation in this chapter).

10.5 DISCUSSION ON RESULTS

In almost all cases, the kNN algorithm yields the worst performance compared to other classifiers. It is interesting to note that the simple term-frequency weights as a feature provide better (or similar) results than TF-IDF. We also found that word-level unigram and bigram for TF-IDF weighting have a minor impact on final performance metrics. Further, the methods using *Word2Vec*-based feature engineering techniques do not provide any significant performance boost. In some instances,

TABLE 10.3

Classification Accuracy (%) Obtained by Classifiers on the *forums.hak5.org* Dataset

Dataset	Classifier	Feature Extraction Method					
		BOW (Binary Term Weights)	BOW (TF Weights)	TF-IDF (Unigram)	TF-IDF (Bigram)	Word2Vec	Doc2Vec
Binary	Naive Bayes	79.96	78.96	75.36	64.65	–	–
	Logistic Regression	93.67	92.75	90.69	68.87	64.64	73.85
	Random Forest	89.88	89.89	88.3	68.55	58.4	62.88
	Decision Tree	93.4	92.8	92.18	67.03	56.5	58.49
	kNN	47.42	57.08	60.01	53.29	59.96	62.61
Multi-class	Naive Bayes	85.67	83.06	70.68	68.4	–	–
	Logistic Regression	97.39	96.09	83.06	75.24	48.86	94.46
	Random Forest	89.25	94.14	91.86	70.03	48.53	63.19
	Decision Tree	97.39	97.07	97.07	64.5	38.76	54.72
	kNN	88.27	85.99	77.2	28.66	49.19	73.62
	Simple Neural Net	86.97	85.34	83.81	78.83	–	–
	Deep Neural Net	77.85	87.3	79.48	77.85	–	–

TABLE 10.4

Classification Accuracy (%) Obtained by Classifiers on the *nulled.io* Dataset

Dataset	Classifier	Feature Extraction Method					
		BOW (Binary Term Weights)	BOW (TF Weights)	TF-IDF (Unigram)	TF-IDF (Bigram)	Word2Vec	Doc2Vec
Binary	Naive Bayes	91.65	90.75	91.61	90.71	–	–
	Logistic Regression	93.26	93.06	92.01	93.71	93.1	92.85
	Random Forest	92.96	93.16	91.76	90.67	9.18	92.5
	Decision Tree	94.62	94.77	92.75	88.73	87.41	85.26
	kNN	47.56	61.44	78.95	90.42	92.95	92.8
Multi-class	Naive Bayes	74.94	75.66	70.01	72.53	–	–
	Logistic Regression	87.06	86.29	87.59	57.95	42.49	88.31
	Random Forest	84.21	85.27	85.9	76.2	41.04	64.8
	Decision Tree	86.96	86.96	87.01	72.86	34.67	52.34
	kNN	67.84	69.34	74.17	46.45	38.88	77.98
	Simple Neural Net	86.53	86.05	85.22	77.84	–	–
	Deep Neural Net	87.11	86.53	85.66	77.64	–	–

specifically for multi-class datasets, they actually seem to generate much less accurate results compared to TF and TF-IDF weights. One reason can be the fact that we only use 100-dimensional vectors to represent the documents for both *Word2Vec* and *Doc2Vec*. Besides, the multi-class dataset has more categories and a small number of samples per class when compared to the binary-class datasets. Increasing the size of the feature vector may result in enhanced performance. However, it will consume significant computing resources to train such a model. For this reason, we refrain from doing so. We can also see that *Doc2Vec*-based models perform remarkably better than *Word2Vec* models. The *Doc2Vec* model learns word-level and document-level embedding simultaneously. Consequently, it outperforms *Word2Vec* where we use a simple word vector averaging technique to derive document-level vectors.

TABLE 10.5

Result after Applying Topic Modeling on *forums.hak5.org* Binary Dataset

Dataset: forums.hak5.org BINARY

Topic Modeling Method: LDA

Topic	Keywords
0	["password" "see" "google" "know" "site" "android" "github" "user" "use" "windows"]
1	["script" "windows" "file" "usb" "payload" "files" "work" "ducky" "run" "drive"]
2	["pineapple" "connect" "wifi" "internet" "get" "connection" "network" "port" "router" "connected"]
3	["email" "antenna" "mail" "darren" "signal" "spam" "cp" "speed" "wait" "antennas"]
4	["http" "www" "php" "hak5" "https" "forums" "code" "like" "index" "page"]

Topic Modeling Method: NMF

Topic	Keywords
0	["like" "get" "know" "use" "something" "work" "time" "want" "see" "really"]
1	["pineapple" "wifi" "internet" "172" "mark" "connect" "firmware" "via" "connected" "connection"]
2	["php" "hak5" "index" "forums" "topic" "https" "showtopic" "http" "forum" "thread"]
3	["usb" "power" "drive" "ducky" "hub" "rubber" "battery" "adapter" "port" "device"]
4	["file" "script" "payload" "ducky" "payloads" "bunny" "run" "exe" "files" "command"]

TABLE 10.6

Result after Applying Topic Modeling on *forums.hak5.org* Multi-class Dataset

Dataset: forums.hak5.org MULTI-CLASS

Category: "Credentials"

Topic Modeling Method: LDA

Topic	Keywords
0	["password" "username" "get" "login" "use" "like" "user" "know" "account" "need"]
1	["password" "username" "file" "pineapple" "root" "http" "use" "using" "get" "ssh"]
2	["hashcat" "use" "john" "using" "txt" "net" "post" "username" "mimikatz" "gmail"]
3	["barcode" "assets" "date" "ass_tag" "echo" "strlen" "mysql_query" "die" "simpleassets" "update"]
4	["network" "wan2" "option" "uci" "set" "172" "etc" "interface" "config" "device"]

Topic Modeling Method: NMF

Topic	Keywords
0	["password" "root" "username" "user" "login" "file" "use" "get" "account" "server"]
1	["wan2" "uci" "network" "set" "12d1" "iptables" "firewall" "sleep" "172" "init"]
2	["wifi" "pineapple" "connect" "network" "wpa2" "enterprise" "connected" "internet" "client" "using"]
3	["php" "splash" "nodogsplash" "portal" "auth" "page" "html" "evil" "call" "www"]
4	["option" "config" "255" "ifname" "proto" "interface" "device" "ht_capab" "netmask" "ipaddr"]

It is also noticeable that none of the neural network-based classifiers (a shallow and a deep neural network) is unable to beat the traditional classifiers we use in this experiment. While neural network algorithms, especially deep neural networks, have made a breakthrough in several computer vision applications and outperform all machine learning-based approaches, their performance for text classification is about the same as the regular classifiers. Recurrent Neural Network (RNN)

TABLE 10.7

Result after Applying Topic Modeling on *nulled.io* Binary Dataset

Dataset: nulled.io BINARY

Topic Modeling Method: LDA

Topic	Keywords
0	["http" "www" "color" "https" "application" "php" "url" "topic" "hack" "nulled"]
1	["bol" "bot" "scripts" "legends" "nulled" "download" "work" "login" "use" "auth"]
2	["php" "gmail" "hotmail" "net" "yahoo" "hide" "http" "index" "html" "server"]
3	["hide" "https" "file" "www" "http" "download" "spoiler" "virus" "link" "virustotal"]
4	["local" "255" "end" "function" "class" "user" "bot" "serverversion" "autoupdatermsg" "true"]

Topic Modeling Method: NMF

Topic	Keywords
0	["like" "script" "get" "use" "know" "thanks" "bot" "crack" "working" "help"]
1	["legends" "bot" "dominate" "login" "download" "available" "enemies" "occur" "monitoring" "freeze"]
2	["hide" "http" "mega" "password" "https" "enjoy" "download" "upvote" "html" "mediafire"]
3	["bol" "scripts" "vip" "studio" "script" "exe" "bolvip" "hijacker" "cracked" "use"]
4	["nulled" "www" "https" "auth" "topic" "php" "http" "access" "need" "cracked"]

TABLE 10.8

Result after Applying Topic Modeling on nulled.io Multi-class Dataset

Dataset: nulled.io MULTI-CLASS

Category: "Credentials"

Topic Modeling Method: LDA

Topic	Keywords
0	["username" "password" "file" "download" "email" "level" "hide" "region" "pages" "summoner"]
1	["user" "password" "username" "php" "http" "admin" "login" "hide" "www" "script"]
2	["var" "steam" "system" "string" "using" "function" "require" "new" "windows" "data"]
3	["user" "action" "password" "email" "login" "https" "host" "username" "agent" "form"]
4	["password" "username" "account" "accounts" "get" "email" "use" "like" "need" "login"]

Topic Modeling Method: NMF

Topic	Keywords
0	["action" "user" "email" "agent" "login" "host" "form" "https" "referer" "http"]
1	["account" "password" "username" "hide" "bol" "accounts" "get" "login" "nulled" "use"]
2	["checker" "check" "delay" "accounts" "killerabgg" "owned" "pvpnetconnect" "library" "last" "current"]
3	["rune" "pages" "rank" "validated" "skins" "champions" "level" "unranked" "summoner" "region"]
4	["user" "export" "scrape" "press" "select" "pick" "number" "page" "list" "check"]

and long-short-term memory (LSTM)-based neural network algorithms specialized in Natural Language Processing perform better in this case [25]. However, we did not use them. There are two reasons for this decision. First, training RNN-based models requires a good amount of computing resources and they are very slow. Second, our basic classifiers have already provided promising results.

Our experimental results show that current topic modeling algorithms on binary dataset do not uncover security-relevant keywords. However, when we apply these algorithms (LDA and NMF) on multi-class datasets as a whole and separately on each security-relevant category, they provide us with topics and keywords that make much sense in the context of their security relevance. Further, we find that the quality of generated keywords for the given topics depends significantly on preprocessing. If we do not preprocess the data properly, the algorithm may get lost in the clutter of words such as stop words and other commonly used words, and the outcome may not represent the latent topics in the dataset.

10.6   CHALLENGES AND FUTURE SCOPES

We face several issues throughout this research work; for example, the first major challenge is the scarcity of open-source threat intelligence datasets for either analysis or benchmarking. We finally manage to get a leaked dataset for *nulled.io* forum from the Internet archive website, *https://archive. org* that we use as a ground truth. The second challenge is to collect data from hacker forums for actual experimentation. We observe that most hacker forums employ some kind of anti-crawling technology, such as CAPTCHA, making it difficult to collect data using an automated program like a web crawler. Further, some forums only allow invite-only registrations. It means we cannot get access to the forum unless we have an account in it and to get an account, we need to get an invitation from a registered member of that forum first. We, however, managed to collect data from a popular forum named *forums.hak5.org* since it does not have such restrictions.

One limitation of our work is that we construct the dataset using a simple keyword searching technique. This process may result in biased measurements. A possible future extension of this work would be using more advanced dataset construction purposes. Alternatively, it is possible to refine and clean the relevant and non-relevant classes using the relevance feedback provided by some human experts in the cybersecurity domain.

10.7   CONCLUSION

Our study collected and analyzed data from a popular hacker forum, forums.hak5.org, for the purpose of identifying and classifying possible CTI. For feature extraction and document vectorization, we employed both traditional and cutting-edge deep learning algorithms. We then developed and utilized several machine learning and neural network models for classifications. Our results showed that it is possible to classify security-relevant posts and irrelevant ones with high accuracy. We further evaluate the performance of these algorithms on the multi-class dataset. Moreover, we utilized two algorithms as an unsupervised learning approach. We were able to extract the top words for a different number of topics on these datasets using LDA and NMF topic modeling algorithms. Further, we applied all these approaches to a labeled dataset named *nulled.io*.

REFERENCES

1. Apurv Singh Gautam, Yamini Gahlot, and Pooja Kamat. Hacker forum exploit and classification for proactive cyber threat intelligence. In *International Conference on Inventive Computation Technologies*, pages 279–285. Springer, 2019.
2. Liang Guo, Senhao Wen, Dewei Wang, Shanbiao Wang, Qianxun Wang, and Hualin Liu. Overview of cyber threat intelligence description. In *International Conference on Applications and Techniques in Cyber Security and Intelligence*, pages 343–350. Springer, 2021.
3. Mauro Conti, Tooska Dargahi, and Ali Dehghantanha. Cyber threat intelligence: Challenges and opportunities. In *Cyber Threat Intelligence*, pages 1–6. Springer, 2018.
4. Ajay Modi, Zhibo Sun, Anupam Panwar, Tejas Khairnar, Ziming Zhao, Adam Doupé, Gail-Joon Ahn, and Paul Black. Towards automated threat intelligence fusion. In *2016 IEEE 2nd International Conference on Collaboration and Internet Computing (CIC)*, pages 408–416. IEEE, 2016.

5. Asif Karim, Sami Azam, Bharanidharan Shanmugam, Krishnan Kannoorpatti, and Mamoun Alazab. A comprehensive survey for intelligent spam email detection. *IEEE Access*, 7:168261–168295, 2019.

6. Asif Karim, Sami Azam, Bharanidharan Shanmugam, and Krishnan Kannoorpatti. Efficient clustering of emails into spam and ham: The foundational study of a comprehensive unsupervised framework. *IEEE Access*, 8:154759–154788, 2020.

7. Char Sample, Jennifer Cowley, Tim Watson, and Carsten Maple. Re-thinking threat intelligence. In *2016 International Conference on Cyber Conflict (CyCon US)*, pages 1–9. IEEE, 2016.

8. Vasileios Mavroeidis and Siri Bromander. Cyber threat intelligence model: An evaluation of taxonomies, sharing standards, and ontologies within cyber threat intelligence. In *2017 European Intelligence and Security Informatics Conference (EISIC)*, pages 91–98. IEEE, 2017.

9. Gartner Inc. Definition: Threat intelligence, May 2013. https://www.gartner.com/en/documents/2487216/definitionthreat-intelligence.

10. Thomas J Holt, Deborah Strumsky, Olga Smirnova, and Max Kilger. Examining the social networks of malware writers and hackers. *International Journal of Cyber Criminology*, 6(1), 2012.

11. Tianyi Wang and Kam Pui Chow. Automatic tagging of cyber threat intelligence unstructured data using semantics extraction. In *2019 IEEE International Conference on Intelligence and Security Informatics (ISI)*, pages 197–199. IEEE, 2019.

12. Brett van Niekerk, Trishana Ramluckan, and Petrus Duvenage. An analysis of selected cyber intelligence texts. In *Proceedings of the 18th European Conference on Cyber Warfare and Security*, pages 554–559, 2019.

13. Ba Dung Le, Guanhua Wang, Mehwish Nasim, and Ali Babar. Gathering cyber threat intelligence from Twitter using novelty classification. arXiv preprint arXiv:1907.01755, 2019.

14. Benjamin Ampel, Sagar Samtani, Hongyi Zhu, Steven Ullman, and Hsinchun Chen. Labeling hacker exploits for proactive cyber threat intelligence: A deep transfer learning approach. In *2020 IEEE International Conference on Intelligence and Security Informatics (ISI)*, pages 1–6. IEEE, 2020.

15. Jun Zhao, Qiben Yan, Jianxin Li, Minglai Shao, Zuti He, and Bo Li. Timiner: Automatically extracting and analyzing categorized cyber threat intelligence from social data. *Computers & Security*, 95:101867, 2020.

16. Paris Koloveas, Thanasis Chantzios, Sofia Alevizopoulou, Spiros Skiadopoulos, and Christos Tryfonopoulos. intime: A machine learning-based framework for gathering and leveraging web data to cyber-threat intelligence. *Electronics*, 10(7):818, 2021.

17. Yali Gao, Li Xiaoyong, Peng Hao, Binxing Fang, and Philip Yu. Hincti: A cyber threat intelligence modeling and identification system based on heterogeneous information network. *IEEE Transactions on Knowledge and Data Engineering*, 2020.

18. Masashi Kadoguchi, Shota Hayashi, Masaki Hashimoto, and Akira Otsuka. Exploring the dark web for cyber threat intelligence using machine learning. In *2019 IEEE International Conference on Intelligence and Security Informatics (ISI)*, pages 200–202. IEEE, 2019.

19. Forums – hak5 forums. https://forums.hak5.org/. (Accessed on 08/15/2021).

20. 0x00sec – the home of the hacker. https://0x00sec.org/. (Accessed on 08/15/2021).

21. Victoria Bobicev. Text classification: The case of multiple labels. In *2016 International Conference on Communications (COMM)*, pages 39–42. IEEE, 2016.

22. Ke Li, Hui Wen, Hong Li, Hongsong Zhu, and Limin Sun. Security osif: Toward automatic discovery and analysis of event based cyber threat intelligence. In *2018 IEEE SmartWorld, Ubiquitous Intelligence & Computing, Advanced & Trusted Computing, Scalable Computing & Communications, Cloud & Big Data Computing, Internet of People and Smart City Innovation*, pages 741–747. IEEE, 2018.

23. Fabrizio Sebastiani. Machine learning in automated text categorization. *ACM Computing Surveys (CSUR)*, 34(1):1–47, 2002.

24. Bofang Li, Aleksandr Drozd, Yuhe Guo, Tao Liu, Satoshi Matsuoka, and Xiaoyong Du. Scaling word-2vec on big corpus. *Data Science and Engineering*, 4(2):157–175, 2019.

25. Md Tazimul Hoque, Ashraful Islam, Eshtiak Ahmed, Khondaker A Mamun, and Mohammad Nurul Huda. Analyzing performance of different machine learning approaches with doc2vec for classifying sentiment of Bengali natural language. In *2019 International Conference on Electrical, Computer and Communication Engineering (ECCE)*, pages 1–5. IEEE, 2019.

26. Qufei Chen and Marina Sokolova. Word2vec and doc2vec in unsupervised sentiment analysis of clinical discharge summaries. arXiv preprint arXiv:1805.00352, 2018.

27. Metin Bilgin and Izzet Fatih S̜entürk. Sentiment analysis on Twitter data with semi-supervised doc2vec. In *2017 International Conference on Computer Science and Engineering (UBMK)*, pages 661–666. IEEE, 2017.

28. Otgonpurev Mendsaikhan, Hirokazu Hasegawa, Yukiko Yamaguchi, and Hajime Shimada. Identification of cybersecurity specific content using the doc2vec language model. In *2019 IEEE 43rd Annual Computer Software and Applications Conference (COMPSAC)*, volume 1, pages 396–401. IEEE, 2019.

29. Jipeng Qiang, Zhenyu Qian, Yun Li, Yunhao Yuan, and Xindong Wu. Short text topic modeling techniques, applications, and performance: A survey. *IEEE Transactions on Knowledge and Data Engineering*, 2020.

30. Tian Shi, Kyeongpil Kang, Jaegul Choo, and Chandan K Reddy. Short-text topic modeling via non-negative matrix factorization enriched with local word-context correlations. In *Proceedings of the 2018 World Wide Web Conference*, pages 1105–1114, 2018.

31. Andrew L Maas, Raymond E Daly, Peter T Pham, Dan Huang, Andrew Y Ng, and Christopher Potts. Learning word vectors for sentiment analysis. In *Proceedings of the 49th Annual Meeting of the Association for Computational Linguistics: Human Language Technologies*, pages 142–150, Portland, OR. Association for Computational Linguistics, June 2011.

32. Uci machine learning repository: Twenty newsgroups data set. https://archive.ics.uci.edu/ml/ datasets/ Twenty%20Newsgroups. (Accessed on 08/15/2021).

11 Cyber-Physical Energy Systems Security

Attacks, Vulnerabilities and Risk Management

Sayada Sonia Akter, Rezwan Ahmed,
Ferdous Hasan Khan, and Mohammad Shahriar Rahman
United International University

CONTENTS

DOI: 10.1201/9781003267812-11

ACRONYMS

EV	Electric Vehicle
ICS	Industrial Control System
MG	Microgrid
EMT	Electromagnetic Transient
CPES	Cyber-Physical Energy System
CB	Circuit Breaker
DCS	Distributed Control Systems
DIA	Data Integrity Attack
RTU	Remote Terminal Unit
SCADA	Supervisory Control and Data Acquisition
TDA	Time-Delay Attack
T&D	Transmission and Distribution

11.1 INTRODUCTION

Experts in power and control systems have been constantly working for the last few years to come up with new tools and strategies that will make it easier to monitor and control the physical electricity systems. Computer science and electronics engineers both are working on the cyber systems at the same time to improve the performance of computer and communication systems. It leads to the development of ubiquitous computing. Every gadget and technological device in our daily lives is connected to low-cost processing and communication networks. It will, without a doubt, have a profound impact on energy systems [1]. The merging of physical and cyber systems leads to the development of a new digital technology known as the Cyber-Physical System (CPS). In CPS, the cyber system gathers the data from the physical system with the help of the sensors and gives back the control signal towards the physical system in order to achieve the goals [2] and [3], as depicted in Figure 11.1.

It is vital to link physical energy systems with cyber systems in order to ensure the efficient and secure operation of the energy systems [4]. A Cyber-Physical Energy System (CPES) integrates and

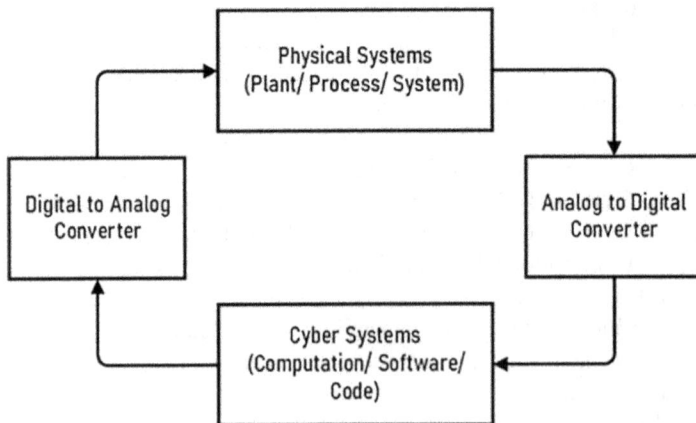

FIGURE 11.1 Flow diagram of cyber-physical system.

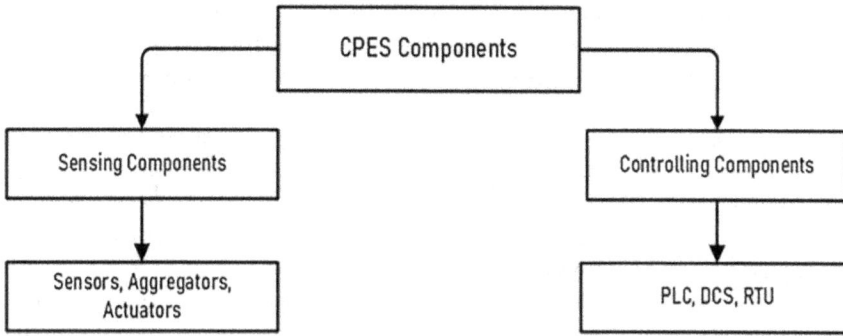

FIGURE 11.2 Components of CPES.

organizes internet and energy system elements; which are developed from control and embedded systems in order to monitor and regulate the physical energy system without any delay. As shown in Figure 11.2, CPES components communicate and organize physical power system actions with the help of embedded computers and networks, receiving feedback on how physical power system events affect computations and vice versa [3]. As a result, cyber security is essential in guaranteeing the reliable and secure functioning of a CPES infrastructure. While cyber-attacks may not directly harm the physical equipment of a smart grid, they can degrade or even destroy regular operations of CPES. This may cause system instabilities, costly operations, and other difficulties related to CPES.

11.1.1 CPES COMPONENTS

There are two types of CPES components as shown in Figure 11.2. Sensing components are used to collect and sense information, whereas controlling components are used to monitor and control the signals.

11.1.1.1 Sensing Components

Sensors that collect data and send it to aggregators are mostly found at the upper layer. This information is sent to the actuators for analysis so that the best possible decisions can be made about what to do next. The following is a list of CPES sensing components [5]:

- **Sensors:** Collect and record real-world data and make sure the data is correct. Therefore, knowing to read the data is very important as the decisions will be made depending on this data.
- **Aggregators:** Aggregators are usually found at the transmission layer (such as routers, switches, and gateways), where they process the data from sensors before making a decision. This data is then gathered and summarized after a statistical analysis.
- **Actuators:** Application layer is where the actuators are found. They make the information visible to the outside world based on the decisions received from the aggregators. In terms of how they work, actuators take in electrical signals as input and generate physical moves as output.

11.1.1.2 Controlling Components

Controlling components are crucial in controlling, monitoring, and managing the signals in order to achieve a higher level of accuracy and protection against malicious attacks or accidents, most notably signal jamming, noise, and interference [5].

- **Programmable Logic Controllers (PLCs):** PLCs were first made to replace hard-wired relays. These are now considered industrial digital computers that control manufacturing processes like robotic device performance and fault diagnosis processing, giving them more flexibility and reliability.
- **Distributed Control Systems (DCS):** DCS are computerized control systems that distribute independent controllers throughout the system under the supervision of a central operator. The remote monitoring and supervision method increases the DCS's dependability while decreasing its installation cost. DCS can be compared to Supervisory Control and Data Acquisition (SCADA) systems in some instances.
- **Remote Terminal Units (RTUs):** RTUs are electronic devices that are controlled by a microprocessor. The primary function of RTU is to connect SCADA to the physical objects via a supervisory messaging system that controls objects through the transfer of telemetry data.

11.1.2   CPES Layers

Individually comprehending the physical and computational components of the power system is not convenient enough; instead, we must comprehend their interrelation as shown in Figure 11.3. Understanding the combined behaviour of software, computers, physical power systems, and networks is required for the design of all such systems [4]. In the CPES, there are three levels of interaction as shown in Figure 11.3.

- **First Level-Physical System:** The transformer, generator, transmission line, dynamic load, and other components of the power system interact with power system controllers. Core component's data of the power system arrives at a power system controller receives and creates a control signal, which is then given back to the core components of the power system for optimal power grid operation.
- **Second Level-Cyber-Physical Interaction:** Power system control and the communication infrastructure have a second degree of interaction. All of the functions of the subsystems (sensors, actuators, interfaces, control, computation, and communication units) in CPES are coordinated by the communication infrastructure.
- **Third Level-Cyber System:** Interaction between the communication infrastructure and the cyber system occurs here. The principal objective of cyber systems is to execute sophisticated grid operations such as state estimate, load forecasting, variable optimization, oscillation monitoring, voltage control, wide-area monitoring and control, model validation, operations planning, and stability analysis [6].

11.1.3   CPES Security Concerns

The secure operation of the power grid depends not only on the flow of power in the physical system but also on the information flow in the cyber system, i.e., Information and Communication Technology (ICT). Despite the fact that the cyber system assures the power grid's efficient, safe, and secure operation, power shortages have happened in the past owing to cyber system failure. When COVID-19 impacted the society for the first time, "working from home" increased the rate of cyber-related incidents from around 5,000 per week in February 2020 to over 200,000 per week in late April 2020 [7]. The malicious exploitation potential of ICT and control interfaces of Internet-of-things (IoT)-controllable loads might, more than ever before, pose a genuine cyber-threat to power grid operations [8]. CPES should have a number of security goals [9,10] such as

- **Safety:** Human, technology, public services, and even the environment should not be at risk because of how the system or the machines work.

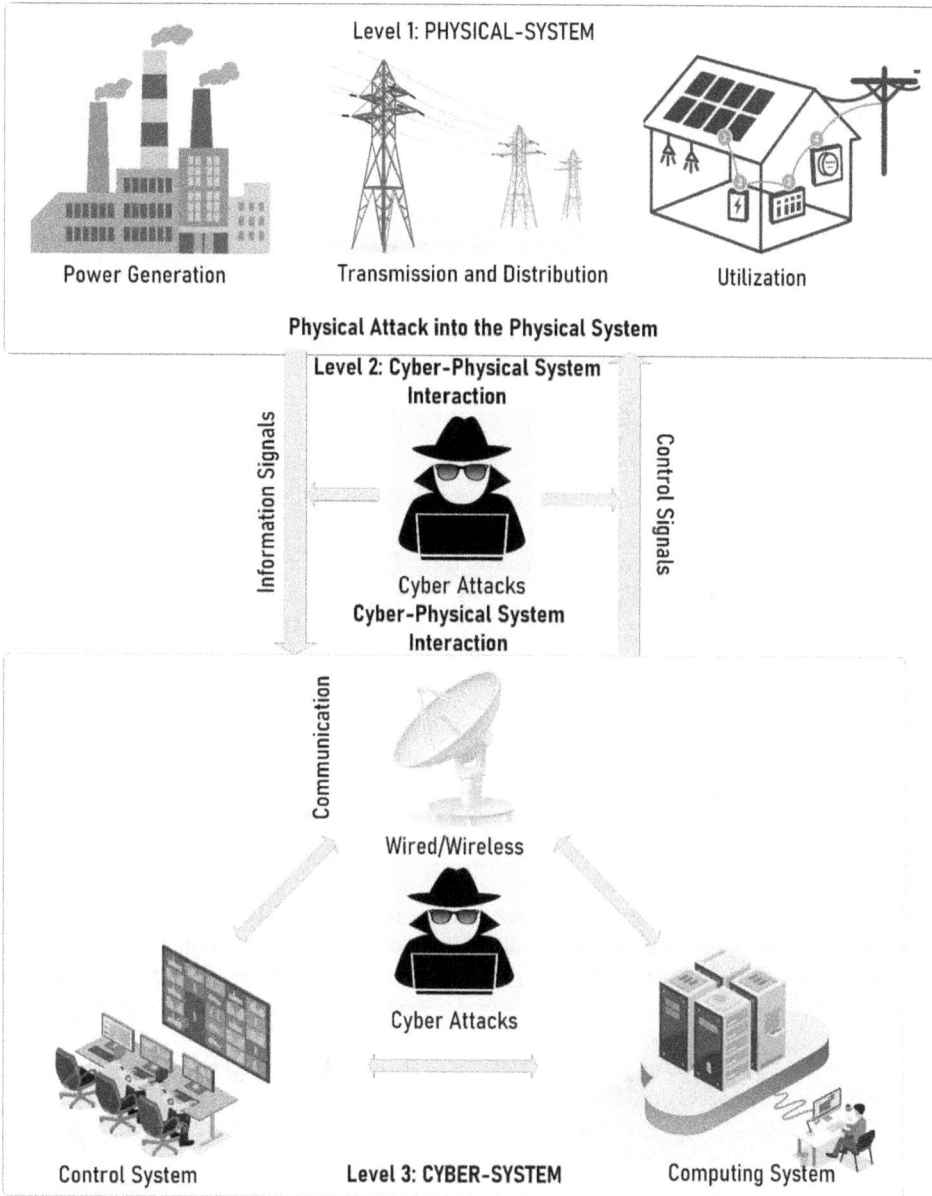

FIGURE 11.3 Layer diagram of CPES.

- **Security:** System and the machines should be protected from any kind of unauthorized or accidental accesses, changes or disruptions due to malware, remote attacks, etc.
- **Availability of Resources:** People who use electricity or gas need to have uninterrupted access to their bills and other information including the control of messages.
- **Privacy:** Data should only be accessible to people who owns that data and others who have been explicitly and legally given permission to see the data, not to the general public.
- **Integrity:** How well the system can keep data and commands from being changed.
- **Resilience:** The ability to stay as normal as possible even during the cyber/physical crisis.

- **Reliability:** Ability of the systems to do what it needs to do under certain circumstances. This includes components like smart metres and other parts of the smart grid.
- **Accuracy:** The system should be able to correctly figure out how much energy has been used and support the effective flow of information.

Therefore, to safeguard CPES's sophisticated power grid control networks from cyber-attacks, risk and vulnerability assessments must be conducted [11]. Based on the above-mentioned papers and the need for a comprehensive study, we present the fundamentals of CPES, as well as related previous works, different types of security threats and attacks throughout the history of CPES, case studies, Cyber-Attack Detection and Mitigation Methods, Cyber-Attack Risk Analysis, and an overall Risk Evaluation Approach for CPES in this chapter.

11.1.4 Contribution of This Chapter

As previously stated, CPES is vulnerable to a wide range of cyber and physical attacks, as well as coordinated cyber-physical attacks when they are combined. While various survey articles have been published in CPES security, most of the studies are dispersed. The purpose of this chapter is to bring together all of the conceivable concerns linked to CPES, in order to provide the readers with a comprehensive understanding of CPES, its security concerns – threats and vulnerabilities associated with case studies along with potential cyber-attacks. Main contributions of this chapter are given below:

- We have reviewed the basics of CPES – components and layers along with security concerns of CPES.
- We looked at the threats that smart grids may face, focusing on threats that are specific to this type of infrastructure, both Physical and Cyber threats along with the vulnerabilities that a CPES may have.
- We studied the loopholes and the types of CPES attacks and presented an attack tree.
- We analysed major cyber-attacks in the history of CPES in detail, with each event being broken down into parts such as attack objectives, types, and impacts of the attacks.
- After reviewing relevant research works, we have elaborated an overall risk evaluation approach that incorporates all the steps that must be taken in order to mitigate zero-day vulnerabilities.
- We have given an overall overview of CPES forensics. At the end of this chapter, we identified a number of high potential areas related to CPES forensics for further investigations, for which the information presented in this chapter can serve as a solid foundation.

The contributions of this chapter are summarized in Table 11.1, which contrasts other literature works with the contributions of this chapter. This table, as well as a review of the literature, will be discussed in further detail later in Section 11.2.

The chapter is structured as follows: Section 11.2 provides related work on today's power grid and its security. Section 11.3 describes CPES threats and vulnerabilities. Section 11.4 contains the taxonomy of CPES attacks, while in Section 11.5 we have discussed some severe historical cyber-attacks on power grids. We have proposed the steps a CPES should follow in order to evaluate the risks. We started the ground for future work in Section 11.7 and concluded the chapter in Section 11.8.

11.2 RELATED WORK

As mentioned earlier, in CPES, there are physical and cyber layers where the physical layer consists of different energy providers and end users and the cyber layer consists of sensors, communication networks, SCADA systems, and control systems. Computers, networked data communications, and

TABLE 11.1

Comparison between Literature

Reference	CPES Structure	CPES Threats	CPES Vulnerabilities	Types of CPES Attacks	Analysis of Attacks	Case Studies	Risk Assessment Approach	Basics of CPES Forensics
[12]	√	√	√	√	√	X	√	X
[13]	√	√	√	X	X	X	√	X
[14]	√	√	√	√	X	X	X	X
[15]	√		√	√	√	X	√	X
[16]	√	√	√	√	√	X	√	X
[17]	X	X	X	√	X	X	√	X
[18]	X	√	X	√	√	X	X	X
[19]	X	X	√	√	√	X	X	X
[20]	√	X	X	X	X	√	√	X
[21]	X	X	√	√	√	√	X	X
[22]	X	X	X	√	X	√	√	X
[23]	√	X	√	√	√	X	√	X
[24]	√	X	X	X	√	X	X	X
[25]	X	√	√	√	√	X	X	X
[26]	X	√	√	X	√	√	X	X
[27]	X	X	X	√	√	√	√	X
[28]	√	X	X	√	√	X	X	X
[29]	X	√	√	√	√	X	√	X
[30]	X	√	√	√	X	√	√	X
[31]	X	√	√	√	√	X	X	X
[32]	√	X	√	X	X	X	X	X
[33]	X	√	√	√	X	X	√	X
This chapter [2022]	√	√	√	√	√	√	√	√

graphical user interfaces make up a SCADA, which lets supervisors monitor the machines and the processes. Proper operation of a CPES is highly dependent on collection, procession, and transmission of data from physical layer to cyber layer [34], which puts the CPES at the risk of cyber-attacks. Power grid's delicate balance could easily lead to a series of protection mechanisms being activated, which could cause cascading failures and large-scale blackouts such as physical attacks, cyber-attacks, and cyber-physical attacks (also called coordinated attacks). These are the three main types of outside attacks on CPES that can happen [35]. The authors [35] surveyed the main CPS security vulnerabilities, threats, and attacks, along with the key issues and challenges; these issues and challenges are the same for CPES as well. Other authors [36] described about a number of attacks on industrial energy networks that have had a big impact on how people think about cyber security.

Wang-Lu [25] looked at the security risks of a typical CPES, which includes transmission and distribution subsystems. They explained the security requirements and thoroughly looked at network threats with case studies. Similar to Wang-Lu's study, He and Yan studied cyber-physical attacks in the smart grids and portrayed what could happen if an intruder tries to attack the energy generation, transmission, distribution, and electricity markets. Real-life cyber-attack incidents are shown in Ref. [37]. The paper discusses attacks on traditional energy networks and those on the smart metering networks. The authors also represent a threat taxonomy taking security threats into account. A comprehensive view of security and privacy concerns of CPES, detailed taxonomy of attacks, a comprehensive study for security and privacy goals and corresponding solutions were presented in Ref. [38]. Kimani et al. evaluated and investigated the primary obstacles and potential vulnerabilities that are hindering the development of IoT-based smart grid networks in Ref. [16]. They reviewed some high-profile cyber-attack examples and divided those attacks into – device attack, data attack, privacy attack, and network availability attack. They also outlined a risk mitigation plan in their paper.

The authors in Ref. [14] provided a comprehensive and methodical overview of denial-of-service (DoS) attack taxonomies as well as a survey of possible solution techniques to this. The authors in Ref. [6] looked at a wide range of modelling, simulation, and analysis methods, as well as different types of cyber-attacks, different ways to detect and stop cyber-attacks on the real power system and cyber security measures for modern CPES. The authors in Ref. [17] gave an overview of the CPS security landscape, with a focus on CPES. They showed a threat model, the possible attack points, and system flaws. The general framework for modelling, simulating, assessing, and mitigating attacks in a CPS was shown by looking at a few attack scenarios that target CPES. A model was proposed in Ref. [39] that can show how various parts of CPSs work together. Wadhawan et al. presented a comprehensive study on smart grid security against cyber-physical attacks in Ref. [40]. They showed how to predict system failures early in order to establish a strong and resilient power system.

Tian et al. looked at two types of DoS attacks and how they affect CPES in Ref. [41]. The first attack is thought to be a stealthy false data injection that is done to hide the attack from detection algorithms. The second, which is thought to be a non-stealthy attack, aims to do the most damage to the power system by targeting the most vulnerable transmission line, preventing power dispatch and initiating load shedding. The abilities of an attacker and the characteristics of the adversary model were discussed in Refs. [42,43]. Conventional grids and smart grids are two different aspects that were discussed in Ref. [44]. They studied the process of using renewable energy in a smart grid system where grid control is important for energy management. They also mentioned how to make sure the grid is reliable and how to control it so that customers do not have to go without electricity. Mathas et al. presented the threats that CPES may face, an assessment of the consequences of each attack type, figuring out features that can be used to detect attacks, and listing ways that can be used to mitigate them [45]. The authors in Ref. [46] discussed some of the problems that may come with integrating CPES. They considered how to improve CPES protection and security, evaluate CPES vulnerabilities, diagnose CPES attacks, and propose some defensive measures.

11.3 CPES THREATS AND VULNERABILITIES

Security services were not designed into CPES, leaving the door open for attackers to exploit numerous vulnerabilities and threats to initiate security attacks. This is due to the heterogeneous nature of CPES devices as they operate in different cyber domains and interact using various protocols and technologies.

11.3.1 CPES SECURITY THREATS

As mentioned below, CPES security threats can be characterized as cyber and physical threats, and when integrated, cyber-physical threats might emerge. Figure 11.4 represents the types of threats.

11.3.1.1 Cyber Threats

For a variety of reasons, the main focus of CPES security is highly dependent on cyber threats rather than physical threats [47]. Aside from SCADA vulnerabilities, the evolution of Advanced Metering Infrastructure has led to the rise of newly unknown cyber threats [48]. Physical attacks require physical presence and actual instruments, whereas electronic attacks can now be launched from any machine. Without appropriate prevention and defensive actions, cyber-attacks are difficult

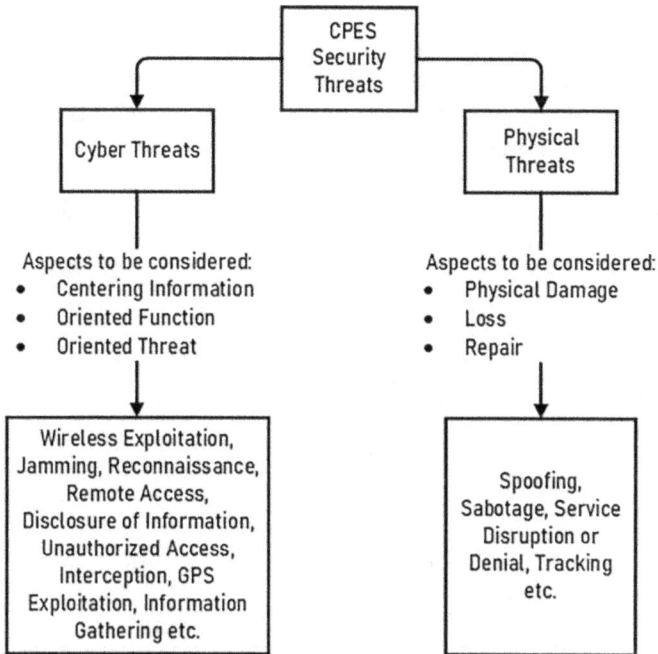

FIGURE 11.4 CPES security threat tree.

to moderate and resist in the sector of CPES. Because CPES security is not confined to a single aspect, it may be seen from a variety of angles, including [33]

- It is required that the flow of data to be protected during the storage phase, transmission phase, and the distribution phase.
- The cyber-physical components must be integrated into an overall CPES.
- The threats impact data confidentiality, integrity, availability, and accountability.

Because of the aforementioned issues, CPES is vulnerable to threats such as

- **Wireless Exploitation:** It needs to know how the system works so that an intruder can use its wireless abilities to get access to or control over a system from afar or even disrupt the system's operations. This leads to a collision and/or a loss of control [21].
- **Jamming:** Jamming intends to disrupt the connection of data between local controllers and smart metres, which is a critical initial step in an intruder's attempt to launch a range of cyber-attacks. To disrupt the demand-response system, an attacker might, for example, delay or stop smart meter reading collection and jam real-time rate signals transmitted in the last mile. Even jamming attacks which are small-scale can result in unavailability of data samples for state estimation [49].
- **Reconnaissance:** An example of this kind of threat is, when intelligence agencies keep running operations against a country's computational Intelligence and Industrial Control System (ICS) mostly using malware. This is because traditional defenses are not strong enough to keep data private [30].
- **Interception:** Hackers can get access into private conversations by exploiting flaws that already exist or that have been found. This leads to another type of privacy and confidentiality breach [50].

- **Disclosure of Information:** Hackers can get any private or personal information through the interception of communication traffic with wireless hacking tools, which means they can break both privacy and confidentiality [50].
- **Information Gathering:** It is illegal for software companies to get files and audit logs from any device and sell this information for marketing and commercial purposes. Such illegal activities are known as "information gathering" [51].
- **Remote Access:** This is mostly accomplished through attempting to gain remote access to the CPES infrastructure, for the purpose of production disruptions, financial losses, blackouts, and data theft. Moreover, Havex trojans are among the most dangerous viruses that can be weaponized and utilized as part of a nation's CPES cyber-warfare campaign management [52].
- **Unauthorized Access:** Attackers try to get into a network without permission through a logical or physical breach in order to get sensitive information, which is a breach of privacy [52].

11.3.1.2 Physical Threats

Physical threats could be categorized by three factors mentioned below [53].

- **Physical Damage:** Due to the fact that transmission lines can be sabotaged and disrupted, the main source of danger comes from power-generating substations that are not as well-protected as transmission lines. Risk assessment can help users lessen the risk of physical tampering or theft by adversaries, such as Advanced Persistent Threats, but it is almost impossible to stop them [54].
- **Loss:** The most frightening thing is when a malicious person causes multiple substation failures. If the smart grid is really damaged, major cities can go completely dark for hours. There was a blackout in the United States on August 14, 2003, because the People Liberation Army took down power lines all over the United States [55].
- **Repair:** If there are problems or disruptions, it can be based on a self-healing process [54]. Self-healing is able to detect and isolate the affected components and send alerts to the control system. The control system then automatically moves backup resources to meet the needs. The goal is to get the service restored as quickly as possible. Thus, self-healing is able to respond quickly to extremely bad damages.

Physical threats that a CPES should be concerned of

- **Spoofing:** It happens when a malicious person or group pretends to be a trusted person or group. This is an opportunity for attackers to spoof sensors; for example, they can send false or misleading data to the control centre.
- **Sabotage:** It is called sabotage when someone steals legal communication traffic and sends it to someone else, or messes with the communication process. Attackers, for example, can damage parts of the CPES that are exposed in the power system. There is a risk that service will be interrupted and/or cut off, which could cause a total or partial blackout.
- **Service Disruption or Denial:** Attackers can physically change the settings on any device to stop a service or stop a service.
- **Tracking:** It is easy for someone to get into a machine and attach a malicious device to it or track the people who use it.

Physical attacks, like damaging power substations or transmission lines, usually cause a lot of damage to the infrastructure [54]. There has been a lot of research done on the risk of a physical attack [12]. There are also cyber-attacks that always target SCADA systems, which can stop or mess with

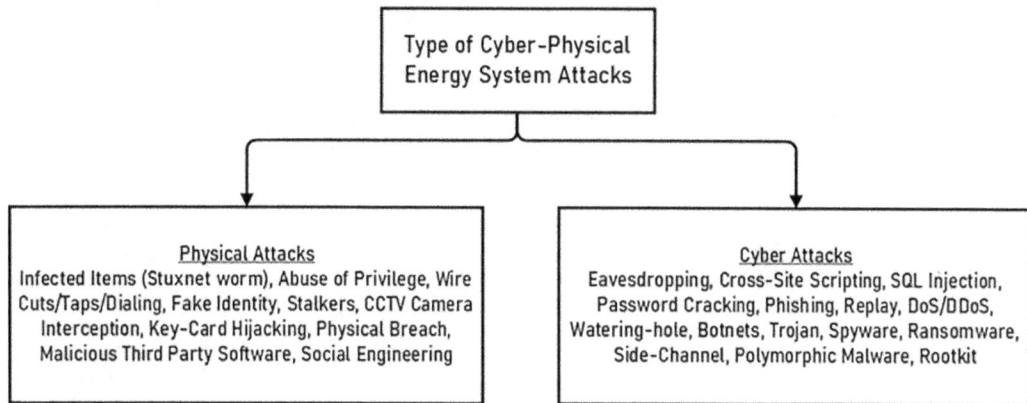

FIGURE 11.5 Cyber-physical attack types.

data transfer or even make it impossible to read [53]. Coordinated attacks can also be made by combining two attacks.

11.3.2 CPES Vulnerabilities

Unpatched security flaws that can be exploited for the purpose of industrial espionage are referred to as vulnerabilities (reconnaissance or active attacks). Therefore, a vulnerability assessment comprises the discovery and analysis of existing CPES weaknesses as well as the determination of necessary remedial and preventative steps to minimize, mitigate, or even completely eliminate any vulnerability [46].

11.3.2.1 Cyber Vulnerabilities

ICS applications are vulnerable to security threats as they rely extensively on open standard protocols such as the widely used Inter-Control Centre Communications Protocol and the Transmission Control Protocol/Internet Protocol. In addition to significant buffer overflow vulnerabilities, these open-source standard protocols frequently lack the most basic security precautions [56]. For example, the Remote Procedure Call protocol and ICS are vulnerable to a variety of attacks, including those associated with the malwares like Stuxnet (1 and 2), Duqu (1.0, 1.5, and 2.0), Gauss, RED October, Shamoon (1, 2, and 3), Mahdi malware, and Slammer Worm attack [57].

11.3.2.2 Physical Vulnerabilities

Physical tampering with cyber-physical components may result in inaccurate data being recorded. A vulnerability is defined as a large number of physical components that are not protected from physical damage, tampering, alteration, modification, or even sabotage. Physical exposure of CPES components is defined as the exposure of a large number of physical components that are not protected from physical damage [39]. Figure 11.5 has examples of varieties of physical and cyber-attacks.

11.4 CYBER-ATTACKS AND CYBER SECURITY IN CPES

In comparison to other cyber-attacks on information technology systems, the cyber-attack on CPES is very severe. Despite the fact that attacker tactics have a lot in common with traditional cyber-attacks, their capacity to cause grid disruption is greatly dependent on the power system applications or control functions that such systems are able to serve. A cyber-attack on the CPES is depicted

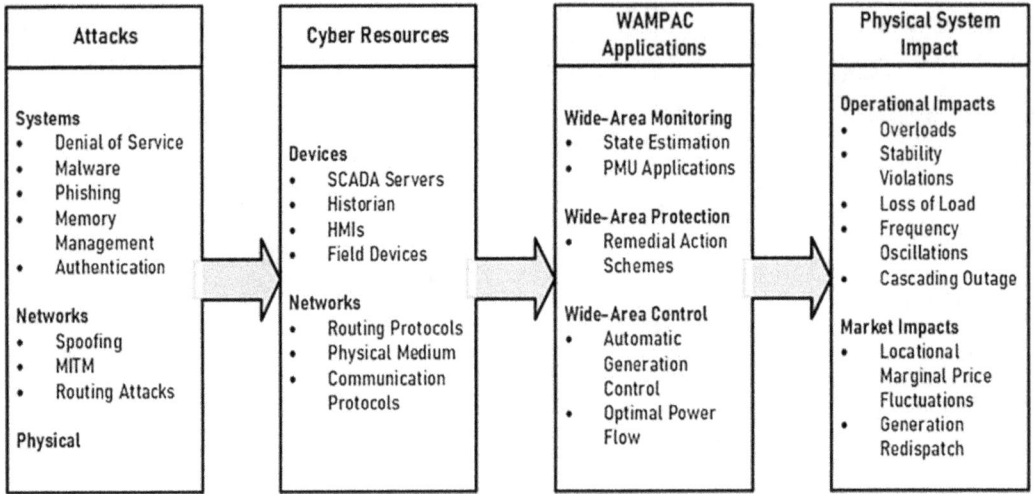

FIGURE 11.6 Mapping from cyber-attacks to control actions to system impacts.

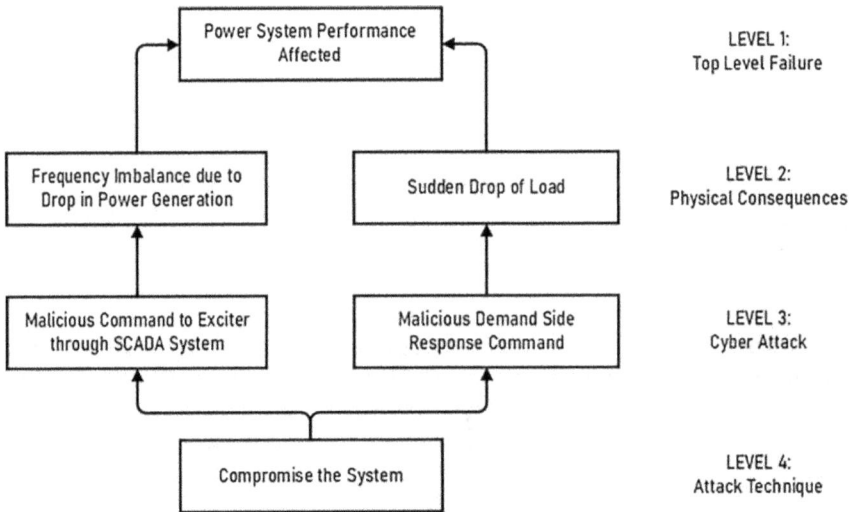

FIGURE 11.7 Cyber network intrusion and attack tree.

in Figure 11.6, which shows how the system would be affected. The first step taken by an attacker is to compromise the availability, integrity, or confidentiality of a component of the cyber system that enables CPES, such as a server or a database. It has an impact on a number of power applications and control activities on the grid as a result of the deterioration. The ability of the attacker to modify the control functions would therefore result in a direct influence on the physical system as a result of the attack.

An alternative method of cyber network intrusion is illustrated in Figure 11.7, which simply describes the cyber-attack procedure presented in Refs. [40,58]. In addition, the attack tree for smart grid applications is also depicted in this diagram. Level 1 power transmission to the consumer is defined as uninterrupted power transmission with no interruptions. Level 2 refers to the physical system factors that cause a power grid blackout to take place. Level 3 cyber-attacks against CPES

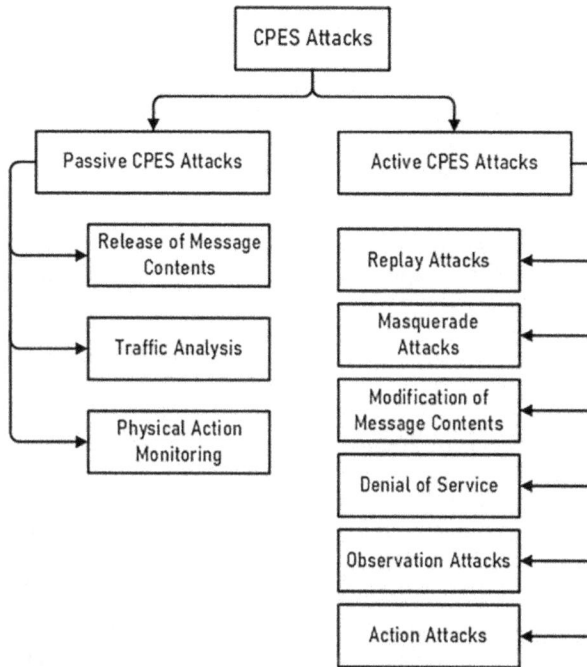

FIGURE 11.8   CPES attack types.

have physical consequences. The attacker has complete control of the SCADA and RTU, causing the power-generating system to go down for a long time. Finally, the attack technique that will be employed to carry out the attack is denoted by the level 4 designation.

CPES attacks can be divided into two categories: passive attacks and active attacks, each with its own set of characteristics and consequences. Methodologies such as sources, intentions, and objectives are all independent of one another. As illustrated in Figure 11.8, each of these classes is further classified into several forms.

11.4.1 PASSIVE ATTACKS

In the field of information security, passive attacks are defined as those that provide access to the CPES in some manner but do not directly impact it. Eavesdropping the information being transferred, as well as accessing data stored in the system and observing CPES behaviour, is an example of such activities.

- The contents of communication messages that contain sensitive or private data, such as system information, control information, internal decisions or other data among CPES components, collaborative CPES, or between a CPES and other supported systems, such as cloud and fog computing, will be made public if the communication contents are publicly disclosed [59].
- If the messages are not encrypted, outsiders can see the contents of the communications in the CPES by monitoring communication traffic. If the monitoring is encrypted, it may not reveal the information; however, it can offer attackers with a pattern of communication, sensing, and information of control among the CPES's many components. The intruders might use this information to figure out what sort of sensing or control signals were being sent, where the CPES components were located, and what type and frequency of the CPES operations were being performed.

- By monitoring physical activities, intruders are able to get an idea about a CPES and the practical operations taking place in the physical environment. All of these have the potential to cause a range of problems and consequences, including endangering the confidentiality of the system, breaching the privacy of consumers, patients, or organizations, as well as allowing industrial and commercial surveillance and espionage.

In order to conceal transmitted information, control signals, and feedback content, appropriate physical security measures as well as powerful encryption/decryption methods for data at rest or in transit can be implemented. Such an approach will eventually prevent many passive cyber-attacks from being launched in the first place.

11.4.2 ACTIVE ATTACKS

The most important characteristic of these attacks is that they cause some level of damage or disruption to the CPES. Obtaining access to the system and altering communication messages, stored data, or planned operations are examples of how intruders might launch an attack. It is possible to detect active attacks much more quickly than passive attacks because the changes or damages they cause are more perceptible.

- Replay attacks are carried out silently by the intruder, who captures messages or action signals that are being exchanged and re-sends them over the computer network in order to cause unwanted effects, such as inaccurate cyber or physical activities, to occur.
- The masquerade attack occurs when an intruder with no or limited privileges in the CPES impersonates entities with higher privileges in order to get unauthorized access to restricted information and resources, as well as the ability to conduct illegal operations.
- Modification of message content attacks cause illegal cyber or physical activities by altering, delaying, or reordering some sensing or control messages.
- When outsiders flood CPES with bogus requests and message exchanges, they are committing a DoS attack. This sort of attack is frequently carried out by flooding some CPES components with messages that exceed their capacity, causing the component, and perhaps the entire CPES, to halt or decrease operational performance.
- Attackers must be able to get physical access to some of the CPES's sensing components in order to conduct observation attacks. The attack is carried out either by disabling the sensors or by making false observations utilizing the sensors that have been disabled; e.g., intentionally raising the temperature near a temperature sensor in order to get it to signal an incorrect state.
- Actuators and action controllers in the CPES are vulnerable to physical attacks, which are also known as action attacks. Intruders have the ability to control the responses of the actuator in order to impact the result of their operations. In the CPES, altering the material type in a 3D printer can result in the created object being faulty or failing to satisfy specifications. Direct attacks are unable to be carried out in the absence of physical access to certain CPES components.

CPES is vulnerable to a variety of passive and active attacks that can compromise different components. Attacks on one or both of these components are possible; both cyber and physical components have consequences on the other. Figure 11.9 demonstrates an example of this.

For example, attacks on the ICS have the potential to alter how a device or machine operates. These attacks can disrupt or manipulate communications resulting in some software modules behaving in an uncontrollable manner. Hacking into a video camera and using the videos to obtain trade secrets is another attack scenario. These attacks are in addition to the more common ones on the software parts.

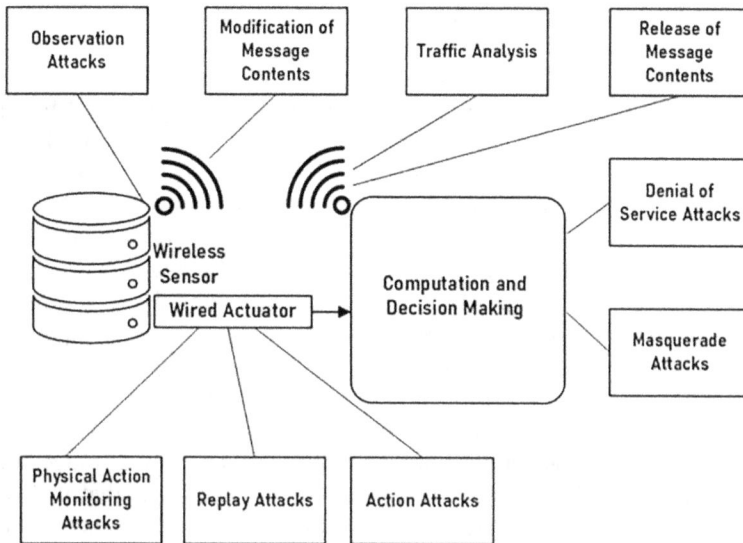

FIGURE 11.9 CPES major components with possible passive and active security attacks.

It is possible to protect the CPES from active attacks in a variety of ways, including increasing physical security at operational sites, implementing strong access control policies, implementing multiple message validation steps in critical parts of the CPES, and incorporating active monitoring techniques into CPES operations.

11.5 CYBER-ATTACK ANALYSIS

Numerous CPES security attacks occurred in the real-world targeted parts of the CPES, such as SCADA systems or the ICS, and disrupted industrial processes by directly attacking these systems. Depending on the conditions, the computing infrastructure that operates and manages the control systems may be targeted for attack. In 2012, Saudi Aramco was the target of an attack that infected over 30,000 workstations with malware, resulting in extensive operational delays [60].

11.5.1 SOME RECENT CYBER-ATTACKS IN CPES

We examine some of the most important cyber-attacks in history in order to have a better understanding of the hazards posed by cyber-attacks on vital infrastructures and other targets.

- In 1982, a number of significant events occurred in the energy industry of SAU. Following the successful exploitation of a fundamental firewall vulnerability, several devices were hacked and reset from a single point of failure. While the impact was minimal in this instance, things could have turned out quite differently [61].
- The MS SQL Server 2000 worm infected the Davis-Besse nuclear power station in Ohio, USA on January 25, 2003. The infection generated a data overload in the site network, preventing computers from communicating with one another [26]. The link was established across an unprotected network, and the Slammer Worm spread from there to the nuclear plant's corporate network. The Davis-Besse business network's performance began to slacken significantly. Microsoft published a software fix for Microsoft SQL Server and Microsoft SQL Server 2000 Desktop Engine around six months before the event (MSDE 2000). Microsoft asked that all of its customers download the patch from the company's

website. The Davis-Besse nuclear power station's system administrator failed to apply the patch to one of the network's servers. The Slammer Worm took advantage of this unpatched server's vulnerability and attacked the whole network [62].

- The Iranian nuclear power grid's SCADA system was compromised by the Stuxnet malware in 2010. The Stuxnet malware was designed to attack and disrupt Iranian infrastructure. A very complex worm travelled on USB flash drives and propagated through Microsoft Windows operating systems. Siemens Step 7 software, which is utilized by industrial computers [63], was the target of this malware attack. There were three phases of this attack [27]. First, it began by analysing and focusing on Windows networks and computer systems. After infecting these devices, the worm proceeded to replicate itself continuously. In the second step, the computer penetrated the Siemens Step 7 application, which runs on Windows operating systems. The worm was ultimately successful in gaining access to the industrial program logic controllers by exploiting the Step 7 software. This last step provided the worm's developers with access to key industrial data, along with the ability to handle a wide range of equipment across a wide range of industrial settings. It was so ubiquitous that if a USB device was plugged into an infected machine, the worm would penetrate the USB device and propagate to all the subsequent computing systems to which the USB was connected [64]. Although Iran did not provide information regarding the attack's implications, it is believed that the Stuxnet worm had compromised 984 nuclear enrichment centrifuges. According to the most recent projections, this resulted in a 30% decrease in enrichment efficiency [27].
- Similar to the above-mentioned Iranian attack, Brazil Power Plant in 2011 caused the plant's management systems to freeze up and not display any data [65].
- Korea Hydro & Nuclear Power (or KHNP) declared in an official statement on December 23, 2014 that its computer systems had been compromised. The organization who hacked KHNP's server acquired access to several plant computers and stole nuclear reactor designs, information on numerous support systems, and personal information on over 10,000 personnel. Additionally, the hackers stated on Twitter that "Unless you stop operating the nuclear power plants until Christmas and give us $10 billion, we will continue to release the secret data related to the facility." According to the South Korean government's investigative report, hackers inserted "Kimsuky" malware in 3,571 KHNP employers' emails using IP addresses in China and Russia [66].
- The Ukraine Cyber Attack in December 2015, which resulted in the loss of electricity for around 225,000 customers, is regarded as the greatest cyber-attack-related blackout in power system history. The lights came back on in three hours in most cases, but because the hackers had sabotaged management systems, workers had to travel to substations to manually close breakers the hackers had remotely opened [67]. Power plant operators fell victim to a complex cyber-attack that included spear phishing emails to obtain early access to their systems, BlackEnergy 3 malware to steal credentials, and other approaches later on. BlackEnergy began as a malware system for executing DoS attacks, which are intended to prohibit legitimate users from reaching a server through a variety of methods. Since then, BlackEnergy has developed into a powerful tool for data exfiltration or the unauthorized movement of information from a computer. Such a transfer might be manual and carried out by someone with access to the computer, or automatically carried out by malicious software installed on the targeted machine [68].
- In 2016, a "severe cyber-attack" on Israel's power grid was launched by a simple ransomware on an industry regulator's office network, paralyzing several of the authority's computer systems. This event happened during two consecutive days of record-breaking winter power usage, with the Israel Electric Corporation reporting a demand of 12,610 MW as temperatures fell below the freezing point [53].

- When mysterious hackers caused a Saudi Arabian oil refinery to shut down in August 2017, an investigation revealed that the malware used in the attack had unparalleled, particularly devastating potential. Triton, also known as Trisis, a family of malware built to compromise industrial safety systems was behind this attack [24].
- In 2017, an Irish power grid EirGrid was compromised by a foreign actor. In April, hackers obtained access to an Irish operator's UK Vodafone network. After the attack, hackers compromised Eirgrid's routers in Wales and Northern Ireland. As a result, they got access to all unencrypted communications sent and received by the firm, leaving the networks exposed to "devious attack" [19].
- Venezuela suffered major power outages after repeated attacks against the central control system of country's main electricity generator, the Simon Bolivar Hydroelectric Plant in Bolivar State, commonly known as the Guri Dam in March 2019. This targeted three of the five generators and forced the dam's turbines to stop.
- On March 5, 2019, electrical utilities in the western United States suffered a cyber-attack that caused operators of the power control centre to lose connection with multiple remote power-producing facilities for periods of several minutes at a time, according to the company. Later, it was discovered that the problem was caused by the fact that internet-facing firewalls were rebooting and going down for no apparent reason. Each reboot caused communications between a controller and a generator site to be disrupted for not more than five minutes, but the pattern continued, and the power system hack lasted over ten hours (Table 11.2) [26].

11.5.2 CPES-Specific Attacks: Case Study

11.5.2.1 Case Study 1: Cross-Layer Firmware Attacks

Cross-layer firmware attacks refer to attacks on the firmware code of embedded devices, which is the read-only inhabitant code that includes microcode and large-scale guidance level routines. These attacks aim to make and spread changes from the device layer to the framework and application layers, respectively. Usually, embedded devices in modern CPS run on metal hardware without an operating system and start up solid single-purpose software. In these kinds of devices, tasks are done on a single-thread infinite loop. If the firmware code for the embedded device is changed in a malicious way, people who do not belong to the embedded device could see it. These attacks can have a wide-ranging effect on different parts and cycles of the CPS. In a CPES, for example, an attacker could change the firmware on the inverters used by battery energy storage systems or electric vehicle chargers. This would make the system's frequency and voltage lists change causing the system becomes unstable. When the Ukrainian force matrix attack happened in 2015 [13], hackers changed the authentic firmware of sequential-to-ethernet converters at power substations and made them inoperable [15].

A cross-layer firmware attack is carried out by an adversary who has the capacity to breach the physical device, proving the cyber system layer. The execution of an over-the-air cross-layer firmware attack that compromises a device through the cyber system layer would likewise require the displaying of the cyber system layer, i.e., the correspondence network that fills in as the medium and section point for the attack [17].

11.5.2.2 Case Study 2: Load-Changing Attacks

An adversary can be either completely or partially uninformed of the framework topology to conduct a load-changing attack. Such constrained information about the CPES may be obtained, for example, using open-source knowledge tactics, and could be used to help coordinate the attack in the best-case scenario. The adversary can play out the attack remotely through the IoT devices controlling the high-wattage loads [22]. Load-changing attacks are planned attempts to disrupt grid through power outages, voltage hitches, and periodic variations.

TABLE 11.2

Major Attacks in the Cyber-Physical Energy Industry

Year	Location	Attack Objects	Type	Impact
1982	Soviet Union (Russia)	Gas pipeline control software	Code manipulation	3 kg TNT equivalent explosion.
1999	Bellingham, USA	Slowdown of SCADA system of a gasoline pipeline	Code manipulation	Huge fireball that killed three people and injured many others.
2003	Ohio, USA	Slammer Worm penetrated the nuclear plant control system	Malware injection	Parameter display system was off for five hours.
2007	Idaho National Laboratory, USA	Aurora attack manipulated a CB of a diesel generator	False data injection	Exploded generator.
2008	Turkey	Attackers manipulated control system parameters of the oil pipeline	False data injection	Oil explosion and 30k barrels are spelled in water.
2010	Iran	Malware looked for traces of Siemens Step 7 software	Code manipulation	Believed to have damaged 984 uranium enrichment centrifuges, resulting in a 30% reduction in enrichment efficiency.
2011	Jirau power plant, Brazil	Malware looked for traces of Siemens Step 7 software	Distributed DoS attack	The plant's management systems got frozen up and did not display data.
2012	Saudi Arabia & Qatar	Malware affected Aramco and RasGas	Malware injection	Generation and delivery of energy have been affected.
2014	Korea Hydro & Nuclear Power, South Korea	Hackers sent emails with "Kimsuky" malware	Malware injection through phishing	Emails hackers gained access to plant computers and released stolen blueprints of nuclear reactor, details on various support systems, and personal data.
2015	Kiev, Ukraine	Attack on the breaker's sittings in three distribution companies	False data injection	The attack was designed to simply destroy data or shut down the plant and to sabotage the firm's operations and trigger an explosion.
2016	Public Utility Authority, Israel	A simple ransomware outbreak on the office network of an industry regulator.	Ransomware malware infection	Resulted in many of the computer systems used by the authority being "paralysed."
2017	Saudi Arabia	Designed to target the Triconex Safety Instrumented System controllers	Malware injection	The attack was intended to cause an explosion that would have killed people.
2019	Western United States	Created DoS attack with the help of a network of hacked computers caused by outdated firmware.	False data injection	A low-impact attack, no blackouts, but left a historical mark on American infrastructure.

It is critical to evaluate the dynamic impact of frequency instabilities generated by load-changing attacks [59] as they have the potential to cause system failure. As a result, in order to concentrate on these recurring dangers, the authors [59] simulate the physical-system layer using an electromagnetic transient (EMT) technique with the use of a real-time simulation environment. The generators are represented as concurrent machines that are concerned with the elements of the stator, the field, and the damper windings. Similar to CPES-specific attack 1, load-changing attacks anticipate admittance to a variety of devices in order to properly design a successful attack strategy.

Load-changing cyber-attacks can result in controlled load shedding [23], but not in cascade failures. Additionally, the authors [69] examined the effects of a dynamic load-changing attack on power system stability, in which a dilettante adversary controls changes in the compromised load depending on the system frequency feedback.

11.5.2.3 Case Study 3: Time-Delay Attacks

Time-delay attacks (TDAs) are designed to disrupt the operation of a compromised control framework by delaying estimates or control instructions from sensors and actuators. TDAs are a type of Data Availability Attack. This type of attack does not need a significant number of attacker assets. For example, it is very possible that it will be carried out through network congestion, which is caused by flooding the organization with a large quantity of data. Thus, it interferes the normal operation of the attacked framework's infrastructure. As a result of their potential to disrupt the strength of isolated microgrids (MGs), or even the overall power grid, TDAs are regarded as a significant threat to CPES [17]. This is due to their potential ability to delay estimations or control orders communicated and received from detecting and control devices (e.g., smart metres and phasor measurement units).

In the framework of the TDA contextual analysis, the researchers acknowledge the presence of an absent opponent who has no knowledge of the system topology. Furthermore, because this type of attack is carried out by providing generous delays, which are mostly at the network level, ownership of the specified device is not necessarily required. Considering that the purpose of TDA planning is to weaken power frameworks by disabling controls that are critical to the framework's resource activity, a TDA may be considered to be a targeted attack. Depending on the scale and complexity of the compromised CPES, the adversary may require a small or large number of capabilities and assets.

The authors in Ref. [70] created and simulated a TDA scenario in order to demonstrate the impact of such an event on a MG CPES. The deliberate islanding order from the MGs regulator is passed to the main grid at time $t = 10$ seconds causing a MG to be disconnected from the main grid. Consequently, due to a lack of available generating capacity in the system, the MG regulator sends an order to reduce the amount of power delivered to a controlled load through a circuit breaker (CB). A TDA by the adversary will postpone the load-shedding command issued from the MG regulator to one of the controlled loads, resulting in severe aggravation at the physical-system layer of CPES. Because the TDA occurs at the cyber system layer of the CPES, models for the cyber system layer and the physical-system layer are necessary for this specific contextual analysis in order to do a real-time co-simulation of the separate levels. In this type of attack, an adversary does not require considerable resources or talents to consider CPES as a viable option, as long as the system has not been maintained with cutting-edge protection components. This "low-bar" need for resources increases the possibility of successfully launching such an attack against a vulnerable CPES.

11.5.2.4 Case Study 4: Propagating Attacks on Integrated
Transmission and Distribution CPES

Real-time simulation inside CPES co-simulation testbeds can result in exhaustive and accurate simulation results that are ready to depict the dynamic conduct of CPES transmission and distribution networks. Integration of the transmission and distribution (T&D) models can be used to holistically evaluate the impact of disruptions (e.g., malicious attacks and faults) in electric power systems (EPS) and can demonstrate how maloperations on the transmission system can extend to and affect the distribution system. Data integrity attacks (DIAs) were examined using integrated T&D simulation models in Ref. [66]. For the development of continuous integrated T&D models, a variety of strategies can be utilized. Different platforms offer a variety of techniques and approaches that allow for the concurrent execution of various systems in escalating EMT situations, depending on the platform. It is possible that the threat model will need to be modified for specific details, depending on the T&D element targeted by an adversary and the sort of attack. For this use case, researchers anticipated an opponent who had a thorough understanding of the system's structure as well as its constituent parts. The attacker also sought to disrupt the T&D system by perniciously influencing switching devices, i.e., the CBs.

A real-time EMT T&D system was modelled in this case study to evaluate the interactions of spreading attacks and disturbances between transmission and an imbalanced distribution system. The authors scaled a part of the systems' characteristics in order to get a match between the power generation and load consumption between the power grid benchmarks.

In order to assess the bi-directional impact of propagation attacks in integrated T&D models of CPES, the researchers designed two attack scenarios for this case study. In the first scenario, it was assumed that the attacker has the capacity of modifying the topology of the EPS. Decoupling the T&D system at the point of common coupling might be accomplished through the use of a DIA on the EPS switch devices (i.e., the distribution feeder CBs). An attacker capable of compromising components on the transmission side of the power system was believed to be present in the second scenario. After conducting tests on such attacks, the authors [28] discovered that they can cause irreparable damage to generators if they persist for three minutes or more.

11.6 CPES RISK EVALUATION

The assessment of cyber security risks prior to putting any CPES into action is critical in order to determine the economic impact of any risk on a country's economy. Such risk management is effective based on the measurement and analysis of risk followed by the deployment of appropriate security measures in line with the degree and impact of that risk. We have summarized the findings of our extensive investigation into the risk assessment procedure through Figure 11.10.

11.6.1 RISK IDENTIFICATION AND MANAGEMENT

Risk Management is implemented in order to detect, analyse, rank, evaluate, plan, and monitor any potential risks that may arise as a result of the risk assessment process.

- **Identifying Risks:** A risk identification process begins with the detection and recognition of risks that may have a negative influence on performance or expected results, and then the characterization of those risks is performed [31].
- **Analysing Risks:** Once a risk has been discovered, it is necessary to analyse the frequency and effect of the risk in order to fully comprehend the nature of the risk. The vulnerability assessment of infrastructure is the first stage in the risk assessment process. Finding cyber vulnerabilities in control system settings presents a number of difficulties because of the high availability requirements and reliance on out-of-date systems and protocols [29]. Vulnerability assessments should begin with a thorough identification of cyber assets. This includes software, hardware, and communication protocols. Following that, activities such as penetration testing and vulnerability scanning can be performed in order to uncover possible security flaws in the network environment. It is also recommended to conduct ongoing investigation of vendor security advisories, system logs, and installed intrusion detection systems to find further system issues.

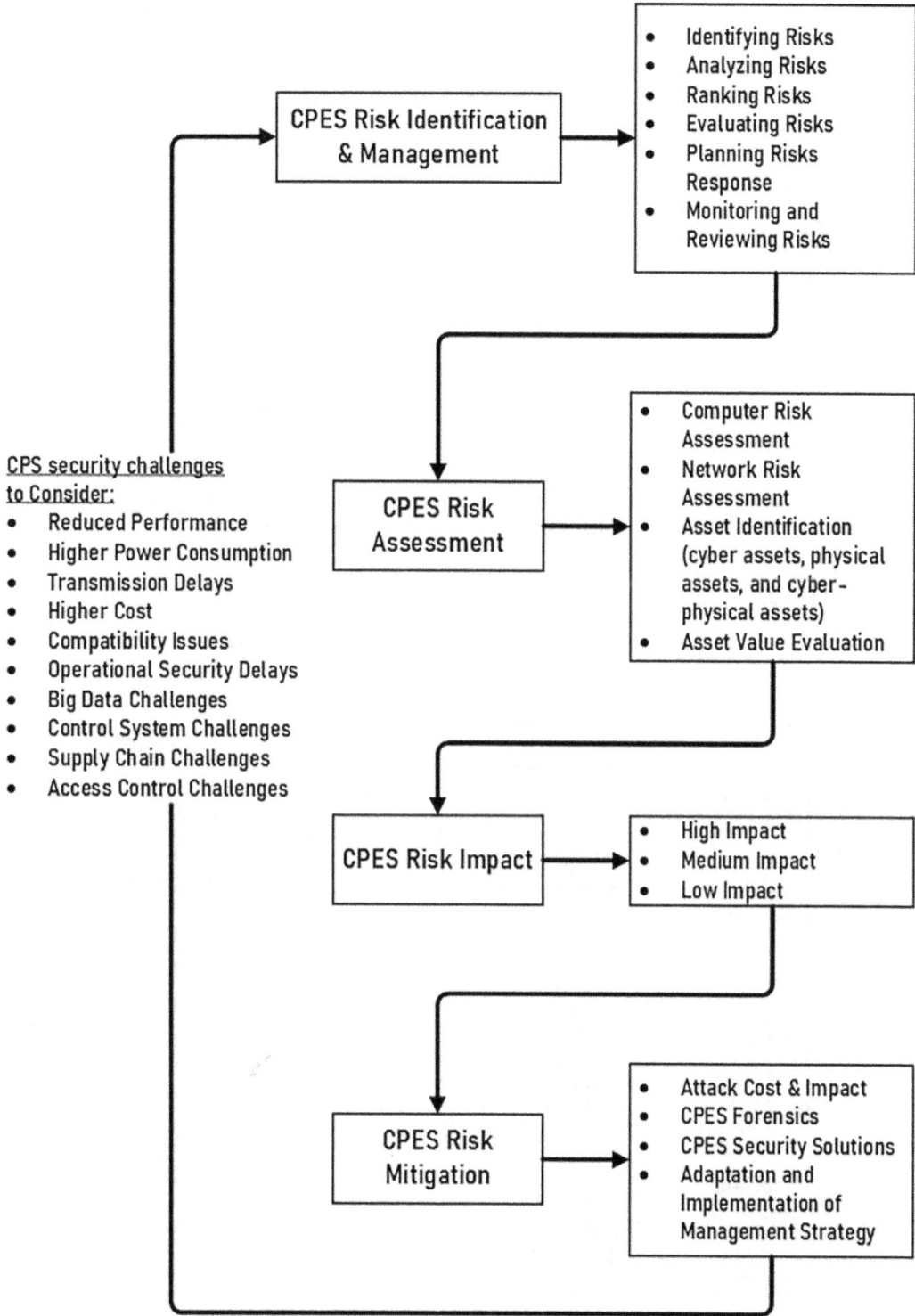

FIGURE 11.10 CPES Risk Evaluation.

An application impact analysis should be carried out in the second phase following the identification of cyber vulnerabilities in order to determine the possible impact on the applications offered by the infrastructure. A physical impact study should be carried out to determine the impact of the attack on electrical applications once the attack's influence on electrical applications has been determined. This study may be carried out by applying power system modelling approaches to examine steady state and transient performances, as well as changes in grid stability parameters such as voltage, frequency, and rotor angle.

- **Ranking Risk:** Risks are graded according to their severity, which is determined by a combination of the effect of the risk and the likelihood that it will occur.
- **Evaluating Risks:** Depending on their categorization, risks are either thought to be tolerable or necessitate substantial therapy and rapid care.
- **Planning Risks Response:** The highest-ranking risks are analysed in order to address, alter, and limit them so that the risk level can bring back to an acceptable level. As a consequence, risk reduction strategies as well as preventative and contingency plans are established.
- **Monitoring and Reviewing Risks:** Risks are monitored, recorded, and analysed on a regular basis. These risks are handled as soon as there is any suspicious activity to ensure that a serious threat does not arise.

11.6.2   RISK ASSESSMENT

Figure 11.11 represents the relationship between power applications and the infrastructure that supports them [71]. Typically, risk is thought of as the frequency of an event multiplied by its possibility

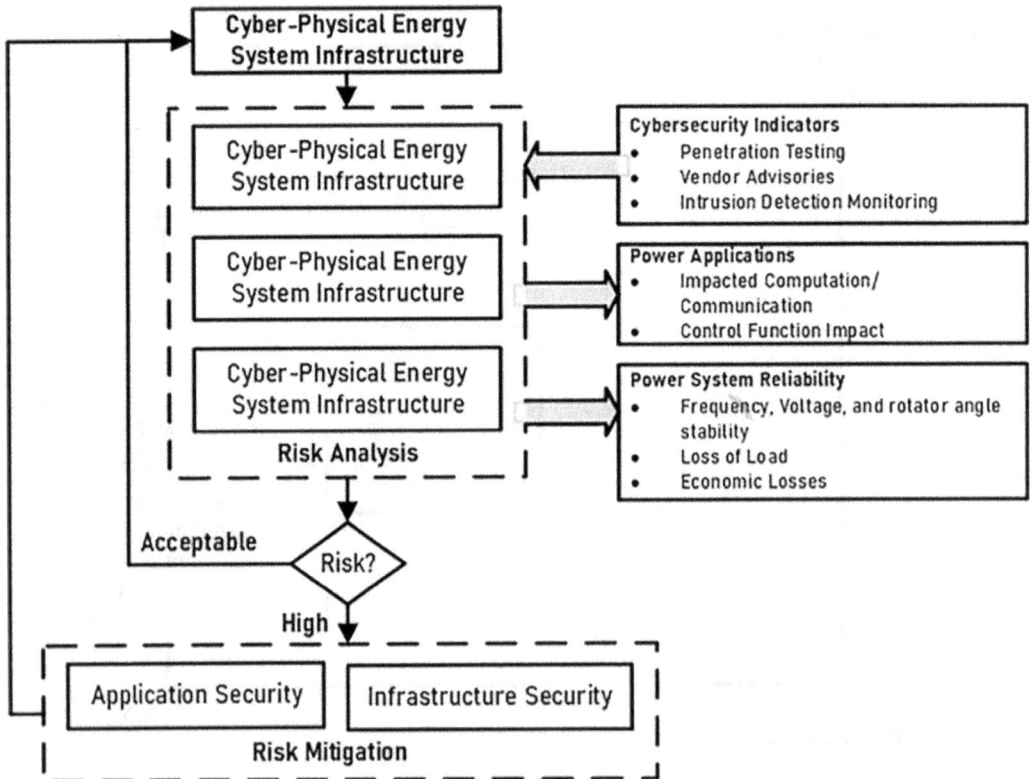

FIGURE 11.11   Risk assessment methodology.

TABLE 11.3

Common Control System Vulnerabilities Weaknesses

Software Weakness	Configuration Weaknesses	Network Security Weaknesses
1. Invalid input verification	1. Illegitimate privileges, permissions, and access controls	1. Common weaknesses related to network design
2. Code with poor quality	2. Lack of authentication and credentials mismanagement	2. Firewall rules that are weak
3. Authorization, responsibilities, and user access	3. Maintenance and configuration of security	3. Configuration vulnerabilities related with network components
4. Incorrect authentication	4. Audit of policy/planning/ procedure and accountability	4. Accountability and audit
5. Inadequacy of data authenticity verification	5. Configuration	
6. Issues related to cryptography		
7. Management of credentials		
8. Maintenance and configuration		

[60]. When it comes to gaining access to important control functions, how probable is it that an attacker will succeed? Is this something that the infrastructure vulnerability analysis stage should take into consideration? Following the identification of possible vulnerabilities, an application impact analysis should be carried out in order to establish which power system control functions are affected and which are not. This information should be used to analyse the impact on the physical system after it has been collected. The common control system vulnerabilities are depicted in Table 11.3.

11.6.3 RISK IMPACT

Risk is evaluated in terms of its potential influence on CPES. We may break it down into three categories:

- **High Impact:** If the danger occurs, it can have severe and damaging consequences for CPES. It is used to assess and neutralize advanced threats.
- **Medium Impact:** If it occurs, the impact will be less severe. It does, however, pose a severe threat to CEPS. It is used to assess and counter emerging threats.
- **Low Impact:** If this risk occurs, it will have a minor impact and will not cause serious harm. This kind of risk can be handled quickly. It is used to assess and prevent basic risks.

A security attack can take different forms, and the most typical of them are detailed below:

- **Service Delays:** CPES is susceptible to service delays, which can compromise their performance and cause them to become offline (blackout) until the problem is remedied by restoration or backup.
- **Affected Performance:** System interruptions due to malicious (cyber-attack) or non-malicious (accident) events can progressively degrade CPES performance and cause it to behave abnormally. This can have a serious influence on the decision process.
- **Cascading Failures:** Sensor failures, software problems, and overheating are all examples of cascading failures that might result in environmental disasters.
- **Financial Losses:** The use of ransomware can result in significant loss of data that is unrecoverable if the backup is not maintained or if the ransom is not paid in full. When this happens, there are enormous financial losses in both the short and long terms, especially if the information is lost. It might take months, if not years, for CPES to regain its previous levels of performance.

- An increase in investment may be necessary to combat sophisticated threats and zero-day attacks, which may necessitate an increase in security spending as part of a defense-in-depth strategy.
- **Deaths:** Deaths caused by dangerous or intentional activities such as flooding, radiation, fire, taking control of safety measure systems, or electric shock.

11.6.4 RISK MITIGATION

This can be done by establishing a more robust supporting infrastructure or by introducing more powerful power applications. Discovering how to concentrate on individual or combination tactics may lead to the development of novel mitigation measures. In addition to cyber and physical security, the formulation and execution of a well-designed risk management plan are required. Anti-counterfeiting measures, supply chain risk management, and data security and protection are all essential components of a successful risk reduction approach. In order to support these models, forensic and recovery strategies should be implemented. In addition to assisting in the analysis of cyber-attacks, this may also be used to coordinate and collaborate with relevant authority in order to detect external cyber-attack vectors. Because of this, it is possible to build logical security measures that are preventive, detective, repressive, and corrective in nature [71].

11.6.5 CPES FORENSICS

When security measures fail and an attack occurs, the importance of forensics investigation cannot be underestimated. Introduction of effective forensics tools and skills will make it easier to investigate situations [32]. Therefore, CPES must have both security and efficient forensics capabilities for accessing and analysing logs of events that occurred before, during, and after the incident. It is also helpful for preventing future incidents.

There are several forms of forensics, including digital forensics and physical evidence forensics, all of which intersect with network forensics. Digital forensics has made significant progress in the recent years. A number of different types of digital forensics have been developed. In computer forensics, there are many different categories to choose from, including network forensics, virtual machine forensics, mobile device, and cloud computing forensics [71]. CPES forensics, in contrast to other types of digital forensics, is a newer type that is still in the early stages of development, hence requires more attention. CPES forensics is a combination of the cyber (software, networks, etc.) and physical (parts, forensics) domains of CPES. It is an extension of the cyber domain. Physical forensics has been practiced and perfected over a long period of time. During this process, certain important ideas have been developed which may be applied in a variety of situations. Network forensics is the systematic tracking, collecting, and analysis of network traffic, as defined by the International Standard Organization [32]. Furthermore, depending on the extent of their deployment and the components that make up their system, CPES forensics incorporates characteristics from all other types of forensic investigation. As an example, if the CPES makes use of cloud-based services, cloud forensics would be necessary. Similarly, if cyber-attacks are launched against mobile devices, mobile forensics will be required. We refer to Figure 11.12 for a representation of the typical digital forensic process flow [18], which can be combined with CPES.

A technique based on a framework for analysing cloud forensics is possible to apply in CPES forensics as shown in Ref. [72]. In this paradigm, there are three variables to consider: technological, organizational, and legal considerations. Each level approaches the subject of forensics from a different perspective. The technical component of forensics in CPES settings refers to the methodologies, techniques, approaches, and tools that are required to conduct forensics in these environments. Technical aspects of forensics include the gathering of information, virtualization/simulation, and the addition of preventative and corrective methods. It is possible that CPES may remain small and self-contained from an organizational sense, or that it could grow to become a worldwide system

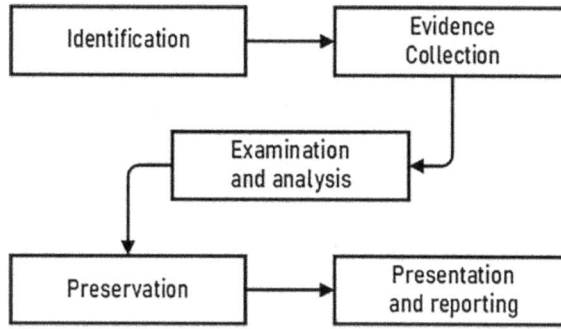

FIGURE 11.12 Traditional digital forensic process flow.

owned and run by a number of different firms. As the CPES expands in size, the forensics processes get more advanced [20]. The legal component of CPES forensics comprises the expansion of current laws and regulations to include CPES forensic acts and evidences as part of its overall framework. Also necessary are the definitions of CPES forensics ideas, methods, and rules, which will help to guarantee that forensics operations do not break any laws or regulations. It is also critical to ensure that these methods and techniques correspond to the evidence collecting, administration, and validation procedures that have been permitted by law.

11.7 GROUND FOR FUTURE WORK

Considering the security requirements of CPES, we need to look into related security solutions. The CPES forensics did not get much deeper in this chapter. However, there is a lot of scope for research on the characteristics of different factors in the context of CPES forensics, its associated challenges, and how they work together. One recent challenge in CPES forensics is forensics-by-design. In this method, forensic efforts can be provided within the CPES as a built-in feature during the development of a CPES environment. When an incident occurs, the built-in CPES forensic capabilities will provide investigators with the information they require for the forensics investigation process in order to figure out where security breaches come from and how they work. In addition, it keeps and examines evidence and draws conclusions. This helps to answer the six most important forensics questions: what, why, how, who, when, and where. There is tremendous potential for further investigation in this area of forensics.

Thus, our future work can be carried out by exploring CPES forensics-by-design, cyber-attack detection methods, CPES security solutions, and post quantum security of CPES. The road towards a stable quantum computer is still uncertain. However, it will be able to solve mathematical problems previously thought to be intractable once one is built, thereby breaking all the existing security mechanisms. This will require the study and development of post quantum security mechanisms. Therefore, we have a huge ground to work on, in order to secure modern and future CPES.

11.8 CONCLUSION

CPES, a next-generation power infrastructure that provides for the two-way flow of electricity and information, establishes a large distributed automated electrical power delivery network. It is one of the most vital infrastructures in today's world. Advanced ICTs comprising the seamless interconnection of computing, communication, control, and human factors are used to monitor and manage the system. It is still possible to hack into the electric power grid, despite the fact that digital technologies are more efficient and dependable in monitoring and controlling capacity. The power grid also involves a complex interdependency between cyber and physical systems. A cyber-attack on the physical power system disrupts the secure operation of power systems by affecting the flow of information through the network. By altering the information flow, a cyber-attack on CPES compromises the secure operation of power system. Once physical attacks on the power system are successfully coordinated with cyber-attacks, the large-scale CPES becomes difficult to operate. Cyber-attacks on critical infrastructure like the EPS are of serious concern. The R&D community throughout the world is placing a high priority on research into CPES security to protect our current and future energy infrastructure.

REFERENCES

1. Orumwense EF, Abo-Al-Ez K. A systematic review to aligning research paths: Energy cyber-physical systems. *Cogent Engineering*. 2019 Jan 1;6(1):1700738.
2. Silva FA. Cyber-physical-social systems and constructs in electric power engineering [Book News]. *IEEE Industrial Electronics Magazine*. 2017 Dec 21;11(4):50–5.
3. Cao Y, Li Y, Liu X, Rehtanz C. *Cyber-Physical Energy and Power Systems*. Springer, Singapore; 2020.
4. Shi L, Dai Q, Ni Y. Cyber–physical interactions in power systems: A review of models, methods, and applications. *Electric Power Systems Research*. 2018 Oct 1;163:396–412.
5. Gubbi J, Buyya R, Marusic S, Palaniswami M. Internet of things (IoT): A vision, architectural elements, and future directions. *Future Generation Computer Systems*. 2013 Sep 1;29(7):1645–60.
6. Yohanandhan RV, Elavarasan RM, Manoharan P, Mihet-Popa L. Cyber-physical power system (CPPS): A review on modeling, simulation, and analysis with cyber security applications. *IEEE Access*. 2020 Aug 14;8:151019–64.
7. Cyber attack trends: 2020 mid-year report. Available on www.isaca.org/\\resources/news-and-trends/industry-news/2020/top-cyberattacks-of-2\\020-and-how-to-build-cyberresiliency.
8. Ospina J, Liu X, Konstantinou C, Dvorkin Y. On the feasibility of load-changing attacks in power systems during the covid-19 pandemic. *IEEE Access*. 2020 Dec 25;9:2545–63.
9. Anderson R, Fuloria S. *Smart Meter Security: A Survey*. University of Cambridge Computer Laboratory, United Kingdom; 2011.
10. Vassilakis C, Kolokotronis N, Limniotis K, Mathas C-M, Grammatikakis K-P, Kavallieros D, Bilali G, Shiaeles S, Ludlow J. Cyber-Trust Project D2.1: Threat landscape: Trends and methods. Technical report. Cyber-Trust Consortium; 2018.
11. Wang Q, Tai W, Tang Y, Ni M. Review of the false data injection attack against the cyber-physical power system. *IET Cyber-Physical Systems: Theory & Applications*. 2019 Jun 27;4(2):101–7.
12. Bilis EI, Kröger W, Nan C. Performance of electric power systems under physical malicious attacks. *IEEE Systems Journal*. 2013 May 1;7(4):854–65.
13. Center A. Analysis of the cyber-attack on the Ukrainian power grid. Technical report; 2016.
14. Huseinović A, Mrdović S, Bicakci K, Uludag S. A survey of denial-of-service attacks and solutions in the smart grid. *IEEE Access*. 2020 Sep 25;8:177447–70.
15. Tian J, Wang B, Li J, Konstantinou C. Adversarial attack and defense methods for neural network-based state estimation in smart grid.
16. Kimani K, Oduol V, Langat K. Cyber security challenges for IoT-based smart grid networks. *International Journal of Critical Infrastructure Protection*. 2019 Jun 1;25:36–49.
17. Zografopoulos I, Ospina J, Liu X, Konstantinou C. Cyber-physical energy systems security: Threat modeling, risk assessment, resources, metrics, and case studies. *IEEE Access*. 2021 Feb 10;9:29775–818.
18. Datta S, Majumder K, De D. Review on cloud forensics: An open discussion on challenges and capabilities. *International Journal of Computer Applications*. 2016;145(1):1–8.
19. EirGrid targeted by 'state sponsored' hackers leaving networks exposed to 'devious attack. www.independent.ie/irish-news/news/exclusive-eirgrid-targeted-by-\\state-sponsored-hackers-leaving-networks-exposed-to-devious-attack-\\36003502.html.
20. Pilli ES, Joshi RC, Niyogi R. Network forensic frameworks: Survey and research challenges. *Digital Investigation*. 2010 Oct 1;7(1–2):14–27.
21. Checkoway M, Kantor A, Shacham S, Koscher C, Roesner K, Checkoway S, InMcCoy D, Kantor B, Anderson D, Shacham H, Savage S, Koscher K, Czeskis A, Roesner F, Kohno T. *Comprehensive Experimental Analyses of Automotive Attack Surfaces, 20th USENIX Security Symposium (USENIX Security 11)*; 2011 (pp. 1–16). USENIX Association, San Francisco, CA.
22. Soltan S, Mittal P, Poor HV. BlackIoT: IoT botnet of high wattage devices can disrupt the power grid. In *27th {USENIX} Security Symposium ({USENIX} Security 18)*; 2018 (pp. 15–32).

23. Huang B, Cardenas AA, Baldick R. Not everything is dark and gloomy: Power grid protections against IoT demand attacks. In *28th {USENIX} Security Symposium ({USENIX} Security 19)*; 2019 (pp. 1115–1132).
24. Xenotime: Hackers behind triton malware turn to power grids; 2020 June 25.
25. Wang W, Lu Z. Cyber security in the smart grid: Survey and challenges. *Computer Networks*. 2013 Apr 7;57(5):1344–71.
26. Data, RISI. Slammer impact on Ohio nuclear plant. https://www.risidata.com/Database/Detail/slammer-impact-on-ohio-nuclear-plant#:~:text=Description%3A,to%20communicate%20with%20each%20other. *Retrieved July* 7 (2003): 2019.
27. John M. Israeli test on worm called crucial in Iran nuclear delay. The New York Times. 2011.
28. Meserve J. Staged cyber attack reveals vulnerability in power grid; 2007. http://edition.cnn.com/2007/US/09/26/power.at.risk/index.html.
29. Singh VK, Sharma R, Govindarasu M. Testbed-based performance evaluation of attack resilient control for wind farm SCADA system. In *2020 IEEE Power & Energy Society General Meeting (PESGM)*; 2020 Aug 2 (pp. 1–5). IEEE.
30. Makrakis GM, Kolias C, Kambourakis G, Rieger C, Benjamin J. Vulnerabilities and attacks against industrial control systems and critical infrastructures. arXiv preprint arXiv:2109.03945; 2021 Sep 8.
31. Stoneburner G, Goguen A, Feringa A. Risk management guide for information technology systems. Nist special publication; 2002 Jul 1;800(30):800–30.
32. Fischer-Hübner S, Alcaraz C, Ferreira A, Fernandez-Gago C, Lopez J, Markatos E, Islami L, Akil M. Stakeholder perspectives and requirements on cybersecurity in Europe. *Journal of Information Security and Applications*. 2021 Sep 1;61:102916.
33. Bou-Harb E. A brief survey of security approaches for cyber-physical systems. In *2016 8th IFIP International Conference on New Technologies, Mobility and Security (NTMS)*; 2016 Nov 21 (pp. 1–5). IEEE.
34. Mazumder SK, Kulkarni A, Sahoo S, Blaabjerg F, Mantooth A, Balda J, Zhao Y, Ramos-Ruiz J, Enjeti P, Kumar PR, Xie L. A review of current research trends in power-electronic innovations in cyber-physical systems. *IEEE Journal of Emerging and Selected Topics in Power Electronics*. 2021 Jan 14.
35. Tu H, Xia Y, Chi KT, Chen X. A hybrid cyber-attack model for cyber-physical power systems. *IEEE Access*. 2020 Jun 18;8:114876–83.
36. Lieskovan T, Hajny J, Cika P. Smart grid security: Survey and challenges. In *2019 11th International Congress on Ultra-Modern Telecommunications and Control Systems and Workshops (ICUMT)*; 2019 Oct 28 (pp. 1–5). IEEE.
37. He H, Yan J. Cyber-physical attacks and defences in the smart grid: A survey. *IET Cyber-Physical Systems: Theory & Applications*. 2016 Dec 1;1(1):13–27.
38. Kumar P, Lin Y, Bai G, Paverd A, Dong JS, Martin A. Smart grid metering networks: A survey on security, privacy and open research issues. *IEEE Communications Surveys & Tutorials*. 2019 Feb 14;21(3):2886–927.
39. Humayed A, Lin J, Li F, Luo B. Cyber-physical systems security – a survey. *IEEE Internet of Things Journal*. 2017;4(6):1802–1831.
40. Wadhawan Y, AlMajali A, Neuman CA. Comprehensive analysis of smart grid systems against cyber-physical attacks. *Electronics*. 2018;7(249).
41. Tian J, Wang B, Li T, Shang F, Cao K. Coordinated cyber-physical attacks considering DoS attacks in power systems. *International Journal of Robust and Nonlinear Control*. 2020 Jul 25;30(11):4345–58.
42. Li H, Li H, Zhang H, Yuan W. Black-box attack against handwritten signature verification with region-restricted adversarial perturbations. *Pattern Recognition*. 2021 Mar 1;111:107689.
43. Ren K, Zheng T, Qin Z, Liu X. Adversarial attacks and defenses in deep learning. *Engineering*. 2020 Mar 1;6(3):346–60.
44. Ourahou M, Ayrir W, Hassouni BE, Haddi A. Review on smart grid control and reliability in presence of renewable energies: Challenges and prospects. *Mathematics and Computers in Simulation*. 2020 Jan 1;167:19–31.
45. Mathas CM, Grammatikakis KP, Vassilakis C, Kolokotronis N, Bilali VG, Kavallieros D. Threat landscape for smart grid systems. In *Proceedings of the 15th International Conference on Availability, Reliability and Security*; 2020 Aug 25 (pp. 1–7).
46. Alrefaei F, Alzahrani A, Song H, Zohdy M. Security of cyber physical systems: Vulnerabilities, attacks and countermeasure. In *2020 IEEE International IOT, Electronics and Mechatronics Conference (IEMTRONICS)*; 2020 Sep 9 (pp. 1–6). IEEE.
47. Alguliyev R, Imamverdiyev Y, Sukhostat L. Cyber-physical systems and their security issues. *Computers in Industry*. 2018 Sep 1;100:212–23.

48. Coffey K, Smith R, Maglaras L, Janicke H. Vulnerability analysis of network scanning on SCADA systems. *Security and Communication Networks*. 2018 Mar 13;2018.

49. Liu H, Chen Y, Chuah MC, Yang J, Poor HV. Enabling self-healing smart grid through jamming resilient local controller switching. *IEEE Transactions on Dependable and Secure Computing*. 2015 Sep 17;14(4):377–91.

50. The Security Ledger. That Israeli grid attack? Just more ransomware. https://securityledger.com/2016/01/that-israeli-grid-attack-just-more-\\ransomware/.

51. Yaacoub JP, Noura H, Salman O, Chehab A. Security analysis of drones' systems: Attacks, limitations, and recommendations. *Internet of Things*; 2020 Sep 1;11:100218.

52. Vávra J, Hromada M. Evaluation of anomaly detection based on classification in relation to SCADA. In *2017 International Conference on Military Technologies (ICMT)*; 2017 May 31 (pp. 330–334). IEEE.

53. Kshetri N, Voas J. Hacking power grids: A current problem. *Computer*. 2017 Dec 18;50(12):91–5.

54. Chen TM, Sanchez-Aarnoutse JC, Buford J. Petri net modeling of cyber-physical attacks on smart grid. *IEEE Transactions on Smart Grid*. 2011 Aug 1;2(4):741–9.

55. Eun Y-S, Aßmann JS. Cyberwar: Taking stock of security and warfare in the digital age. *International Studies Perspectives*. 2016.

56. Bencsáth B, Ács-Kurucz G, Molnár G, Vaspöri G, Buttyán L, Kamarás R. Duqu 2.0: A comparison to duqu. Budapest; 2015 Feb;27:2016.

57. Gaietta M. *The Trajectory of Iran's Nuclear Program*; 2016. Springer.

58. Davis KR, Davis CM, Zonouz SA, Bobba RB, Berthier R, Garcia L, Sauer PW. A cyber-physical modeling and assessment framework for power grid infrastructures. *IEEE Transactions on Smart Grid*. 2015 May 7;6(5):2464–75.

59. Shekari T, Gholami A, Aminifar F, Sanaye-Pasand M. An adaptive wide-area load shedding scheme incorporating power system real-time limitations. *IEEE Systems Journal*. 2016 Apr 14;12(1):759–67.

60. Bronk C, Tikk-Ringas E. The cyber-attack on Saudi Aramco. *Survival*. 2013 May 1;55(2):81–96.

61. Musleh AS, Chen G, Dong ZY. A survey on the detection algorithms for false data injection attacks in smart grids. *IEEE Transactions on Smart Grid*. 2019 Oct 30;11(3):2218–34.

62. Poulsen K. Slammer worm crashed Ohio nuke plant network. http://www. Security focus. com/news/6767; 2003.

63. Stuxnet work attack on Iranian nuclear facilities; 2020 June 25.

64. Kushner D. The real story of stuxnet. *IEEE Spectrum*. 2013 Mar 7;50(3):48–53.

65. EirGrid targeted by 'state sponsored' hackers leaving networks exposed to 'devious attack'. www.independent.ie/irish-news/news/exclusive-eirgrid-targeted-by-\\state-sponsored-hackers-leaving-networks-exposed-to-devious-attack-\\36003502.html.

66. Min J. North Korea's asymmetric attack on South Korea's nuclear power plants. http://large.stanford.edu/courses/2017/ph241/min1/Journal.

67. Analysis of the cyber attack on the Ukrainian power grid. *Electricity Information Sharing and Analysis Center*. 2016.

68. Anton C. BlackEnergy by the SSHBearDoor: Attacks against Ukrainian news media and electric industry. WeLiveSecurity. COM; 2016.

69. Amini S, Pasqualetti F, Mohsenian-Rad H. Dynamic load altering attacks against power system stability: Attack models and protection schemes. *IEEE Transactions on Smart Grid*. 2016 Oct 27;9(4):2862–72.

70. Ospina J, Zografopoulos I, Liu X, Konstantinou C. Trustworthy cyberphysical energy systems: Time-delay attacks in a real-time co-simulation environment. In *Proceedings of the 2020 Joint Workshop on CPS\&IoT Security and Privacy*; 2020 Nov 9 (pp. 69–69).

71. Sridhar S, Hahn A, Govindarasu M. Cyber–physical system security for the electric power grid. *Proceedings of the IEEE*. 2011 Oct 3;100(1):210–24.

72. Nelson B, Phillips A, Steuart C. Guide to computer forensics and investigations. *Cengage Learning*; 2014 Nov 7.

12 Intrusion Detection Using Machine Learning

Rijwan Khan, Aditi Tiwari,
Aashna Kapoor, and Abhyudaya Mittal
ABES Institute of Technology

CONTENTS

12.1 INTRODUCTION

Intrusion detection system (IDS) is a critical virtual tool of safety, which monitors as well as detects intrusion attacks, also it keeps a check on the networks for apprehensive actions, and they also are prone to false alarm. Consequently, when organizations first created the IDS products then they will need to tweak them. It involves configuring these systems properly to differentiate between authentic traffic of network and mischievous actions. To make sure secure and dependable facts go with the drift throughout numerous businesses, contemporary-day networked enterprise settings necessitate an excessive degree of protection. After conventional protection technology fails, an intrusion detection device works as an adaptable protect tool for device protection. Because cyber-assaults are handiest going to get greater complex, it is vital that protective generation maintain up so it is important to have such a system that can detect the intrusions and this is the reason for which there is a requirement for an IDS. An IDS can help you organize crucial network data as well as increase network security. Every day, your network generates a data which is in large amount, and an IDS can help you find the difference between the vital activities and the less important data. An IDS can help you to save your amount of searching time through hundreds of system logs for crucial information by assisting you in determining which data to pay attention to. When it comes to intrusion detection, this can save you time, save manual effort, and reduce human mistakes.

DOI: 10.1201/9781003267812-12

An IDS is a system that is considered to be a crucial and most important part of a cybersecurity strategy. Although an easy firewall is the inspiration for community protection, many superior assaults can get around it. An IDS provides any other layer of protection and thus making it extra tough for an attacker to get admission to a company's community. It's vital to reflect on consideration of the deployment situation even when selecting an IDS system. In a few circumstances, an IDS can be the pleasant alternative for the job, even as in others, an IPS's included safety can be a higher alternative. An included answer is furnished through the use of an NGFW with integrated IDS/IPS capabilities, which simplifies risk detection and protection administration.

Intruders get access to systems, which leads to intrusions. Authorized users who strive to get extra privileges for which they're now no longer authorized, and licensed customers who abuse the privileges which have been granted to them. IDS relies on a few techniques to determine whether or not an intrusion assault has happened. The following are some of the methodologies:

- **Signature-Based Method:** A method in which a well-known intrusion assault signature is maintained in an IDS database and compared to the current system data. When the IDS detects a match, it considers it an intrusion [1]. Signatures are the styles that the IDS detects and the IDS which might be primarily based totally on signature can effortlessly come across assaults whose sample is already gift with inside the system; however, new malware assaults are tough to come across given that their sample (signature) is not known.
- **Anomaly-Based Method:** A new type of malware is being created at a rapid rate; anomaly-based IDS is designed to identify unknown malware threats. Machine learning (ML) comes into play in anomaly-based IDS to build a reliable business model and everything in between is compared against that model and it is considered suspicious if not found in paradigm. This method can identify both known and undiscovered threats. The disadvantage of this strategy is that it has low accuracy and a high false alarm rate.
- **Hybrid-Based Method:** Hybrid-primarily based total detection is the employment of or extra intrusion detection techniques so as to conquer the restrictions of an unmarried approach at the same time as gaining the advantages of or extra techniques [2].

In today's world, detection of intrusion is considered to be an important component of network security. It is now easier than ever to build a model to detect intrusions because of advances in ML technologies. While present models have a high level of accuracy, another component of intrusion detection is the model's computation time. The IDS should be able to keep up with the high influx of network connections, and with them, possible attacks, as the network speed increases. ML learns historical data patterns through statistical modeling and then predicts the most likely conclusion using new data. As a result, an anomaly-based technique was used to apply the ML algorithm to IDS. As previously stated, the goal is to create a model having higher accuracy and a lower false alarm rate.

12.1.1 IDS CLASSIFICATION

The system used to detect the intrusion can be classified among the two divisions which are as follows:

- **Host-Based IDS:** Host-primarily built IDS is hooked up on the selected endpoint for guarding it against each inner and outside dangers. An IDS with this functionality is probably capable of revealing community visitors to and from the machine, see energetic processes, and taking a look at the system's logs. The visibility of a host-primarily based totally IDS is restricted to its host machine, restricting the context to be had for decision-making; however, it has vast visibility into the host computer's internals.

- **Network-Based IDS (NIDS):** An IDS system that monitors the entire protected network is called an NIDS. Full visibility into all network traffic makes decisions based on meta-data and packet content. While this broader perspective provides more information and the ability to detect multiple threats these systems lack insight into the endpoints they protect.

These systems provide a variety of benefits to businesses starting with their ability to detect security incidents. An IDS can help determine the amount and pattern of abuse. This information is used to improve security systems or create simpler controls. Enterprises can also use IDSs to check for vulnerabilities or problems with their network equipment configuration. These metrics can then be used to assess future threats.

IDSs can also help in assisting industries in gathering supervisory necessities. Corporations can achieve much prominence on networks using a system like this which can detect intrusions, making it easier to obey security criteria.

12.1.2   Why IDS?

Nowadays, the cyber-attack is now becoming extremely sophisticated, posing higher challenges in sensing intrusions accurately. Attempts to prevent the intrusions could jeopardize the credibility of security services, including data availability, confidentiality, and integrity.

An IDS is one of the significant parts of any cybersecurity strategy [3]. Although a basic firewall serves as the base of network security, numerous forward-looking attacks can get through it. By adding another layer of defense, an IDS makes it become more problematic for an attacker to gain unobserved access to a company's system.

It's essential to think about the deployment scenario when choosing an IDS solution. In those instances, an IDS might be the best option, while in others, an IPS's integrated security might be the better choice. An integrated solution can be achieved by using an NGFW with built-in IDS/IPS functionality, which makes it easier for threat detection and safety management [4].

12.2   RELATED WORK

Several studies have been conducted on IDSs. With the introduction of Big Data, traditional techniques become more complicated. Many of today's IDS techniques are incapable of handling cyber-attacks on computer networks, which are dynamic and complex [5]. In this way, artificial intelligence techniques, such as ML, can result in higher rates of tracking, fewer incorrect alarms, and lower costs of communication and computation. Due to this, many researchers are developing quick and accurate IDSs using ML techniques. Here, we discuss some studies that utilized ML algorithms for IDS.

ML is a field that includes the application of mathematical models to bring out beneficial information from huge datasets through the use of methods and algorithms.

There are several ML algorithms (also known as shallow learning) used in IDS, including Decision Trees, K-Nearest Neighbor Networks (KNNs), Artificial Neural Networks (ANNs), Support Vector Machines (SVMs), K-Mean Clustering, Fast Learning Networks, and Ensemble Methods [6,7].

- **Decision Tree:** With supervised ML, a Decision Tree can be used to classify and predict the outcome of a dataset based on a set of decisions. The model has a tree-like structure with nodes, branches, and leaves. In decision trees, each leaf represents an outcome or class label, and the branches represent a decision. Each node represents an attribute or feature. The most common model types are CART, C4.5, and ID3 [8]. For example, Random Forest and XGBoost algorithms build their models from multiple decision trees [9]. To avoid overfitting, the Decision Tree algorithm makes selection to build its model based on

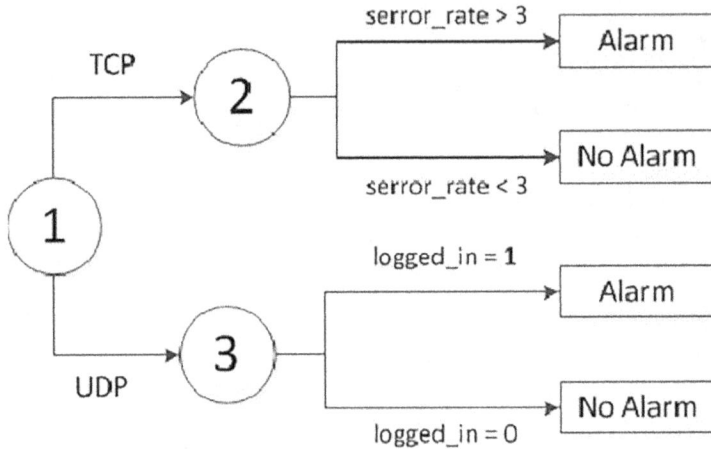

FIGURE 12.1   Decision Tree classification example for intrusion detection.

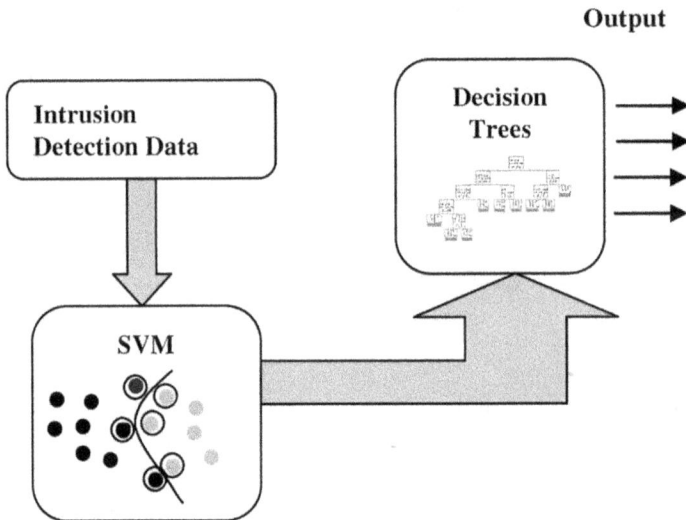

FIGURE 12.2   IDS using SVM and decision tree.

the best features and then performs pruning operations to delete any branches from the tree that aren't needed.

- **SVM:** In ML, SVMs are algorithms based on a hyperplane based on maximum margin in n-dimensional feature space. SVMs have become widely used in IDSs due to their good generalization as well as the ability to overcome the computational complexity [10,11]. In this approach, one uses a kernel function to translate a low-dimensional input vector into a high-dimensional feature space. Then, using the support vectors, we obtain the maximum marginal hyperplane, which serves as a decision boundary (Figures 12.1 and 12.2) [12,13].

- **ANN:** ANN is a supervised ML algorithm. It is based on the behavior of biological neurons in the brain and central nervous system [14,15]. It is made up of processing elements known as neurons (nodes) and the connections that connect them. ANN inputs are typically fed to the artificial neurons in one or more hidden layers, where they are weighted and processed to determine the output to the next layer [16].

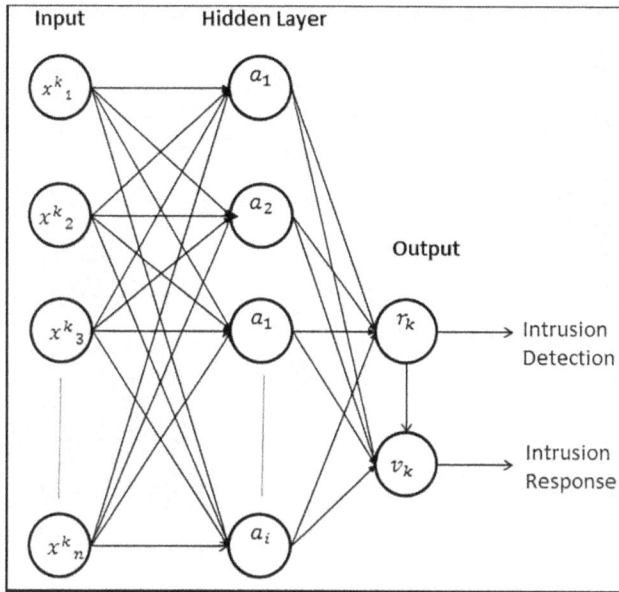

FIGURE 12.3   ANN model of intrusion detection and response.

The set of weights and biases for the hidden layer and output layer neurons can be adaptively tuned using a "learning rule" (often gradient descent-based back-propagation of errors) in ANNs. The ability to perform nonlinear modeling by learning from larger datasets is the main benefit of using an ANN technique and the main problem with training the ANN model is that it takes a long time due to its complexity, which slows down the learning process and leads to a suboptimal solution. Huang et al. proposed a new ANN called an extreme learning machine (ELM) to overcome the limitations of ANN. The ELM is a single hidden layer feed-forward neural network that determines output weights analytically by randomly using the input weights and hidden layer bias [17]. ANN has been used in a variety of computer security domains, including the analysis of software design flaws and the detection of computer viruses (Figure 12.3) [18].

* **K-Means Clustering:** K-Mean clustering is a popular iterative centroid-based ML algorithm that learns unsupervised [19]. Clustering is the process of dividing data into meaningful clusters (or groups) by grouping data that is highly similar. In k-means clustering the number of centroids in a dataset is denoted by the letter K. Normally, distance is used to assign specific data points to a cluster. Within a cluster, the main goal is to reduce the sum of the distances between data points and their respective centroids [20].

  To address IDS, Yao proposed a multilevel semi-supervised ML model framework. The clustering concept is used in conjunction with the RF model. Pure cluster extraction, pattern discovery, fine-grained classification, and model updating were the four modules in the proposed solution [21].

  The idea is that if an attack isn't labeled in one module, it'll be passed on to the next for detection. Even with low instances in the dataset, experimental results showed that the model was superior at detecting attacks (Figure 12.4) [18]

* **KNN:** The KNN algorithm is a low-complexity data mining algorithm that is theoretically mature. It's one of the most basic supervised ML algorithms, and it works by using the concept of "feature similarity" to predict the class of a data sample [20]. If the majority of a sample's k-nearest neighbor samples belong to the same category in a sample space, then the sample belongs to the same category as well (Figure 12.5).

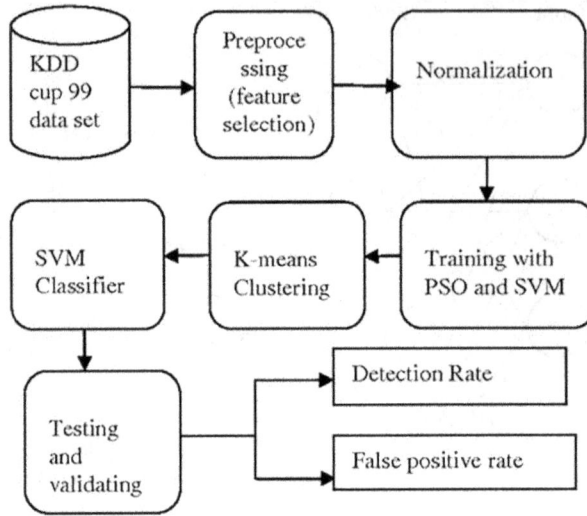

FIGURE 12.4   IDS using k-means clustering.

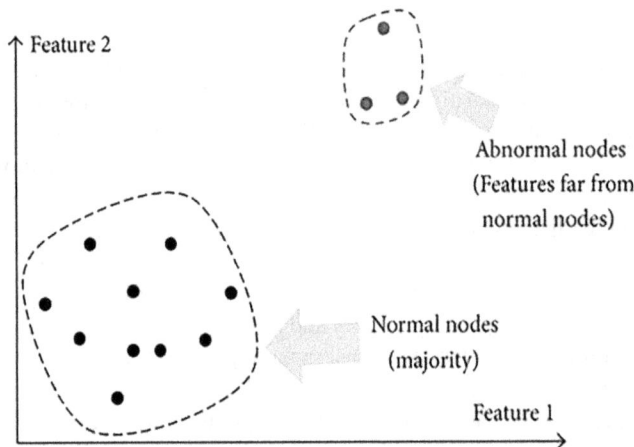

FIGURE 12.5   The schematic diagram of KNN intrusion detection algorithm.

We use a-dimensional vector to represent nodes in the intrusion detection algorithm, such as $a1$, $a2$, an. These dimensions can include the number of routing messages that can be sent in a given amount of time, the number of nodes in the sending routing packets with different destinations, the number of nodes in the receiving routing packets with the same source node, and so on [11,19]. The parameter k in the KNN algorithm influences the model's performance. If the value of k is extremely small, the model may be prone to overfitting. Karats used the CSE-CIC-IDS2018 benchmark dataset to compare the performance of various ML algorithms [5,22]. They solved the dataset imbalance problem by using Synthetic Minority Oversampling Technique to reduce the imbalance ratio, which improved the detection rate for minority class attacks [18].

- **Ensemble Methods:** Shen proposed an IDS based on ensemble methods, using ELM as the base classifier. Ensemble methods are based on the idea of learning at a group level while making use of various classifiers [23,24]. During the ensemble pruning phase, a BAT

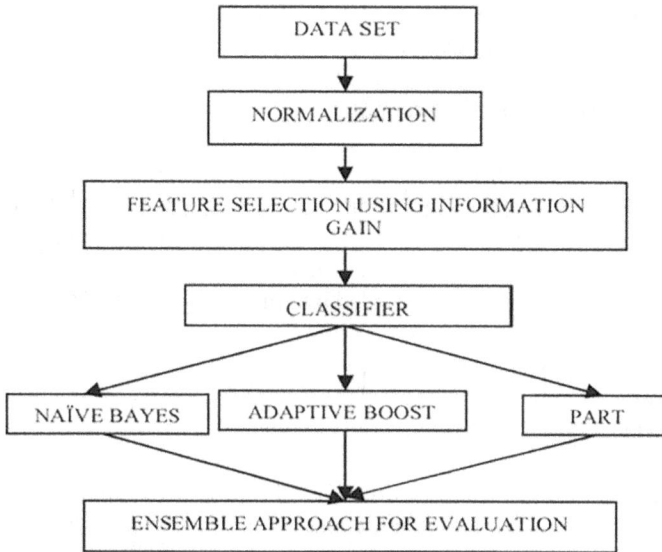

FIGURE 12.6 The use of ML algorithms for an ensemble IDS.

optimization technique is used. The model was tested using KDD Cup'99, NSL-KDD, and Kyoto datasets. Experiments revealed that many ELMs working together in an ensemble perform much better than individual ELMs in terms of performance.

Gao's study presented an adaptive ensemble model that utilizes several base classifiers with an adaptive voting algorithm such as DT, RF, KNN, and Deep Neural Network [24]. Experiments on the NSL-KDD dataset were used to validate the proposed methodology. By comparing other models, the results of experimental analysis demonstrated the effectiveness of the performance. The technique that was put forward produced insufficient results for the weaker attack classes (Figure 12.6).

It is found that a few comparative research studies were achieved in this discipline, however, exhaustive take a look at remains now no longer achieved. This chapter is concerned with designing an IDS for community by analyzing the combinations of maximum famous function choice strategies and classifiers in order to design a system with a good mixture of function choice method and classifier. That analysis will deliver us a concept about which function choice method must be blended with which classifier to construct correct community intrusion detection.

12.3 EXPERIMENT

The basic task of an IDS is to build a classifier model, which is capable of differentiating between the attacks/intrusions and normal connections [21,25]. In order to make an effective IDS, after analyzing the related works done in this field and analyzing the previous research works, some conclusions were drawn [26,27].

Different categories of attacks in an intrusion are as follows [28]:

- **Denial of Service Attack:** The systems traffic and network usage is increased to affect the availability of the services to valid users. For example, seen flood.
- **User to Root Attack:** When someone obtained unwanted access to the admin services. For example, load module.

- **Remote to Local Attack:** It is the unwanted and unverified access to the system from a distant machine. For example, guessing password.
- **Probe Attack:** By scanning the network for misuse, one can gather information about network weaknesses, such as port scanning.

The new malware attacks are difficult to detect since their signatures are unknown and the further disadvantages of the existing strategies are having less efficiency and a high false alarm rate.

Using the hybrid approach, we have achieved great accuracy and a limited intrusion, which is considered to be one of the most efficient methods [29].

The proposed work/experiment can be divided into three major parts [30]:

- Preprocessing of dataset
- Training the model
- Testing of the model

An intrusion detection device has been presented in this section. To start, information preprocessing of the NSL-KDD dataset is done so that it could be used for device schooling and testing. The dataset contains the following files of data [31]:

- **KDDTrain+.ARFF:** The complete teach set of the NSL-KDD dataset with binary labels in ARFF format
- **KDDTrain+.TXT:** The complete NSL-KDD teach set, which includes attack-kind labels and trouble degree in CSV format
- **KDDTrain+_20Percent.ARFF:** This is the 20% subset of KDDTrain+.arff document
- **KDDTrain+_20Percent.TXT:** A 20% subset of the KDDTrain+.txt document
- **KDDTest+.ARFF:** The complete NSL-KDD take a look at set with binary labels in ARFF format
- **KDDTest+.TXT:** The complete NSL-KDD take a look at set, which includes attack-kind labels and trouble degree in CSV format
- **KDDTest-21. ARFF:** A subset of the KDDTest+.arff document which does now no longer encompass statistics with trouble degree of 21 out of 21
- **KDDTest-21.TXT:** A subset of the KDDTest+.txt document which does now no longer encompass statistics with trouble degree of 21 out of 21 (Figure 12.7).

With the wide variety of statistics within the NSL-KDD teach and assessment units, it is low priced to run experiments at the whole set without having to choose a small sample. Therefore, the results of various studies of paintings could be comparable and continuous. The 20% NSL-KDD train+ dataset was used for the analysis. Various properties of the dataset NSL-KDD are displayed in Table 12.1.

After those operations, function choice processes have executed the usage of specific function choice strategies and new methods [9]. At the end, the proposed layered hybrid shape is added as an entire that is tested.

12.3.1 Data Preprocessing

In the dataset, NSL-KDD, transformation and normalization operations have been performed to ensure it is clean and produces better performance. Here, a more efficient dataset is obtained [22].

12.3.1.1 Transformation Operation

Among the elements within the NSL-KDD dataset, the nominal values are transformed to values of numeric type, and the numeric type values of protocol type, provider, and flag features are

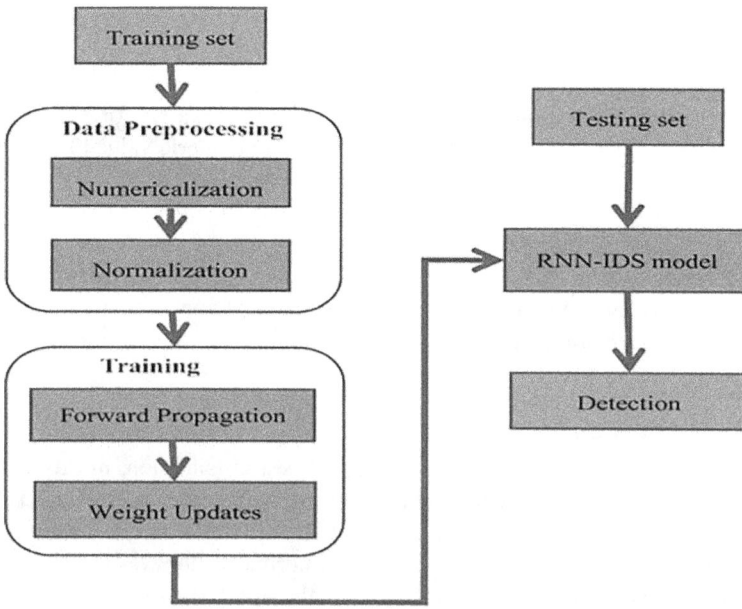

FIGURE 12.7 Model proposed.

TABLE 12.1
Features of NSL-KDD Dataset

No.	Feature Name	No.	Feature Name
1	Duration	22	is_guest_login
2	protocol type	23	**count**
3	service	24	srv_count
4	**flag**	25	**serror_rate**
5	src_bytes	26	**srv_serror_rate**
6	**dst_bytes**	27	**rerror_rate**
7	land	28	srv_rerror_rate
8	wrong_fragment	29	**same_srv_rate**
9	urgent	30	diff_srv_rate
10	hot	31	srv_diff_host_rate
11	num_f ailed_logins	32	dst_host_count
12	logged_in	33	**dst_host_srv_count**
13	num_compromised	34	dst_host_same_srv_rate
14	root_shell	35	dst_host_diff_srv_rate
15	su_attempted	36	dst_host_same_src_port_rate
16	num root	37	dst_host_srv_diff_host_rate
17	num_file_creations	38	**dst_host_serror_rate**
18	num_shells	39	**dst_host_srv_serror_rate**
19	num_access_files	40	dst_host_rerror_rate
20	num_outbound_cmds	41	dst_host_srv_rerror_rate
21	is_host_login	42	

converted [32,33]. The system is designed in such a way to obtain a dataset containing entirely of numerical values [34,35] and to process those values as numerical values throughout the type operations. Additionally, the system indicates the nominal values and numerical equivalents used inside the process of conversion. Each kind of protocol is transformed into a 1, 2, or 3 numeric value, while the attributes of the provider and the flag are a corresponding numeric value [31,36].

The qualities which are transformed are given below:

- **Protocol Type:** This information gives details about the information of protocol used in the connection.
- **Flag:** This quality talks about the information of status of connection.
- **Service:** This information describes the network service information used during the connection.

12.3.1.2 Normalization Operation

As part of preprocessing, normalization of the dataset is a crucial step, mostly in classification. Normalization uses the features of the dataset to convert them into the values similar to each other. By normalizing the dataset, we are able to boost up the operation at the dataset and convey hit results at a faster pace [37]. In this study, the dataset for normalization was used after min-max normalization. There is a reason right here for normalizing the smallest cost to zero and the biggest cost to one, and for unfolding the dataset samples to this zero-1 range [38]. The system used to calculate the brand new cost is given.

12.3.2 PROPOSED FEATURE SELECTION METHODS

This section describes a feature choice algorithm that removes useless facts from datasets, reduces their size, and increases their efficiency [39]. The NSL-KDD dataset is used to experiment with this function choice algorithm. Two unique function choice techniques are supplied and datasets are created with the use of those techniques. The proposed gadget uses these datasets for instruction and testing [40]. By mixing attributes with different function choice strategies, attributes with high deterministic properties have been selected, included within the generated dataset, and avoided neglecting them. In the function choice strategies, which might be extensively used with inside the literature and received excessive overall performance due to the checks made are covered with inside the proposed gadget in acquiring new datasets [41,42].

12.3.2.1 The Technique for the Combination of Various Algorithms for Selecting Features

Figure 12.8 illustrates a block diagram of the proposed technique as well as the relevant characteristic choice algorithm. Since we are using CfsSubsetEval and WrapperSubsetEval as characteristic choice algorithms, this technique has great potential to be used globally. Using the BestFirst seek

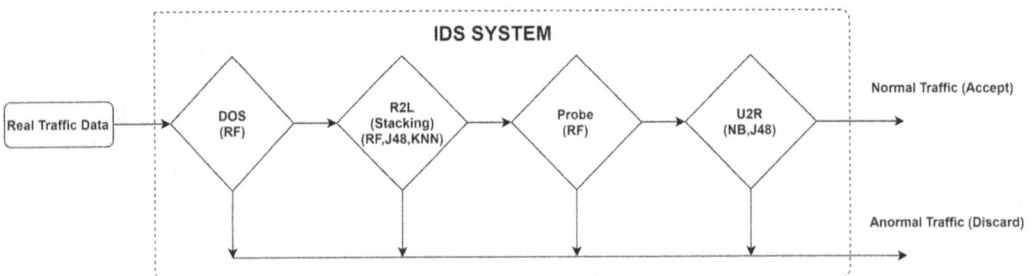

FIGURE 12.8   Block diagram of the layered architecture.

technique, CfsSubsetEval is used in destiny selection on the 20% NSL-KDD education dataset having attributes around 41 [43]. To carry out characteristic choice, WrapperSubsetEval uses a set of rules that identifies one-of-a-kind classifiers and suggests attributes via those algorithms. Following the choice of the one-of-a-kind characteristic choice algorithms, a brand new dataset with 25 attributes is received through merging the chosen features. By combining the attributes chosen through a unique algorithm, this technique is meant to produce a dataset that can support powerful category operations. Then, numbers of attribute in NSL-KDD dataset have decreased from 41 to 25 as a result of the operations performed.

12.3.2.2 The Combining Technique of Various Characteristic Choice Set of Rules Consistent with Protocol Type

In this technique, the characteristic choice technique is accomplished consistent with the protocol kind statistics that's a totally critical characteristic in figuring out the contents of facts site visitors in pc networks. Unlike the opposite characteristic choice technique, this technique makes use of subdatasets, which can be created consistent with the protocol kind [6,7]. The purpose for selecting consistent with the protocol kind is that the protocol statistics is one of the maximum critical additives in phrases of site visitors inside the attributes [26,39]. Figure 12.8 shows the block diagram of the proposed technique with excessive priority given to site visitors. In this way, its miles targets selected attributes with a great preference for site visitors. NSL-KDD's 20% education dataset is split up into subsets based on protocol types before the characteristic choice operation [38].

By examining the site visitors' information within the layering system, it is now possible to determine whether or not they come from everyday users or from an attack [3]. Therefore, information about site visitors is examined with a high-overall performance set of rules that are relevant to all types of attacks, and the danger of dangerous visitors is detected.

12.3.3 Evaluation

Values within a complexity matrix are used to calculate evaluation standards. The elaboration inside a complexity matrix of values is as follows [44,45]:

- **TP (True-Positive):** The wide variety of data samples which can be with inside the intrusions magnificence with inside the dataset and are effectively expected with inside the intrusions magnificence.
- **TN (True-Negative):** The wide variety of data samples which can be with inside the ordinary magnificence with inside the dataset and are effectively expected with inside the ordinary magnificence.
- **FN (False-Negative):** The range of data sample which with inside the intrusions magnificence with inside the dataset and are wrongly anticipated with inside the regular magnificence.
- **FP (False-Positive):** Incorrectly anticipated data sample inside intrusions magnificence that are within the regular magnificence within the dataset.

Following the confusion matrix, these values are then used to calculate the assessment standards [46]:

Accuracy
True-Positive Rate
False-Positive Rate
F-Measure
Matthews Correlation Coefficients
Detection Rate
Time (Figure 12.9)

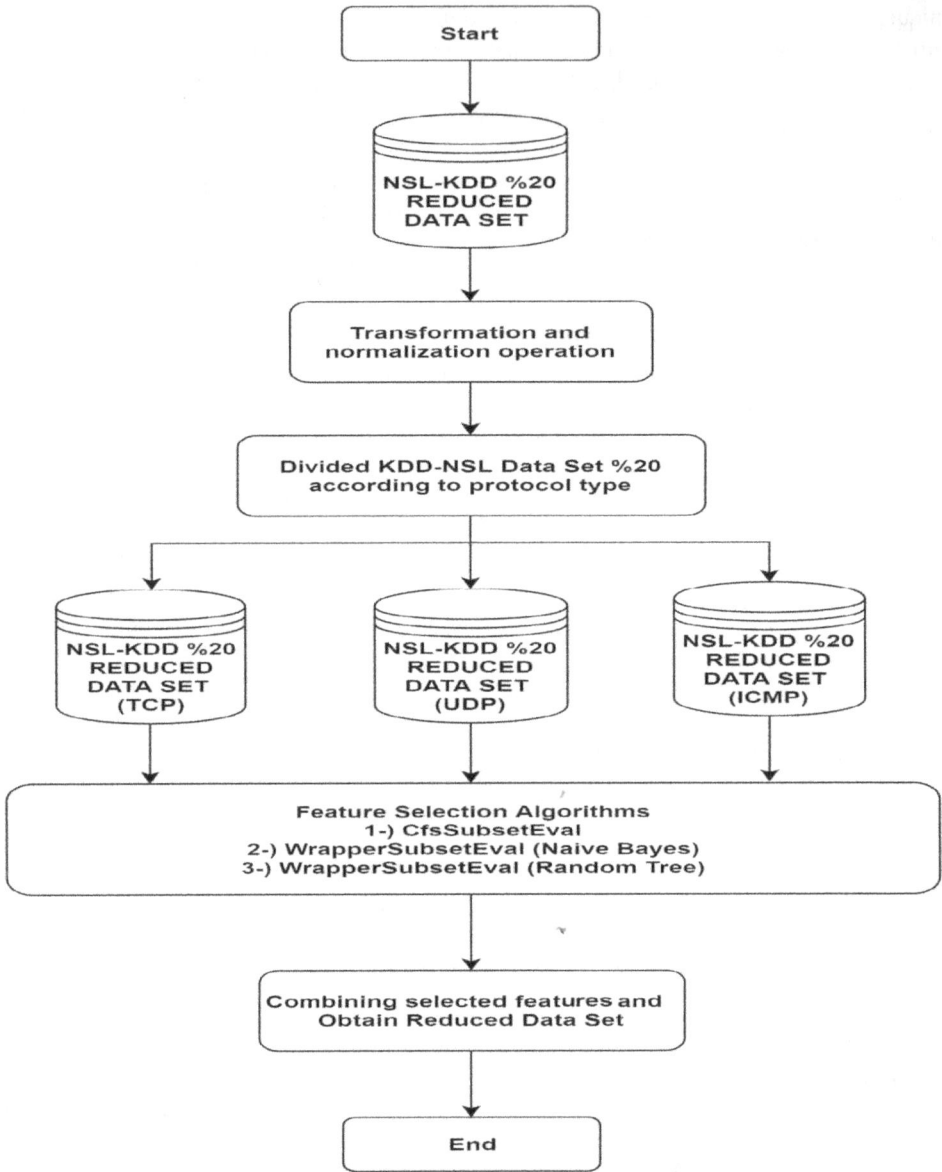

FIGURE 12.9 Hybrid intrusion detection system.

12.4 RESULT

In the chapter, the study that has been done clearly explains the feature figures of dataset counts are decreased by about half using the proposed feature selection methods, and extremely effective results are obtained with these datasets in the testing [36,47]. The training and testing operations are agreed upon using the NSL-KDD percent 20 training dataset. Following the feature collection procedure, tests will be run to establish the most relevant ML technique for the attack sort, as well as the procedures to be deployed. The suggested system's performance was evaluated using several tests like accuracy, DR, TP Rate, FP Rate, F-Measure, MCC, and time. The values obtained for the planned model in the R2L attack type in the DR and TPR criteria are 0.94 and 0.91, accordingly [47].

In comparison to other research in the literature, the DR and TPR criterion values obtained in all assault types are extremely high. In the F-Measure and MCC criterion, the R2L attack type has the lowest value of 0.92. These criteria have excellent values in various assault types. When the runtimes for the various types of attack are compared, it is discovered that the handling time for the U2R attack type is relatively low, while the computation for the R2L type of attack is lengthier than the additional attack categories owing to the usage of a heaping architecture.

The observations show that the k-NN classifier outperforms the others, and that among the feature selection approaches, the IGR feature selection technique outperforms the others and the CFS method outperforms the others [28,30,41].

12.5 CONCLUSION

We looked at recent research on an algorithm for the detection of intrusion based on ML. In addition, throughout the review, we take into account a wide range of ML approaches employed in the intrusion detection area, including single, hybrid, and ensemble classifiers. In terms of the fair outcomes of linked work, more study into creating IDSs utilizing ML approaches is needed. The following topics may be of interest for future investigation:

1. Classifiers that are used as a starting point. The solitary classifier chosen for model assessment and assessment may no longer be a good option for baseline classification. Comparisons of a number of observations and hybrid classifiers in relation of precision would be useful.
2. Selection of features. Because there are a variety of feature selection methods, the evaluated studies that examine feature selection only use one way, it is unclear which method performs the best, especially when it comes to intrusion detection classification techniques.

The RNN-IDS model provides high accurateness in binary and multiclass classification, as well as great modeling capacity for intrusion detection. When compared to traditional classification approaches like J48, naive Bayesian, and random forest, the performance achieves a higher accurateness and detection rate with a lower false-positive rate, particularly when performing multiclass classification on the NSL-KDD dataset.

Furthermore, because all of the combinations provide considerable accuracy, the investigational consequences suggest that ML may be employed in detection of any intrusion. The k-NN classifier outperforms the others, while among the approaches of selection for features, the IGR feature selection method outperforms the other while CFS falls short. IGR feature selection with k-NN has the highest accuracy of all the combinations. As a result of this research, so in conclusion we can say that using a grouping of IGR feature selection and k-NN to create an actual system which can detect intrusions is possible.

REFERENCES

1. Kok SH, Abdullah A, Jhanjhi NZ, Supramaniam M (2019) A review of intrusion detection system using machine learning approach. *Int J Eng Res Technol* 12(1):8–15.
2. Othman SM, Ba-Alwi FM, Alsohybe NT, Al-Hashida AY (2018) Intrusion detection model using machine learning algorithm on Big Data environment. *J Big Data* 5(1):1–2.
3. Bishop CM (1995) *Neural Networks for Pattern Recognition.* Oxford University, England.
4. Al-Yaseen WL, Othman ZA, Nazri MZA (2017) Multi-level hybrid support vector machine and extreme learning machine based on modified K-means for intrusion detection system. *Expert Syst Appl* 67:296–303.
5. Harb HM, Desuky AS (2011) Adaboost ensemble with genetic algorithm post optimization for intrusion detection. *Int J Comput Sci Issues (IJCSI)* 8(5):28.
6. Anderson J (1995) *An Introduction to Neural Networks.* MIT Press, Cambridge.

7. Balajinath B, Raghavan SV (2000) Intrusion detection through behavior model. *Comp Commun* 24:1202–1212.

8. Kuang F, Xu W, Zhang S (2014) A novel hybrid KPCA and SVM with GA model for intrusion detection. *Appl Soft Comput* 18:178–184.

9. Agarwal R, Joshi MV (2000) *A New Framework for Learning Classifier Models in Data Mining.* Department of Computer Science, University of Minnesota.

10. Manickam M, Rajagopalan SP (2018) A hybrid multi-layer intrusion detection system in cloud. *Clust Comput*:1–9. https://doi.org/10.1007/s10586-018-2557-5.

11. Malekipirbazari M, Aksakalli V (2015) Risk assessment in social lending via random forests. *Expert Syst Appl* 42(10):4621–4631.

12. Deng R, Zhuang P, Liang H (2017) CCPA: Coordinated Cyber physical attacks and countermeasures in smart grid. *IEEE Trans Smart Grid* 8(5):2420–2430.

13. Qi L, Dou W, Zhou Y, Yu J, Hu C (2015) A context-aware service evaluation approach over big data for cloud applications. *IEEE Trans Cloud Comput.* https://doi.org/10.1109/TCC.2015.2511764.

14. Wazid M, Das AK (2016) An efficient hybrid anomaly detection scheme using K-means clustering for wireless sensor networks. *Wirel Pers Commun* 90(4):1971–2000.

15. Aljawarneh S, Yassein MB, Aljundi M (2017) An enhanced j48 classification algorithm for the anomaly intrusion detection systems. *Clust Comput*:1–17. https://doi.org/10.1007/s10586-017-1109-8.

16. Depren O, Topallar M, Anarim E, Ciliz MK (2005) An intelligent intrusion detection system (IDS) for anomaly and misuse detection in computer networks. *Expert Syst Appl* 29(4):713–722.

17. Denning DE (1987) An intrusion-detection model. *IEEE Trans Softw Eng* SE-13(2):222–232.

18. Aslahi-Shahri BM, Rahmani R, Chizari M, Maralani A, Eslami M, Golkar MJ, Ebrahimi A (2016) A hybrid method consisting of GA and SVM for intrusion detection system. *Neural Comput Appl* 27(6):1669–1676.

19. Rodriguez-Galiano VF, Ghimire B, Rogan J, Chica-Olmo M, Rigol-Sanchez JP (2012) An assessment of the effectiveness of a random forest classifier for land-cover classification. *ISPRS J Photogramm Remote Sens* 67:93–104.

20. Milenkoski A, Vieira M, Kounev S, Avritzer A, Payne BD (2015) Evaluating computer intrusion detection systems: A survey of common practices. *ACM Comput Surv (CSUR)* 48(1):1–41.

21. Abadeh MS, Habibi J, Barzegar Z, Sergi M (2007) A parallel genetic local search algorithm for intrusion detection in computer networks. *Eng Appl Artif Intel* 20:1058–1069.

22. Alpaydin E (2014) *Introduction to Machine Learning.* MIT Press, Cambridge.

23. Chandola V, Banerjee A, Kumar V (2009) Anomaly detection: A survey. *ACM Comput Surv (CSUR)* 41(3):15.

24. Ertoz L, Kumar V, Lazarevic A, Srivastava J, Tan PN (2002) Data mining for network intrusion detection. In: *Proceedings NSF Workshop on Next Generation Data Mining*, pp 21–30.

25. Liao HJ, Lin CHR, Lin YC, Tung KY (2013) Intrusion detection system: A comprehensive review. *J Netw Comput Appl* 36(1):16–24.

26. Hartigan JA, Wong MA (1979) Algorithm AS 136: A k-means clustering algorithm. *J Royal Stat Soc Ser C (Appl Stat)* 28(1):100–108.

27. https://www.google.com/url?sa=i&url=https%3A%2F%2Fwww.researchgate.net%2Ffigure%2FList-of-features-of-NSL-KDD-dataset_tbl1_329081841&psig=AOvVaw0vJv-D9BfqMS1hGdKPakLn&ust=1637743795816000&source=images&cd=vfe&ved=0CAsQjRxqFwoTCID_z6CNrvQCFQAAAAAdAAAAABAI.

28. Horng SJ, Su MY, Chen YH, Kao TW, Chen RJ, Lai JL, Perkasa CD (2011) A novel intrusion detection system based on hierarchical clustering and support vector machines. *Expert Syst Appl* 38(1):306–313. 102.

29. Alsubhi K, Aib I, Boutaba R (2012) FuzMet: A fuzzy-logic based alert prioritization engine for intrusion detection systems. *Int J Netw Manag* 22(4):263–284.

30. Hwang TS, Lee TJ, Lee YJ (2007) A three-tier IDS via data mining approach. In *Proceedings of the 3rd Annual ACM Workshop on Mining Network Data* (pp. 1–6).

31. https://www.unb.ca/cic/datasets/nsl.html.

32. Breiman L (2001) Random forests. *Mach L* 45(1):5–32.

33. Quinlan RC (1993) *4.5: Programs for Machine Learning.* Morgan Kaufmann Publishers Inc, San Francisco.

34. Guo C, Ping Y, Liu N, Luo SS (2016) A two-level hybrid approach for intrusion detection. *Neurocomputing* 214:391–400.

35. Singh R, Kumar H, Singla RK (2015) An intrusion detection system using network traffic profiling and online sequential extreme learning machine. *Expert Syst Appl* 42(22):8609–8624.

36. Breiman L, Friedman JH, Olshen RA, Stone PJ (1984) *Classification and Regressing Trees.* Wadsworth International Group, California.
37. Cannady J (1998) Artificial neural networks for misuse detection. In: *National Information Systems Security Conference*, vol 26, pp 368–381.
38. An anomaly intrusion detection method using the CSI-KNN algorithm. In: *Proceedings of the 2008 ACM Symposium on Applied Computing.* ACM, pp 921–926.
39. Han J, Pei J, Kamber M (2011) *Data Mining: Concepts and Techniques.* Elsevier, New York.
40. Denoeux T (1995) A k-nearest neighbor classification rule based on Dempster-Shafer theory. *IEEE Trans Syst Man Cybern* 25(5):804–813.
41. Bridges SM, Vaughn RB (2000) Intrusion detection via fuzzy data mining. In *12th Annual Canadian Information Technology Security Symposium* (pp. 109-122), Ottawa.
42. Chavan S, Shah KDN, Mukherjee S (2004) Adaptive neuro-fuzzy intrusion detection systems. In: *Paper Presented at the in Proceedings of the International Conference on Information Technology: Coding and Computing (ITCC'04).*
43. Zhang Z, Shen H (2005) Application of online-training SVMs for real-time intrusion detection with different considerations. *Comput Commun* 28(12):1428–1442.
44. Scarfone K, Mell P Guide to intrusion detection and prevention systems (IDPS). National Institute of Standards and Technology. Special Publication February–2007.
45. Sommer R, Paxson V (2010) Outside the closed world: On using machine learning for network intrusion detection. In: *IEEE Symposium on Security and Privacy*, pp 305–316.
46. Tavallaee M, Bagheri E, Lu W, Ghorbani A (2009) A detailed analysis of the KDD CUP 99 data set. In: *Proceedings of the Second IEEE Symposium on Computational Intelligence for Security and Defense Applications (CISDA)*, pp 53–58.
47. Bouzida Y, Cuppens F, Cuppens-Boulahia N, Gombault S (2004) Efficient intrusion detection using principal component analysis. In: *Paper Presented at the Proceedings of the 3eme conference surla securite et architectures reseaux (SAR)*, Orlando, FL.

13 Network Forensics

Sandeep Kaur, Manjit Sandhu,
Sandeep Sharma, and Ravinder Singh Sawhney
Guru Nanak Dev University

CONTENTS

DOI: 10.1201/9781003267812-13

13.1 INTRODUCTION

Network forensics is affiliated to the digital forensics where the data is analyzed through the network traffic. Basically, it is required to collect the information, get the legal evidences, find the root cause analysis of network, and scrutinize malware behavior. Cybercrime and similar emergency will be responded by the Digital Forensics and Incident Response along with the computer emergency response teams (CERT) or computer security incident response teams (CSIRT) [1]. For the sake of criminal reconstruction, DFIR relies on the proofs which are detected in the file systems, operating systems, and information system hardware [2]. While CERT/CSIRT still take over most event reply placement functions, their advanced tools and techniques are increasingly being incorporated into everyday proactive security practices, such remote forensic triage, in order to level the playing field with sophisticated cybercriminals.

Forensics includes hardware and software that can analyze the data later [3]. Network forensic is required to avoid the hackers to not hack the network so that a particular network provides the security to the systems. With the help of analysis, we can collect the actions that are performed by the hackers and attackers.

In general, the bulk acute attacks, such as Advanced Package Tool, ransomware, espionage, and others, start from a single instance of an unauthorized entry into a network and then evolve into a long-term project for the attackers until the day their goals are met; however, the information which is flowing from one device to another device in and out of the network goes through many different networking and internetworking devices, like firewalls, routers, switches, hubs, web proxies, and others. Our aim is to recognize and investigate all these dissimilar traces [4].

13.2 METHODOLOGY FOR NETWORK FORENSICS

To satisfy clear-cut and significant solution at the end of a network forensic you, as a forensic agent, must follow through a methodological framework. Figure 13.1 represents the steps of Network Forensic Methodology. This track comprises the following steps. Seven steps for examining Network Forensics are Identification, Preservation, Collection, Examination, Analysis, Presentation, and Incident Response.

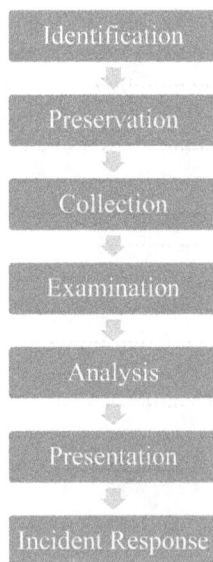

```
Identification
    ↓
Preservation
    ↓
Collection
    ↓
Examination
    ↓
Analysis
    ↓
Presentation
    ↓
Incident Response
```

FIGURE 13.1 Steps of network forensics methodology.

13.2.1 IDENTIFICATION

Identification is the first step toward the process of capturing and analyzing the data on the network. This step leads all other steps to reach the destination. Incidents which are based on network indicators include the recognition and determination of the steps.

13.2.2 PRESERVATION

Isolation of data takes place in the second step of the network forensics methodology for the security and preservation purpose to avoid community from using the gadgets. They should be used in such a way that digital evidences will not be lost [5]. Autopsy and encase like software tools are present for preservation.

13.2.3 COLLECTION

Collection takes place at the third step process. It is responsible for recording the actual scenes and then made a copy of the recorded data by using the procedures and methods for standardization.

13.2.4 EXAMINATION

Record of the visible data will be kept in the examination which is the fourth step of the network forensics methodology [6]. Metadata from data of many pieces will be collected and then send to the court for the purpose of the evidence.

13.2.5 ANALYSIS

The agents of investigation will combine pieces of the fragmented data after identification and preservation step [7,8]. Conclusion will be withdrawn on the basis of the evidences and the data analyses will be taken care. Without data analyses, the conclusion will not be formed. Security Information and Event Management (SIEM) software gives a track record of the activities with the IT environment. The SIEM tools analyze log and event data in real time to provide threat monitoring, event correlation, and incident response – with security information management which collects, analyzes, and reports on log data.

13.2.6 PRESENTATION

The denotation for forensic denote to brought up to the court. The process of sum up and clarification of wind up is done. This should be written down in layman's term using theoretical expressions and all the abstract terminologies should instance of the particular feature.

13.2.7 INCIDENT RESPONSE

The information gathering to authenticate and evaluate the incident purely relies on the intrusion detection. Thus, above are the Network Forensics steps that fulfill the whole process of capturing and analyzing the data. Analysis tools which will be investigated by a network investigator of network forensics are responsible for monitoring information on the network, information which is gathered about the anomalous or malicious traffic [9].

13.3 SOURCES OF EVIDENCE

Network evidences shall be collected from different sources and in this section, we are throwing light on the sources which helps us to collect the data from the network [10]. The basic and some advanced sources are as follows:

1. Spout the wire and the air (TAPs – Test Access Points or Terminal Access Points)
2. Content Addressable Memory (CAM) table working with a network switch
3. Routing tables function for routers
4. Dynamic Host Configuration Protocol (DHCP) record
5. Domain Name System (DNS) server record
6. Domain controller/authentication servers/system record
7. Intrusion detection/penetration (IDS/IPS) records
8. Firewall records
9. Proxy Server records

13.3.1 SPOUT THE WIRE AND THE AIR (TAPs)

The basic source and the simplest form of knowledge contagious is through the TAPs. TAPs will be placed on the network and fiber optic cables to inquire on traffic. TAPs are the basic footstep in the procedure of as long as penetrating clarity over your physical, non-existence, and cloud architecture, retrieving 100% of the network fright on each tapped link [11–13]. In physical networks, a network TAP is typically deployed in a network link connected between the interfaces of two network elements, such as switches or routers, and is treated as part of the network infrastructure, similar to a patch panel. In no real networks, ways that are used for deploying the TAP on network may vary, and variation depends on the interface and environment. Figure 13.2 shows how TAP is connected to router, analysis workstation, and switch to form a network.

13.3.2 CAM TABLE ON A NETWORK SWITCH

Figure 13.3 shows how to store data in CAM table. CAM tables are used to store and map the physical address (MAC, i.e., Media Access Control) and their respective ports. Entries could be manual or automatic; it depends on the network and interconnected devices that are attached to it. Basically, this table is used in a large extent because it can add an MAC address on the network to a virtual system. The concept of mapping is also used to map the MAC address to the physical port. Switches are networking device, which is used in the CAM table and provides us the mirroring also. Mirroring helps the network to inform about the data which are from other networks or they are from the virtual local area networks and systems.

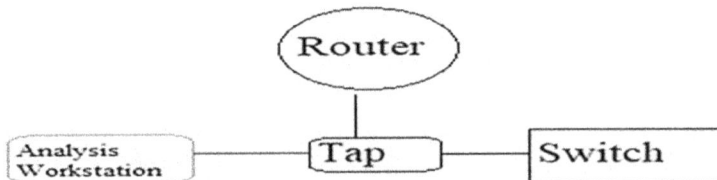

FIGURE 13.2 How TAP works on the network [14].

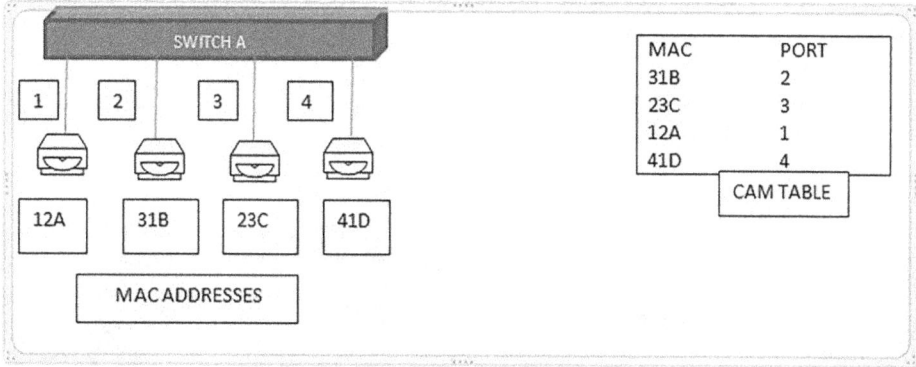

FIGURE 13.3 Entries in the CAM table [14].

13.3.3 ROUTING TABLES FUNCTION FOR ROUTERS

Router includes routing table, which maps ports on the different network through router. Figure 13.4 shows the routing table. The routing table helps us to find the path of the destination network, next hop, and interface while moving through different networking and internetworking devices [15]. Routers have the capability of inbuilt filtering the packets and firewall possibilities. Routing tables are of two types, i.e., static routing table and dynamic table. Static routing table is handled by network administrator whereas dynamic routing table is handled by the software automatically. The record has been made of the traffic that has been traveling through the particular network. Figure 13.4 displays the entries in routing table from network.

DHCP Record: When a specified Internet Protocol address is assigned to the MAC address, then DHCP servers, in general, make the record entries. DHCP worked on the leased system. When they will be changed for the network, time tamp will be changed. They are having an important value in

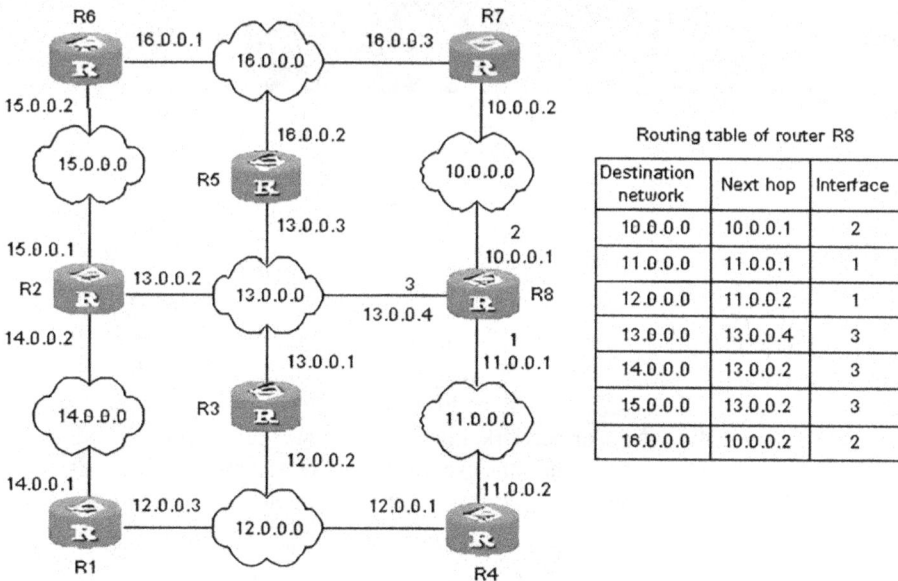

Routing table of router R8

Destination network	Next hop	Interface
10.0.0.0	10.0.0.1	2
11.0.0.0	11.0.0.1	1
12.0.0.0	11.0.0.2	1
13.0.0.0	13.0.0.4	3
14.0.0.0	13.0.0.2	3
15.0.0.0	13.0.0.2	3
16.0.0.0	10.0.0.2	2

FIGURE 13.4 Entries in routing table from a network [14].

LOG Format	ID	Date	Time	Description	IP Address	Host Name	MAC Address

FIGURE 13.5 DHCP log fields [14].

```
Command Prompt

C:\Users\Dell>ipconfig/displaydns

Windows IP Configuration

    safebrowsing.googleapis.com
    ----------------------------------------
    Record Name . . . . . : safebrowsing.googleapis.com
    Record Type . . . . . : 1
    Time To Live . . . . : 112
    Data Length . . . . . : 4
    Section . . . . . . . : Answer
    A (Host) Record . . . : 216.58.203.10


    ogs.google.com
    ----------------------------------------
    Record Name . . . . . : ogs.google.com
    Record Type . . . . . : 5
    Time To Live . . . . : 88
    Data Length . . . . . : 8
    Section . . . . . . . : Answer
    CNAME Record  . . . . : www3.l.google.com


    Record Name . . . . . : www3.l.google.com
    Record Type . . . . . : 1
    Time To Live . . . . : 88
    Data Length . . . . . : 4
    Section . . . . . . . : Answer
    A (Host) Record . . . : 142.250.183.46


    array506.prod.do.dsp.mp.microsoft.com
    ----------------------------------------
    Record Name . . . . . : array506.prod.do.dsp.mp.microsoft.com
    Record Type . . . . . : 1
    Time To Live . . . . : 1448
    Data Length         : 4
```

FIGURE 13.6 DNS server log [14].

the network forensics. The DHCP log fields include the log format, ID, Date, Time, Description, IP Address, Host Name, and MAC Address. Figure 13.5 shows the DHCP log fields, which include the log format, ID, Date, Time, Description, IP Address, Host Name, and MAC Address.

DNS server's Record: If we want to understand IP-to-Hostname resolve at a particular duration of time, then we can focus on the DNS server query records. This DNS server helps to retrieve the previous domain by using the command line mode in the case when the system gets corrupted with the harmful malware on the particular network. Figure 13.6 shows the working of DNS server log in Command prompt and Figure 13.7 displays the event viewer which enables the administrator or user to view the log events such as application, security, setup, system, etc. of windows operating system.

13.3.4 DOMAIN CONTROLLER/AUTHENTICATION SERVERS/SYSTEM RECORDS

The activities that are performed by authentication servers are login time, login attempts to check how many times login has been made, and many login activities within the whole network. The

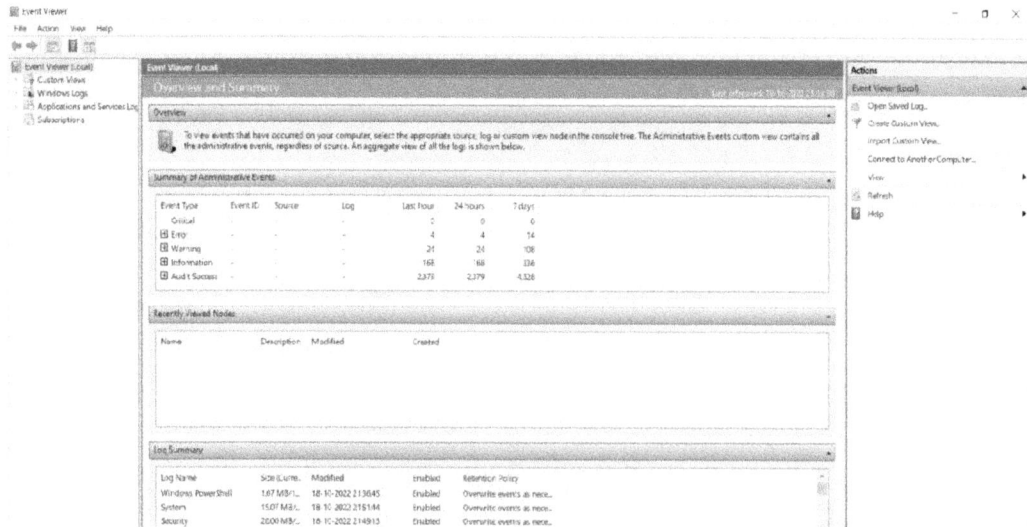

FIGURE 13.7 Event viewer.

compromised machine is used as a launchpad (pivoting) where the number of attackers reaches to meet each other half way host to login to the server of database. How many times the system is failed to logout and passed to login attempts from the system to the database server; moreover, if we want to quickly give away the infected systems then we have to rely on the authentication logs.

13.3.5 IDS/IPS RECORDS

When we deal with the data from the investigation point of view, IDS/IPS logs are the most helpful. The Internet Protocol addresses, signature matched, attacks which are going on, presence of malware, command-and-control servers, source and destination systems for the IP and the Port, timelines, etc. are provided by the IDS/IPS records. Figure 13.8 represent IDS/IPS logs.

Firewall Records: The detailed view of network activities are provided by the firewall records. Firewall solutions not only protect the network on the server or unwanted connections from a network, but it also provides the identification of the type of traffic, the trust score to the outbound endpoint, unwanted ports and also blocks the connection attempts. Figure 13.9 displays firewall records.

13.3.6 PROXY SERVER RECORDS

Figure 13.10 represents Proxy Server Logs. The most powerful feature for the investigator forensic is the web proxies. Records help uncover internal threats and provide explicit details on events such as surfing, the malware based on the source of the web, and the user's behavior on a particular network.

13.4 TOOLS IN DIGITAL FORENSICS

Dumpcap, tcpdump, Xplico, and Network Miner are the tools for general purposes, which are helpful in capturing and analyzing the data. Some tools are used for a particular task such as **snort** which is used for intrusion detection; **ngrep** which is used for Match regular expression; ntop, tstat, and tcpstat are used for the print network. The above-written and many other tools using different

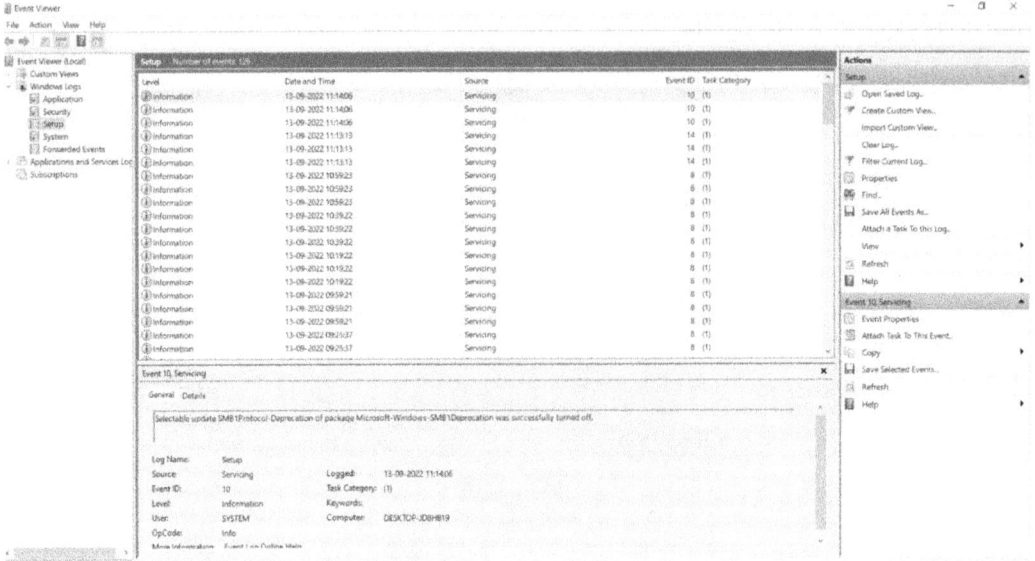

FIGURE 13.8   IDS/IPS logs.

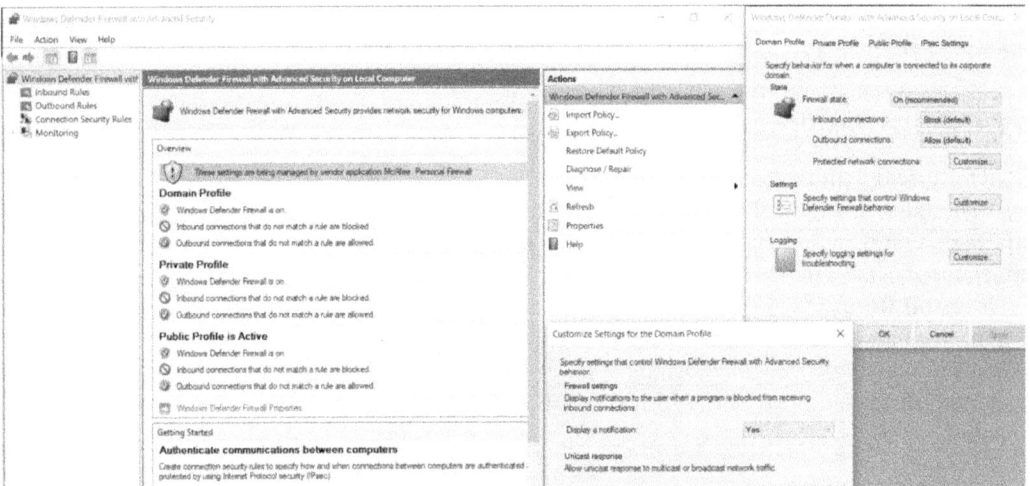

FIGURE 13.9   Firewall logs.

frameworks and libraries like python library (Scapy, Libpcap) are used for the network forensics process [16].

For capturing and analyzing the network traffic, evaluating the performance of network, and finding problems, network protocol determination, privacy investigation and response incident and legal protection, the digital and network forensics tools are used [17]. Make sure you are updated with the latest updates on the softwares and the tools and the technology that you are using. You have to be up to date with the latest versions of the system as well as application softwares.

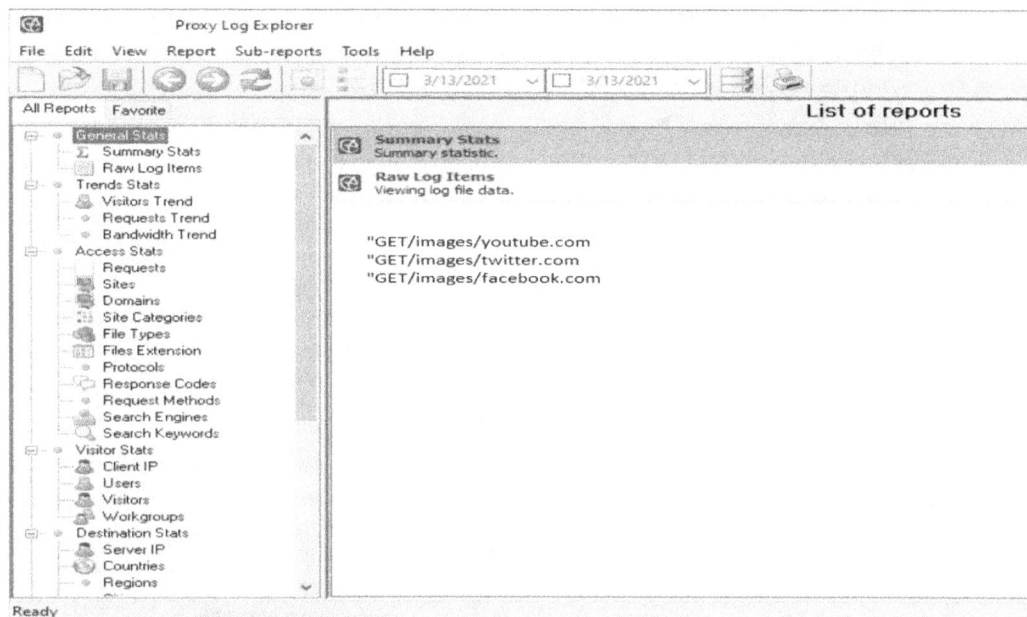

FIGURE 13.10 Proxy server logs [14].

Network attacks can be investigated by a variety of tools. There are two types of tools that are available in online and offline modes. One is open-source software which is available to everyone and another is the paid tools. Tools are basically the softwares that helps us to investigate the attacks on the network [18]. We will discuss the tools that are free and easily available to everyone.

First, we start with the tcpdump and Wireshark that how we can capture and analyze the network traffic. Second, the tool named Network Miner will be discussed. Lastly, we had a look on the snort and Splunk, which are famous tools for the network traffic indexing and analysis [19].

13.4.1 Tcpdump (Command Line)

Tcpdump is data packet network analyzer software, which is purely in command line mode. The packets that are transmitted or received over some network will allow the user to display the tcp/ip and other packets. Traffic will be controlled on the Unix-based systems. The data that is captured is stored in a file, which is compatible with the tools like Wireshark for future analysis. Tcpdump is also used for troubleshooting or for capturing the traffic continuously in large volumes for future analysis. Data link layer and network layer data are captured by the tcpdump. The major problem that can arise is the disk space, while capturing the data for analysis purpose a large amount of data will be grown which leads to network traffic. Tcpdump uses the filters, which helps in capturing unnecessary data. By applying filters, some important data and evidences may lose. So, this software recommends to capture whole data first and then uses the filter to remove unwanted data. Table 13.1 represents the syntax and description of Tcpdump commands correspondingly.

13.4.2 Wireshark (Graphical User Interface)

Wireshark is a tool used in the field of Cyber Security [20,21]. This is the open-source tool used for capturing and analyzing traffic and, moreover, they are used for applying the filters by using the GUI. One can choose interface where someone needs to capture the traffic, on the system [22,23]. Figure 13.11 displays how Wireshark interface looks.

TABLE 13.1

Syntax and Description of Tcpdump Commands

Description	Syntax
Everything on an interface	tcpdump **-i eth0**
Find traffic by IP	tcpdump **host 1.1.1.1**
Filtering by source and/or destination	tcpdump **src 1.1.1.1**
	tcpdump **dst 1.0.0.1**
Finding Packets by Network	tcpdump **net 1.2.3.0/24**
Get packet contents with HEX output	tcpdump **-c 1-X icmp**
Show traffic related to a specific port	tcpdump **port 3389**
	tcpdump **src port 1025**
Show traffic on one protocol	tcpdump **icmp**
Show only IP6 traffic	tcpdump **ip6**
Find traffic using port ranges	tcpdump **portrange 21–23**
Find traffic based on packet size	tcpdump **less 32**
	tcpdump **greater 64**
	tcpdump **<= 128**
Reading/writing captures to a file (CAP)	tcpdump **port 80-w** capture_file
	tcpdump **-r capture_file**

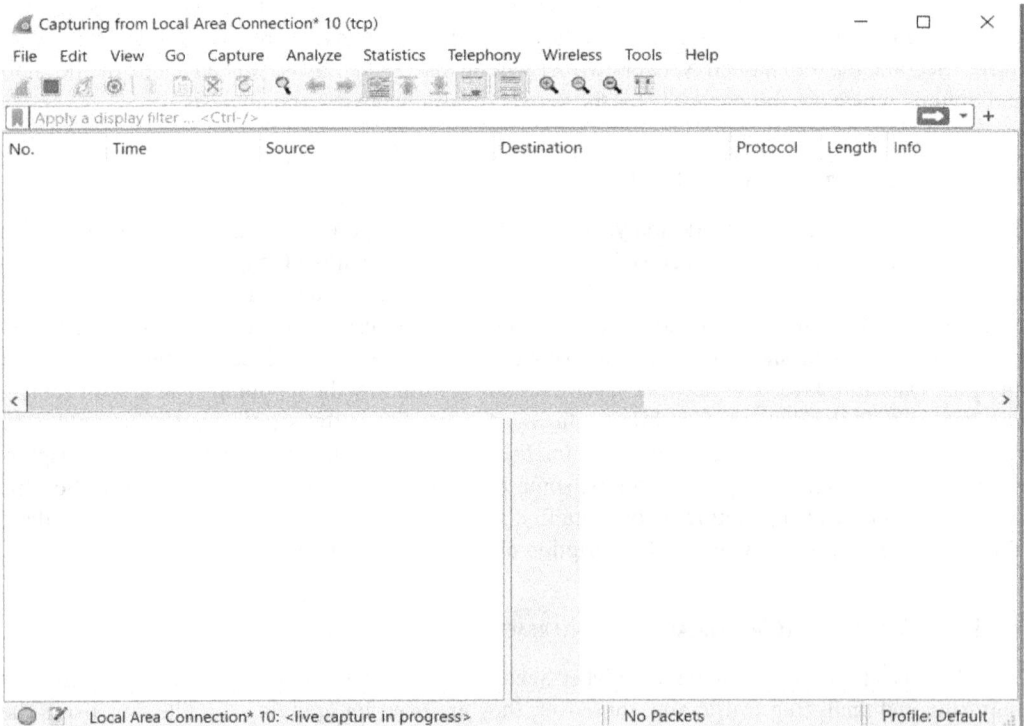

FIGURE 13.11   Wireshark interface.

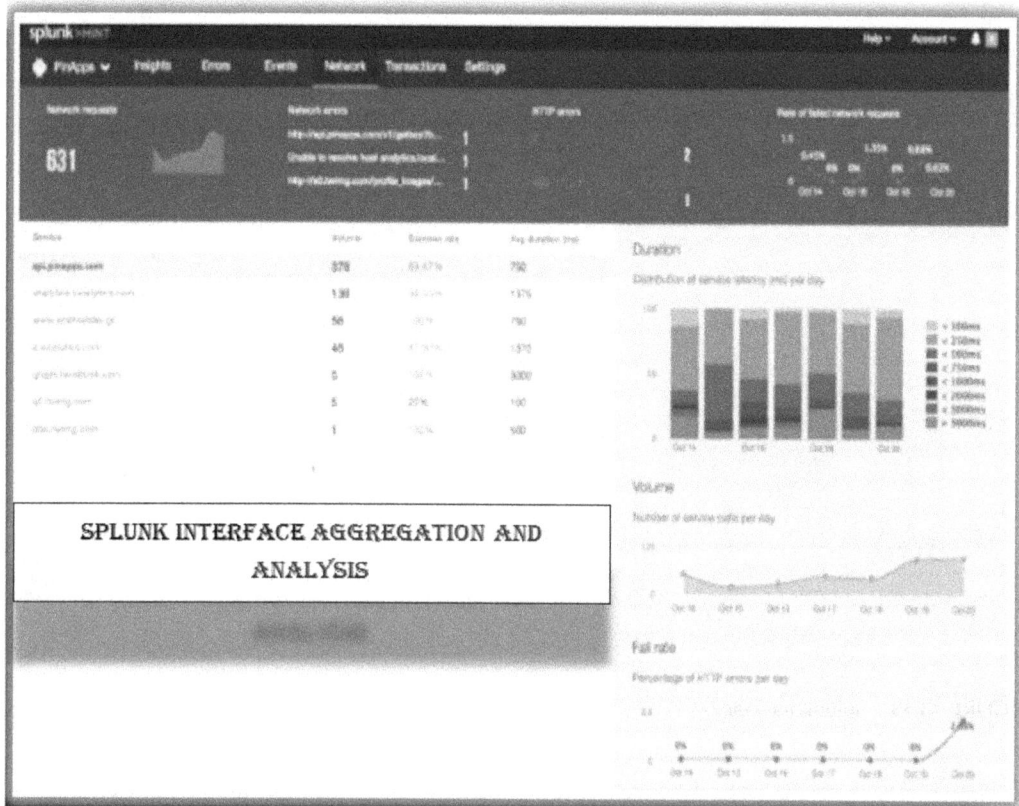

FIGURE 13.12   Network miner interface.

13.4.3   NETWORK MINER

The website netresec.com which is the official website of "Network Miner" is an open-source Network Forensic Analysis Tool for Windows (but also works in Linux/Mac OS X/FreeBSD). Network Miner can be used as a passive network sniffer/packet capturing tool in order to detect operating systems, sessions, hostnames, open ports, etc. without putting any traffic on the network [24]. Network Miner can also parse PCAP files for off-line analysis and to regenerate/reassemble transmitted files and certificates from PCAP files.

Network Miner makes it easy to perform advanced Network Traffic Analysis by providing extracted artifacts in an intuitive user interface. The way data is presented not only makes the analysis simpler, but also saves valuable time for the analyst or forensic investigator [25].

Network Miner has, since its first release in 2007, become a popular tool among incident response teams as well as law enforcement. Network Miner is used today by companies and organizations all over the world. The professional version of the Network Miner also exists. Figure 13.12 displays the Network Miner Interface.

13.4.4   SPLUNK

Splunk is a tool for aggregation and analysis and it is proprietary, highly extensible, and, moreover, it is portable. Splunk operations are capturing, indexing, and correlating the real-time data for a searchable container and production is graphs, alerts, dashboards, and visualizations. During the time of investigation, Splunk plays a role for as long as verification from different sources. The most

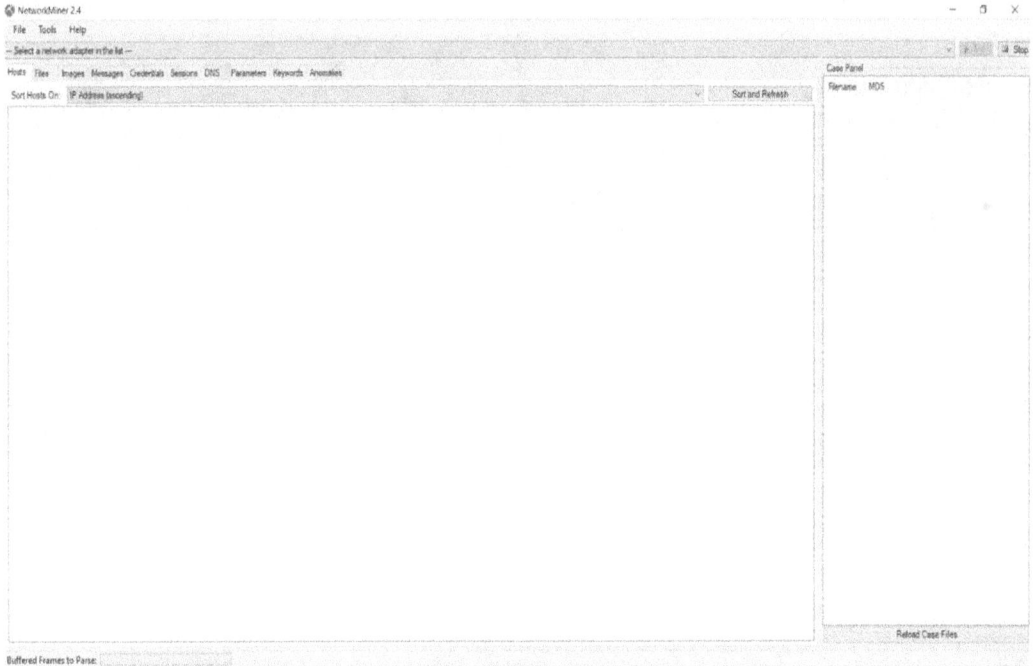

FIGURE 13.13   Splunk interface.

popular and commercial tool is the Splunk; limited features are offered in the free version. This is purely graphical user interface mode and everyone can use this. Figure 13.13 represents the Splunk Interface.

13.4.5   SNORT

Snort is an open-source, free network intrusion detection system and is the bulk in demand. The commercial version of the snort is also available and is presently offered by the Cisco. Snort is customizable this permit the users to add the add-on which are known as preprocessors. Furthermore, it proceeds with the big set of the output alternatives. The fundamental, snort provides the needful protocols. Snort administrator requires to cater for the protocols as the delinquency doesn't come with any protocols at the beginning. A set of rules is already defined on the website of the snort that can be fed into the snort application [26,27]. Moreover, the rules can be written as the custom alert protocols. Figure 13.14 shows the snort interface.

13.4.6   THE SLEUTH KIT

The user working on the hard drive can stores the data permanently on it, for later use. The Sleuth Kit is concentrating on the hard drive only. Delicate information from an inspection point of view is stored on the RAM. That's why the system analyst first collects the artifacts from the virtual memory. It is very useful from an inspection point of view. This is purely command based and different commands are used for the system inspection, they will be explored through the logs from the harmful and delicate system without changing any data from it. The deleted and hidden files will also be shown side by side by this. It has a great look at the partitions too. The biggest drawback of this tool is that user have to recall all the commands which is time consuming process. Figure 13.15 represents the Sleuthkit Interface.

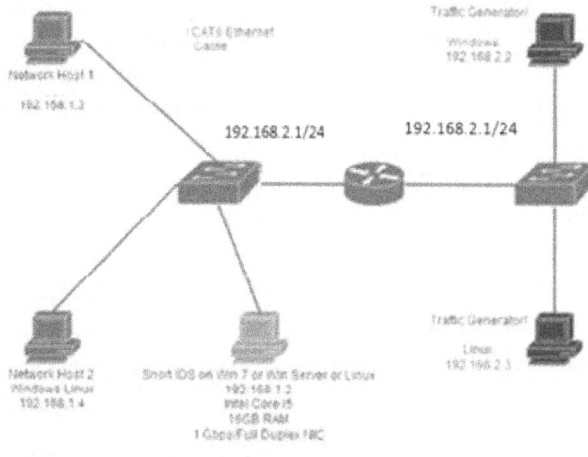

FIGURE 13.14   Snort interface [20].

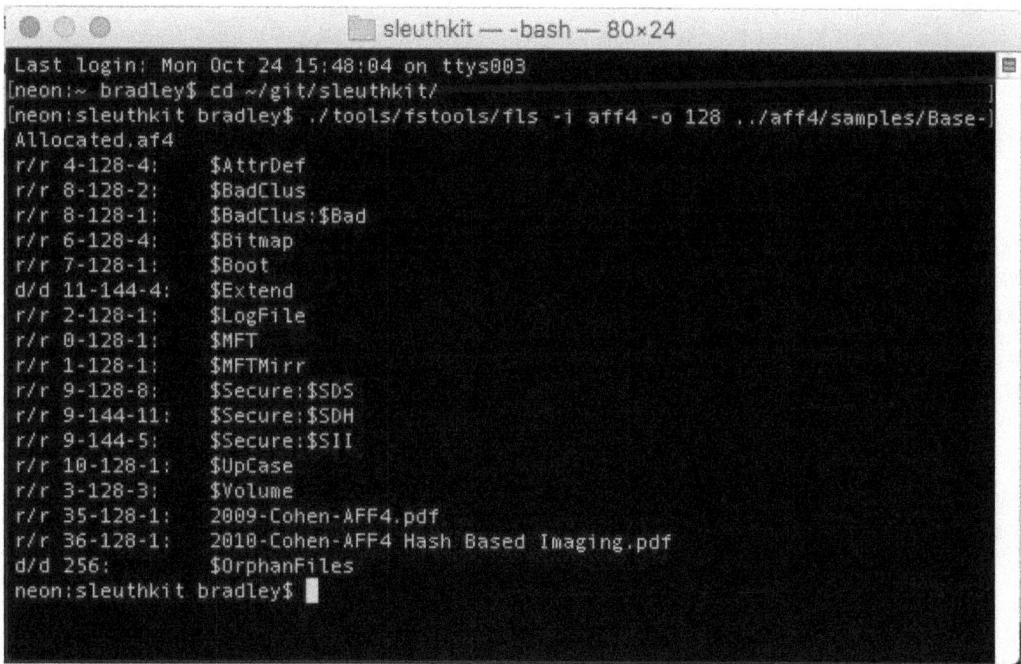

FIGURE 13.15   Sleuth Kit interface.

13.4.7 AUTOPSY

Autopsy is an open-source digital investigation software and used to put the immortal for bring out the hard drive inspection. For conducting digital investigations, the government agencies and corporate investigators use the Autopsy software. This tool is used by the military as well as by the law enforcement. This tool supports both the Linux and the Windows Operating System, Autopsy is pre-installed in the Kali Linux. Figure 13.16 displays the Autopsy interface and this tool comes in the graphical user interface mode and the browser. It's not a standalone tool; it works with the Sleuth Kit as it is a command line.

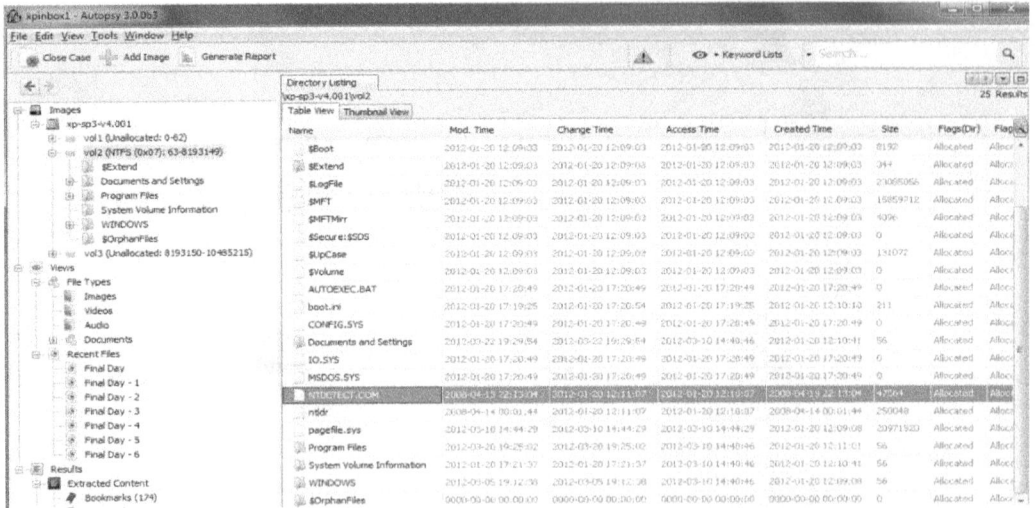

FIGURE 13.16   Autopsy interface.

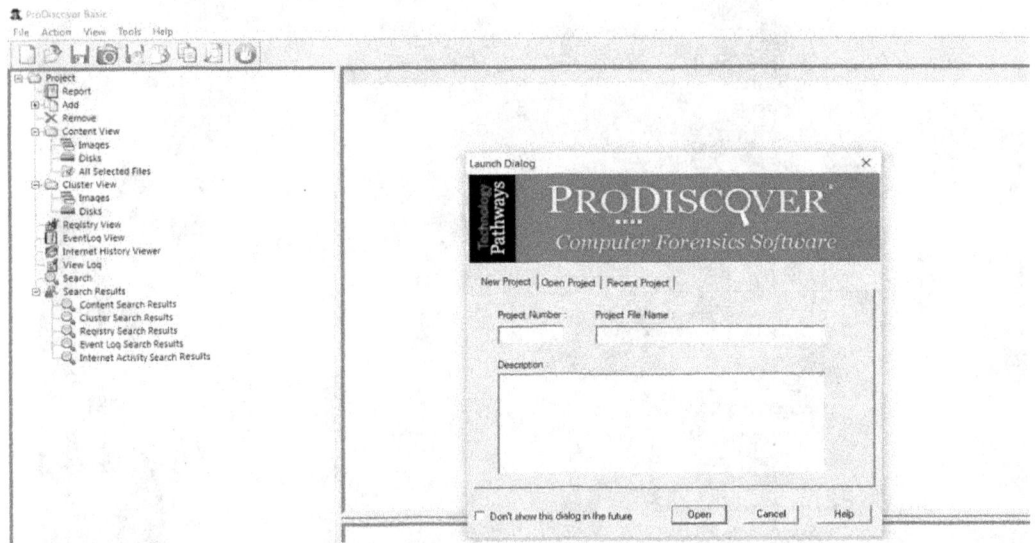

FIGURE 13.17   PRODiscover interface [25].

13.4.8   PRODISCOVER BASIC

ProDiscover comes in the graphical user interface mode as well like the Autopsy. This tool is available for free of cost for investigating the data; this tool makes the duplicate copy of the hard drive. This allows to search and preview the corrupted files, reading the whole data byte-by-byte without changing the data and the meta data. It allows for making the images of the memory, BIOS image, RAM Image, and hard drive images. Duplicate images will be made of it. Data analysis will be done in detail as soon as the image is prepared [28–30]. Figure 13.17 represents the ProDiscover Interface, which shows the users how it looks to the user.

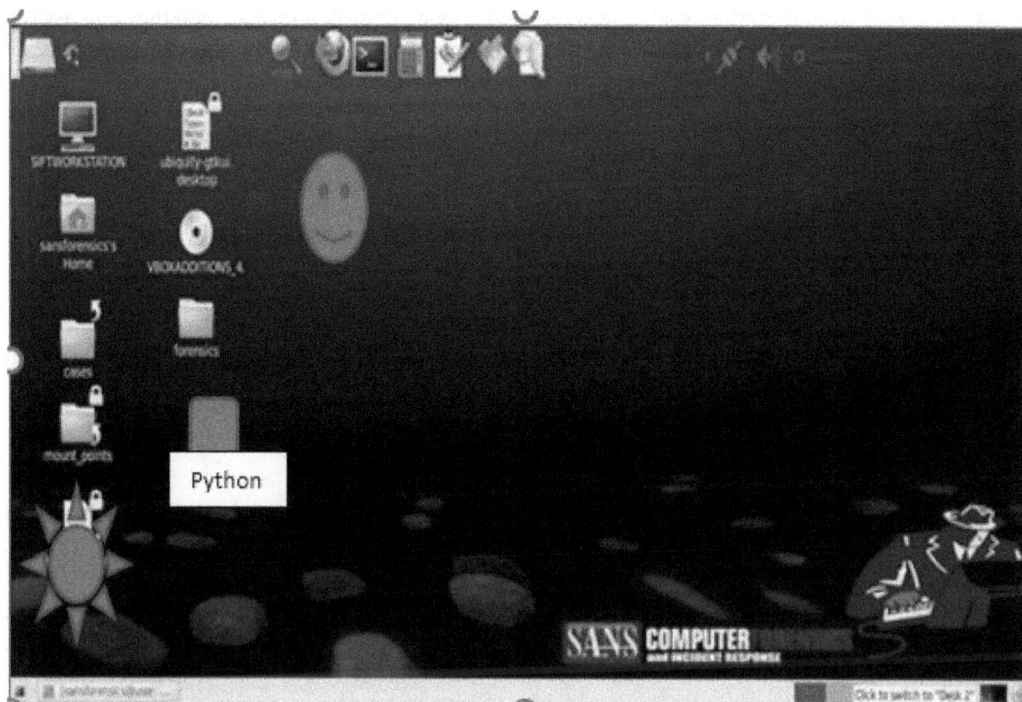

FIGURE 13.18 SANS SIFT interface.

13.4.9 SANS SIFT

The ubuntu-based live CD is the SANS SIFT (Investigative Forensic Toolkit). The in-depth incident response investigation is required to carry the tools for the inspection purpose. It supports Advanced Forensic Format (AFF), RAW (dd) evidence formats, and Expert Witness Format (E01) for analysis. SIFT failure tools like log2timeline etc. timeline will be generated from the system records, recycle bin will be examined by the Rifiuti, data file carving for Scalpel, and many more. Figure 13.18 displays the SANS SIFT interface.

13.4.10 VOLATILENESS

As long as the investigation of volatile memory, Volatileness is the bulk popular tool. Similar to the Sleuth Kit, Volatility is furthermore open-source, free, and supports third-party add-on. The Volatility base acts annually match to expand the bulk imaginative and utility addition to the users to users' structure.

13.5 METHODOLOGY IN DIGITAL FORENSICS

Following are some different digital forensics investigation techniques. Have a look at these forensics techniques, which lead us to capture and analyze the data on the network.

13.5.1 PRESERVING THE EVIDENCE

Write blocker tool can be placed for making a similar copy of the pure data by the Forensic analysts. In nature, if we want to avoid some program as well device from making changes to original data by using the writ blocker. If the user wants to create bit by bit copy of the disc, then they can use the hardware write blockers. If hardware write blockers are unavailable, or addition to it is there, then we can rely on the forensic software write blockers for a similar motive.

13.5.2   WEB SCHEME RECONSTRUCTION

If the user wants to recover the browsing history, files which are temporary, and acquire the cookies then we can rely on this technique. This is very useful in order to remove them conceivable.

13.5.3   FILE SIGNATURE ATTESTATION

This technique is used to differentiate the header and the footer data of unsecured files with familiar(secured) files and compared result is displayed.

13.5.4   NETWORK DEVICE INSPECTION

By the way, this kind of digital forensic investigation technique is intricate. It usually gets employed when logs of servers are unavailable for some reason.

Network keeps all the records that are involved in administering the inspection. At the same time, the inquiry cover routers, switches, and firewalls for all internetworking devices to inspect harmful DNS requests; the unknown IPS leads to connections or unexpected rise in the network scheme [31,32]. Incidentally, here digital inspection methodology is complex. It is generally come by professionals when the records of the servers are unreachable for a little cause.

13.5.5   RECOVERING INVISIBLE FILES

Investigation observers use dissimilar techniques of decipher, cryptography, and drive or image analysis to assiduously focus on invisible data and files. The final point is to obtain a way in to them.

13.6   CONCLUSION

The book chapter Network forensics concludes that the security at every point is needed in the network so that we can preserve the data. For preserving the data, we have to rely on the different hardware and software specifications. Hardware includes all the internetworking devices and the softwares are used to handle the networks virtually on the network. In this chapter, we have discussed the network forensics methodology that which are the different steps that leads to completing the process of investigation. Then, we have discussed the different tools and techniques that capture the evidences and that make analyses accordingly. These tools are software based. While sitting on the computer we have made the results and then we can submit it to the court for the further necessary action. This book chapter can further be expanded by using at least one tool and we can perform the live implementations and the results will be generated.

REFERENCES

1. K. Jiang and R. Xuan, "Book review: guide to computer forensics and investigations," *Journal of Digital Forensics, Security and Law*, vol. 3, no. 5, p. 467, 2008.
2. E. S. Pilli, R. C. Joshi, and R. Niyogi, "Network forensic frameworks: survey and research challenges," *Digital Investigation*, vol. 7, no. 1–2, pp. 14–27, 2010.
3. R. Bejtlich, *The Practice of Network Security Monitoring: Understanding Incident Detection and Response*, No Starch Press, San Francisco, CA, 2013.
4. M. Rasmi and A. Jantan, "A new algorithm to estimate the similarity between the intentions of the cyber crimes for network forensics," in *Proceedings of the 4th International Conference on Electrical Engineering and Informatics (Iceei 2013)*, vol. 11, pp. 540–547, Malaysia, Malaysia, June 2013.

5. B. Cusack and M. Alqahtani, "Acquisition of evidence from network intrusion detection systems," in *Proceedings of the 11th Australian Digital Forensics Conference*, Perth, Western Australia, December 2013.

6. B.-C. Cheng, G.-T. Liao, H.-C. Huang, and P.-H. Hsu, "Cheetah: a space-efficient HNB-based NFAT approach to supporting network forensics," *Annals of Telecommunications – annales des télécommunications*, vol. 69, no. 7–8, pp. 379–389, 2014.

7. D. Wang, T. Li, S. Liu, J. Zhang, and C. Liu, "Dynamical network forensics based on immune agent," in *Proceedings of the Third International Conference on Natural Computation (ICNC 2007)*, vol. 3, pp. 651–656, IEEE, Haikou, China, August 2007.

8. M. Ibrahim, M. T. Abdullah, and A. Dehghantanha, "VoIP evidence model: a new forensic method for investigating VoIP malicious attacks," in *Proceedings of the 2012 International Conference on Cyber Security, Cyber Warfare and Digital Forensic (CyberSec)*, pp. 201–206, IEEE, Kuala Lumpur, Malaysia, June 2012.

9. L. M. Chen, M. C. Chen, W. Liao, and Y. S. Sun, "A scalable network forensics mechanism for stealthy self-propagating attacks," *Computer Communications*, vol. 36, no. 13, pp. 1471–1484, 2013.

10. I. L. Lin, Y. S. Yen, B. L. Wu, and H. Y. Wang, "VoIP network forensic analysis with digital evidence procedure," in *Proceedings of the 485 the 6th International Conference on Networked Computing and Advanced Information Management*, pp. 236–241, IEEE, Seoul, January 2010.

11. W. Ren and H. Jin, "Distributed agent-based real time network intrusion forensics system architecture design," in *Proceedings of the 19th International Conference on Advanced Information Networking and Applications (AINA'05)*, vol. 1, pp. 177–182, IEEE, Taipei, Taiwan, March 2005.

12. E. Jeong and B. Lee, "An IP traceback protocol using a compressed hash table, a sinkhole router and data mining based on network forensics against network attacks," *Future Generation Computer Systems*, vol. 33, pp. 42–52, 2014.

13. Y. Zhu, "Attack pattern discovery in forensic investigation of network attacks," *IEEE Journal on Selected Areas in Communications*, vol. 29, no. 7, pp. 1349–1357, 2011.

14. https://1.bp.blogspot.com.

15. https://i.stack.imgur.com/tuunC.gif.

16. https://arxiv.org/ftp/arxiv/papers/1004/1004.0570.pdf.

17. S. Zander, G. Armitage, and P. Branch, "A survey of covert channels and countermeasures in computer network protocols," *IEEE Communications Surveys & Tutorials*, vol. 9, no. 3, pp. 44–57, 2007.

18. J. Li, D. Zhou, W. Qiu et al., "Application of weighted gene co-expression network analysis for data from paired design," *Scientific Reports*, vol. 8, pp. 622–628, 2018.

19. https://i1.wp.com/ipwithease.com.

20. https://resources.infosecinstitute.com.

21. https://paper.bobylive.com/Security/Hands-On-Network-Forensics.pdf.

22. S. Perry, "Network forensics and the inside job," *Network Security*, vol. 2006, no. 12, p. 13, 2006.

23. D. M. White, "The federal information security management act of 2002: a Potemkin village," *Fordham Law Review*, vol. 497, pp. 79–369, 2010.

24. C. Wang, T. Feng, J. Kim, G. Wang, and W. Zhang, "Catching packet droppers and modifiers in wireless sensor networks," in *Proceedings of the 2009 6th Annual IEEE Communications Society Conference on Sensor, Mesh and Ad 500 Hoc Communications and Networks*, pp. 1–9, IEEE, Rome, Italy, June 2009.

25. https://lifars.com/ 2020/06/the-basics-of-network-forensics/.

26. https://www.researchgate.net/publication/329217872_network_forensics.

27. https://www.rcciit.org/students_projects/projects/it/ 2018/GR7.pdf.

28. F. Akhtar, J. Li, Y. Pei et al., "Diagnosis and prediction of large-for-gestational-age fetus using the stacked generalization method," *Applied Sciences*, vol. 9, no. 20, p. 4317, 2019.

29. A. Imran, J. Li, Y. Pei, J.-J. Yang, and Q. Wang, "Comparative analysis of vessel segmentation techniques in retinal images," *IEEE Access*, vol. 7, p. 114862–114887, 2019.

30. J. Li, L. Liu, J. Sun et al., "Comparison of different machine learning approaches to predict small for gestational age infants," *IEEE Transactions on Big Data*, vol. 6, no. 2, 2016.

31. https://kneda.net/documentos/Learning%20Network%20Forensics.pdf.

32. http://csrc.nist.gov/publications/nistpubs/800-86/SP800-86.pdf.

14 A Deep Neural Network-Based Biometric Random Key Generator for Security Enhancement

Sannidhan M. S., Jason Elroy Martis, and Sudeepa K. B.
NMAM Institute of Technology

CONTENTS

DOI: 10.1201/9781003267812-14

14.1 INTRODUCTION

Preservation of information security is one of the primary challenges in today's contemporary world. Information is very vastly spread across the globe in physical and digital forms. Any physical data can be secured to the fullest with the incorporation of tangible security tools. However, preserving the security of the digital information appearing virtually over electronic devices has turned out to be one of the significant issues and concerns over time. The digital world has even grown to replace day-to-day activities that include a varied range of offline and online applications. A few critical applications include communication, marketing, finance, law enforcement systems, money transactions, cryptocurrencies, etc. With the comprehensive coverage and increasing popularity of its applications, digital information is equally facing constant security threats and spoofing. Hence, the primary obligation of any digital application is to safeguard the information and ensure that it is not stolen or tampered with in any case (Zolotar et al. 2021; Korać et al. 2021).

Cryptography has taken giant steps to deal with information security problems over the digital world and is also striving constantly to overcome the pitfalls of the existing system exposed to a few types of threats (Wang 2021). The strength of any cryptographic algorithm depends on its effectiveness in encoding the information that becomes impossible for the hackers or the hacking tools to decode. With the consideration of the fact mentioned earlier, advancements in modern cryptography have even implemented various intelligent algorithms based on artificial intelligence to enhance its strength (El-Zoghabi et al. 2013). However, with the equal advent of intelligent spoofing and hacking tools, it was possible to suppress the algorithm's ability to decode the information (Maghrebi et al. 2016; Yu & Chen 2018). Deep research has discovered that cryptographic algorithms can attain a better strength with the application of random keys and thus rely heavily on the strength of the random keys utilized. Studies have also proved that the utilization of Pseudo-Random Number Generator (PRNG) is the most effective method to generate a series of truly random sequences (Hu et al. 2020; Blackledge et al. 2015). Despite the utilization of PRNGs, intelligent tools could crack the keys generated due to the loophole in the generation method itself. Hence, designing a smart random number generator that relies on intelligent tools to generate random keys that are practically impossible to break is crucial.

Advancements in biometric technology opened a broader scope for incorporating fingerprints to generate a key sequence that is further utilized for encoding the information to preserve confidentiality (Hao et al. 2006; Dutta et al. 2008; Xi & Hu 2010). Many cryptographic applications came up with the incorporation of fingerprint technology to encode information by utilizing biometric features to generate the key sequence. Nevertheless, there were a couple of fallacies identified in the system. One of them is that it depended on installing additional hardware to collect fingerprints, making it expensive. Another issue is that the existing research exposed that the fingerprint device itself is not very secure as the features can be easily spoofed (Karimovich & Turakulovich 2016). To overcome all the issues of the previously implemented systems, we recommend an intelligent key generation technique based on the artificial intelligence approach. The proposed technique generates a random key sequence by utilizing the seed values obtained from the biometric features of human faces called facial landmarks. We have extracted the facial landmarks from the intended image by training multiple deep neural networks with gigantic data of facial images representing variable races and geographical domains. We have also designed a novel Linear Feedback Shift Register (LFSR) to strengthen the random key sequence further (Sannidhan et al. 2020). The designed LFSR also enhances the length and randomness property of the key series.

14.1.1 SYSTEM CONTRIBUTIONS

Following are the notable contributions of the proposed research work to achieve our objectives

1. Implementation of bounding box using Multi-Task Cascaded Convolution Neural Network (MTCNN) to identify the facial structure of an individual.

2. A Deep Neural Network to extract high-level facial landmarks known as facial embeddings.
3. Design of novel LFSR to generate random subsequences using facial embeddings as seed values.
4. Encryption and decryption of image data using the key sequence generated from random subsequences.
5. Utilization of standard statistical tests to evaluate the randomness quality of key sequence generated.

14.1.2 Chapter Organization

The remainder of this chapter is organized into five unique sections. The upcoming section, i.e., Section 14.2, deliberates the background and literature adopted to carry out the work. Further, Section 14.3 discusses the proposed methodology covering all the aspects of different working modules involved in the system. Section 14.4 debates the necessary implementation procedures concerning incorporating working modules. The Penultimate section, i.e., Section 14.5, describes and validates the results achieved. Finally, Section 14.6 presents the conclusion and future scope of the research work.

14.2 REVIEW OF LITERATURE

We have conducted a comprehensive review of various research articles to gather information pertaining to our research objectives. In that direction, we have referred few notable research ailments related to information security systems using random key generators and different approaches to design LFSRs. We have also equally focused on the recent trends in Deep Learning systems supporting random key generation to strengthen cryptographic systems further. Parallelly, we have also referred to some web repositories to gather valuable information corresponding to the implementation of the work. The following paragraphs present the outcomes of useful research articles that support the implementation of the proposed system.

Zhao et al. (2020) proposed an intelligent cloud security system based on Deep Learning techniques. In their research work, they have encrypted the facial images using key points of the facial image. To derive the key points, researchers have incorporated MTCNN. Later, the derived points are fed into standard encryption techniques to encode the image and derive the password in the form of random key generation. Experimental findings proved that the technique is highly efficient in face detection. Still, the technique does not generate a random key of enough strength when it comes to security concern, as it uses only five key points to derive a random key. In a research article, Kalsi et al. (2017) proposed a method of key generation technique via DNA Deep Learning cryptography powered by the Genetic Algorithm (GA) approach. The authors have successfully generated a total random key by using GA. The generated key is then coupled with a DNA algorithm via trained convolution neural networks to encode the information by concealing the data. The generated key passed the required tests to prove the ability of its randomness. However, the seed values used to generate random keys were not that powerful and appeared to be a breakable sequence.

Research published by Quinga-Socasi et al. (2020) introduced the DL approach of cryptography system using the symmetric key technique. In their research exertion, the authors have proposed the method of deriving a symmetric key from the password entered by the user. To generate the required key from the entered password, they have incorporated an autoencoder neural network system. A generated symmetric key is then utilized for the purpose of encryption. Experimental observations derived from the proposed system successfully uphold the encryption and decryption process. Even though the researchers have introduced a modern Deep Learning approach to replace the traditional symmetric key cryptosystems, the method fails to effectively encrypt the larger size data. Panchal et al. (2017) discussed the utilization of fingerprint features to implement biometric-based cryptography. In their article, the researchers have recommended the method of key generation via features extracted from fingerprints. To facilitate the implementation of the proposed methodology, the authors have incorporated fingerprint sensors to collect the features, and features are then encoded into an inimitable code using convolution neural networks. Investigations performed in the article reveal the key generation's efficient and novel technique. Nevertheless, the system failed to prove the randomness nature of the generated key, and also, the biometric sensors themselves are highly prone to hacking.

In their article, Abed et al. (2018) implemented the concept of securing cloud platforms via biometric systems. In their research work, they have proposed a novel system of generating a 256-bit key length on the fly via the classification of trained biometric samples. To train the biometric samples, researchers have considered fingerprint, face, and keystrokes as the sample. The implemented technique attained the primary objective of generating the key sequence of the intended length. On investigating through the experimental observations, it is evident from the implemented system that the usage of transparent modalities itself is a backlog as they are highly vulnerable to easy attacks due to the exposure. In the article published by Cotrina et al. (2021), the authors have presented a new approach for generating random numbers using the Gaussian distribution approach. In the article, they have chosen LFSR as an ideal generator to generate a sequence of random numbers. As part of their experimental investigation, researchers have even compared LFSR with other standard generators, and they have evidently proved that LFSRs performed ably better than any other generators.

Dhanda et al. (2020) proposed a lightweight cryptosystem to uphold the Internet of Things security. Considering the computational capability of the IoT devices, researchers have concentrated on the implication of a lightweight cryptographic system that can generate a key with reasonable length to satisfy the security ailments. Hence, the authors have carried out an exhaustive analysis of different lightweight cryptographic systems to choose the better one that suits the computational capacity of any IoT device. On performing the analysis, as an outcome, they have identified that the LFSR technique outperforms any other generators in terms of key generation and the involvement of computational overhead. In one of the research papers, Sannidhan et al. (2020) proposed a novel key generation methodology for the stream cipher system using the facial features of a human image. In the proposed system, the authors have used the Speed Up Robust Feature extraction technique to extract the features. The extracted features are later fed as seed values into a novel LFSR to generate a pseudo-random number sequence. Experimental investigations evidently passed standard statistical tests for randomness of generated keys upholding the success of LFSR to generate key sequences from facial features.

14.3 PROPOSED SYSTEM

To overcome the fallacies as well as uphold the literature, we have proposed our system, as shown in Figure 14.1.

As seen in Figure 14.1, our proposed system is broadly considered into two central units: (a) Key generation unit and (b) Encryption/Decryption process. The detailed version of our units is elucidated in subsequent sections.

14.3.1 Key Generation Unit

The key generation unit of our system is the central head that generates key subsequences from a standard image. The familiar image must be of a human face or a set of faces. The system functions as an auto-encoding transformer that converts the image domain into a textual domain. We accomplish this feat with the help of the following processes.

14.3.1.1 Neural Network

Our system is modulated to extract fixed-way points on the utterly unique face. In other terms, the neural network acts as a biometric extractor that accepts a person's image tags to generate

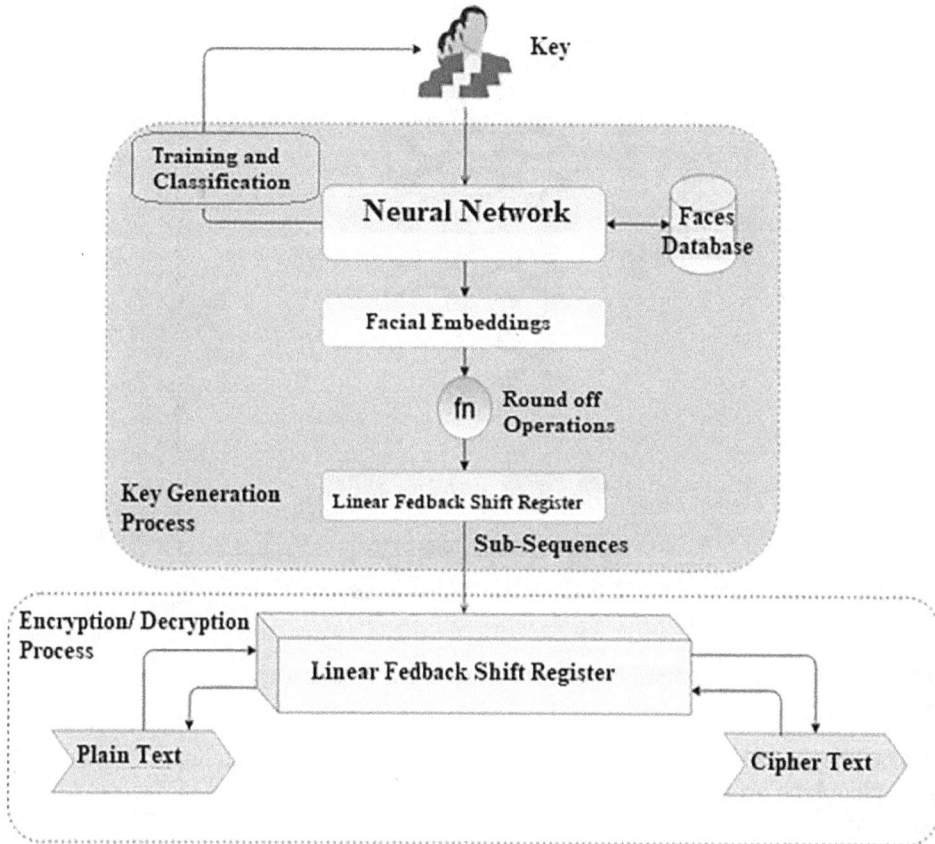

FIGURE 14.1 Block diagram depicting the process involved in our proposed system.

facial waypoints called landmarks. These so-called landmarks must be unique to a specific person. Hence, they are rigorously trained with a vast number of faces from different races and gender. We construct our neural network using a concept of one-shot learning that eases the training speed without compromising the accuracy of the network. Our neural unit consists of two neural networks working conjointly to deduce facial embeddings. The initial network is called the MTCNNs, which detects human faces and points out facial landmarks. The latter utilizes these landmarks to engender a fixed set of facial embeddings unique to that person.

14.3.1.2 Multi-Task Cascaded Convolutional Neural Networks

Abbreviated as MTCNNs, this network is a one-stop solution explicitly designed to identify faces from still images or video clip recordings. The background of this network works on the architecture of Convolutional Neural Networks (CNNs), which is a feature extractor. It is a specialty network because it employs three levels of CNN cascaded together in a pyramid-like fashion, gaining the output in terms of facial landmarks. The three CNNs are termed as P-Net, R-Net, and O-net, respectively.

14.3.1.3 Facenet

The Facenet is a specially designed network to recognize human faces by categorizing them based on a set of fixed feature points similar to a hash sequence generated by any hashing algorithm. The network is a domain transformer that accepts an image input on one end and categorizes faces based on previous training experiences. The network structure is associated in such a way that it compares

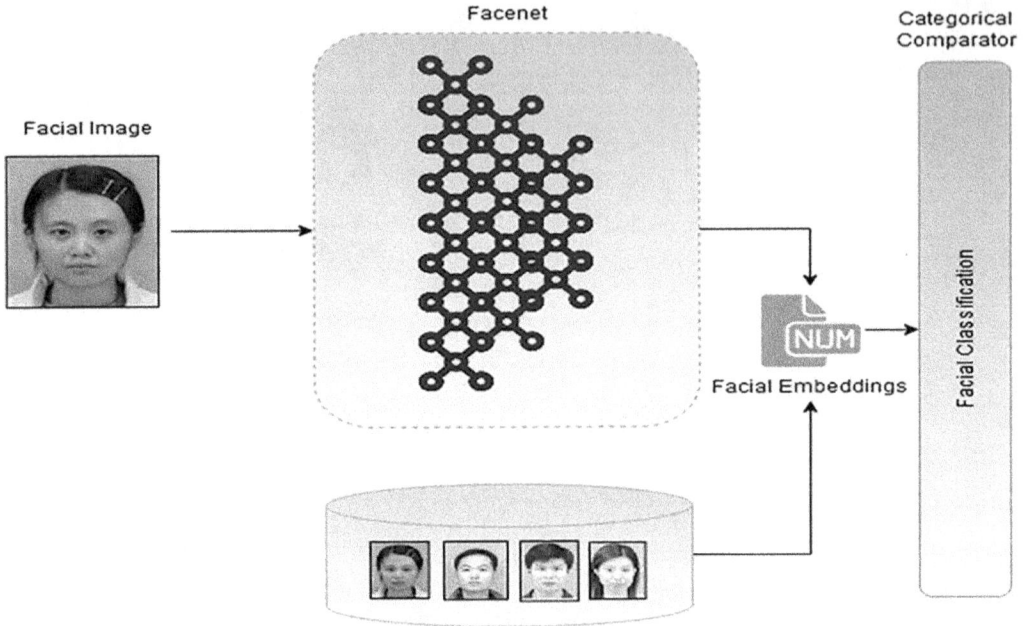

FIGURE 14.2   Block diagram depicting the architecture of our Facenet neural network.

facial embeddings in a straightforward pattern, making the network execution performance sleeker and impending high accuracy. Figure 14.2 illustrates the overall structure of Facenet (Sannidhan et al. 2019; Martis et al. 2020; Pallavi et al. 2021).

14.3.1.4 Round Off Operations

The facial embeddings generated by the Facenet network are 512 in number. Having this entire set of numbers makes the computation too complicated and intensive. To avoid this, we round off the embeddings to 64. Another reason for performing this step is to generate a limited number of subsequences at the computation limit and confuse the attacker. The embeddings obtained from our Facenet include a normalized entry of numbers ranging from 0 to 1. Our proposed LFSR cannot accept floating-point numbers. As a turnaround, we denormalize the numbers to suit our preferences by multiplying all 512 embeddings by 256. Choose 256 because every image in a single channel is of 8 bits leading to 256 grayscale values. The process of round-off is explained in detail in Equations 14.1–14.3.

$$\langle \mathrm{em}_1|...|\mathrm{em}_{512} \rangle = 256 \times \begin{bmatrix} \mathrm{em}_1 \\ \vdots \\ \mathrm{em}_{512} \end{bmatrix} \tag{14.1}$$

$$\langle e_1|...|e_{64} \rangle = T \langle \mathrm{em}_1|...|\mathrm{em}_{512} \rangle \tag{14.2}$$

$$e_1 = \langle \mathrm{em}_1|...|\mathrm{em}_8 \rangle \oplus_{256}, \ e_2 = \langle \mathrm{em}_9|...|\mathrm{em}_{16} \rangle \oplus_{256},... \tag{14.3}$$

From Equations 14.1–14.3, $\langle e_1|...|e_{64} \rangle$ forms the embedding values ranging from 1 to 64, $\langle \mathrm{em}_1|...|\mathrm{em}_{512} \rangle$ are the 512 facial embeddings generated from facenet, and $e_1 = \langle \mathrm{em}_1|...|\mathrm{em}_8 \rangle \oplus_{256}$ is

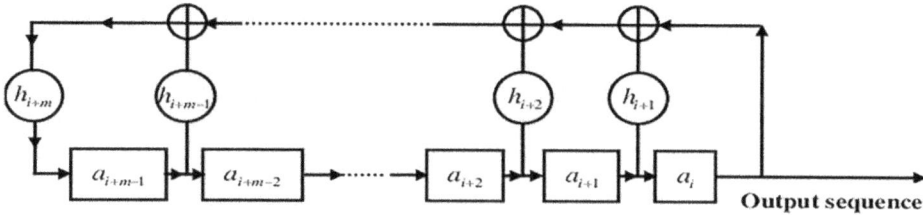

FIGURE 14.3 Figure presenting the internal running of the LFSR.

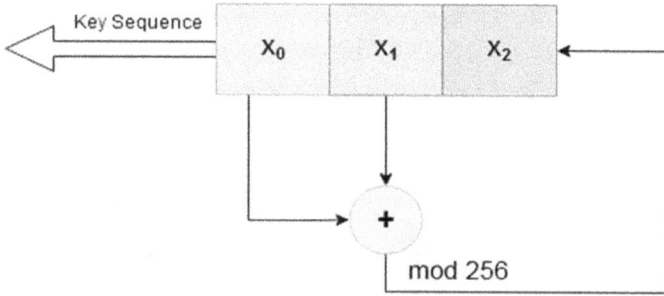

FIGURE 14.4 Figure displaying the implementation of the mathematical equation used in the projected model.

the transformation function T applied to group the eight embeddings together. Since there are eight groupings formed from the 512 embeddings, 64 transformed embeddings are created, as depicted in Equation 14.14. In summary, the round-off operation takes 512 embeddings and converts them to 64 embeddings.

14.3.2 DESIGN OF LFSR

Here, we deliberate the proposal of LFSR. Overall, an LFSR incorporates multiple register entities added in a strict sequence called LFSR stages. Each phase shows the movement station of the catalog at a fixed point in space-time. The standing will be reorganized based on the current positional value added to a predefined confusion operation 'F.' The following sequence series of numbers will predict the functional sequence conforming to the F's predefined feedback function. As an example, we postulate that the attained segmental line of the register to be 'm' and the overall sum of magnitudes to be 'n.' Considering them, the uppermost series of the arrangement of the key is characterized as m^{n-1}. As an example, the wide-ranging premeditated LFSR is represented in Figure 14.3 (Sudeepa et al. 2017, 2020).

A general working design of the LFSR is portrayed in Figure 14.4. It describes the method of a modest LFSR conforming to the actual requirements of our aforementioned projected system. Here, the LFSR consists of three separate stages added in a sequence denoted by Y_0, Y_1 and Y_2. The feedback function 'F' is denoted mathematically by Equation 14.4.

$$F_i \equiv \left(F_{i-1} \oplus_n F_{i-2}\right) \tag{14.4}$$

In order to implement F_i suitable to our constructs, the most convenient operation drills down to Equation 14.5.

$$X_n = \left(X_{n-1} \oplus_{256} X_{n-2}\right) \tag{14.5}$$

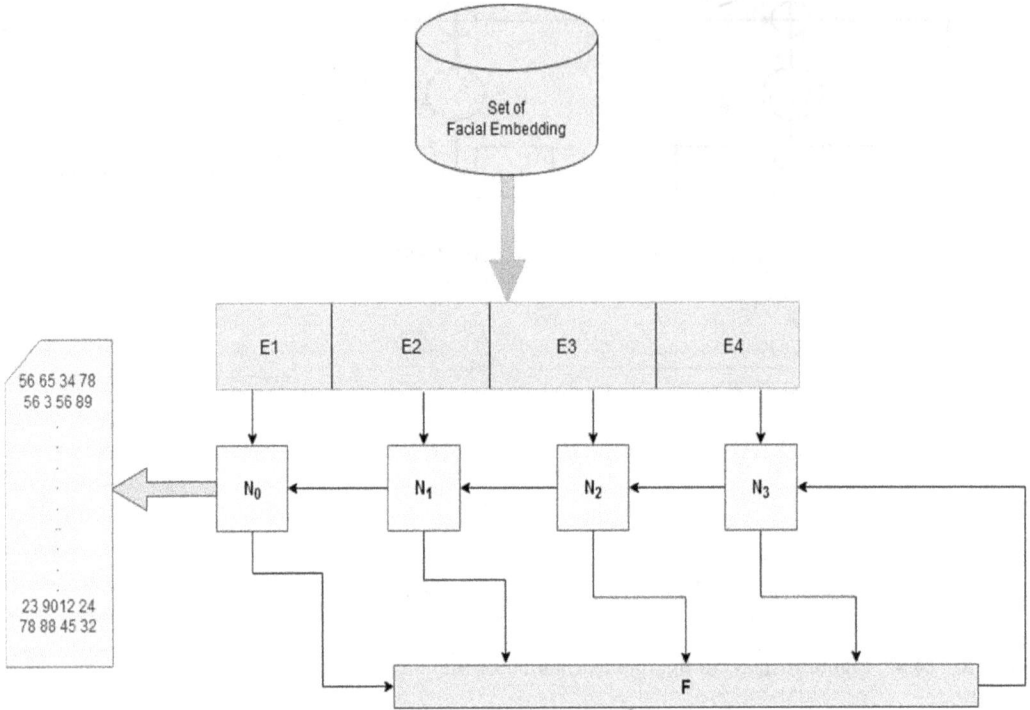

FIGURE 14.5 Design of LFSR to extrude pseudo-random sequences from facial features.

14.3.2.1 Pseudo-Random Number Generator

Figure 14.5 depicts the process of generating a pseudo-random number sequence from extracted facial embeddings. As presented in Figure 14.5, an LFSR operates on a sequence of four at a time facial features, and the individual module is warehoused in a discrete register N_0 to N_3 sequentially. Each sequence of four structures represents the kernel value for a long series of the pseudo-random key cohort. The working concept of the model begins by adding the contents of the register Y_0 and Y_2.

The working sum of N_0 and N is treated to a modulo operation of 256 to add confusion. The new result obtained from the resulting sum is termed as N_n. This N_n is then pushed to N_3 performing a shifting operation in a left-oriented fashion. We repeat this process indefinably until there is a repetition of subsequences. This entire stream of non-repeating subsequences clubbed together is termed the long pseudo-random key. Equations 14.6–14.10 summarize the mathematical layout behind the operation (Pushpalatha et al. 2021).

$$N_n = \left(N_0 \oplus_{256} N_2 \right) \tag{14.6}$$

$$N_3 = N_n \tag{14.7}$$

$$N_2 = N_3 \tag{14.8}$$

$$N_1 = N_2 \tag{14.9}$$

$$N_0 = N_1 \tag{14.10}$$

14.3.3 ENCRYPTION AND DECRYPTION PROCESS

The primary task involves generating the key to carry out the encryption/decryption process. As already mentioned, our project's core idea consists of creating the key using the LFSR discussed in the previous section. The critical sequence becomes more substantial if the arrangement is random. The fact of nature is that there is no utterly accurate system to create total random numbers. Our job here is to mimic at least the true randomness of a pseudo-random number. We aim to achieve this role by using the LFSR.

14.3.3.1 Encryption Unit

The encryption unit takes the keystream file, as shown in Figure 14.6, and forms a modulus 256 to the input to create a cipher. The information must be stream data extracted from an image. We take a bitmap image of three different channels added in a sequence as shown in Equations 14.11 and 14.12, respectively.

$$\text{Stream}_{(i,j)} = \left\{ \begin{array}{c} R_{(i,j)} \\ G_{(i,j)} \\ B_{(i,j)} \end{array} \right\} \text{ for } \left\{ \left(0 < i < I_{\text{width}}\right), \left(0 < j < I_{\text{height}}\right) \right\} \tag{14.11}$$

$$\text{KeyStream}_i = \left[\text{Stream}_{(i,j)}^R \middle| \text{Stream}_{(i,j)}^G \middle| \text{Stream}_{(i,j)}^B \right] \tag{14.12}$$

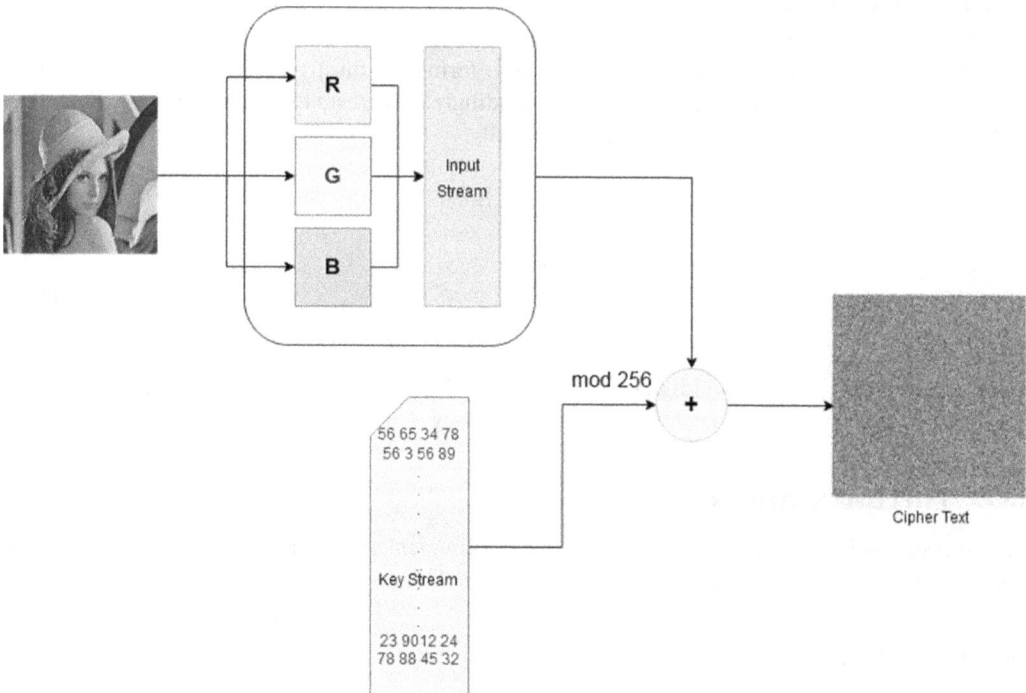

FIGURE 14.6 General process of encryption of an image.

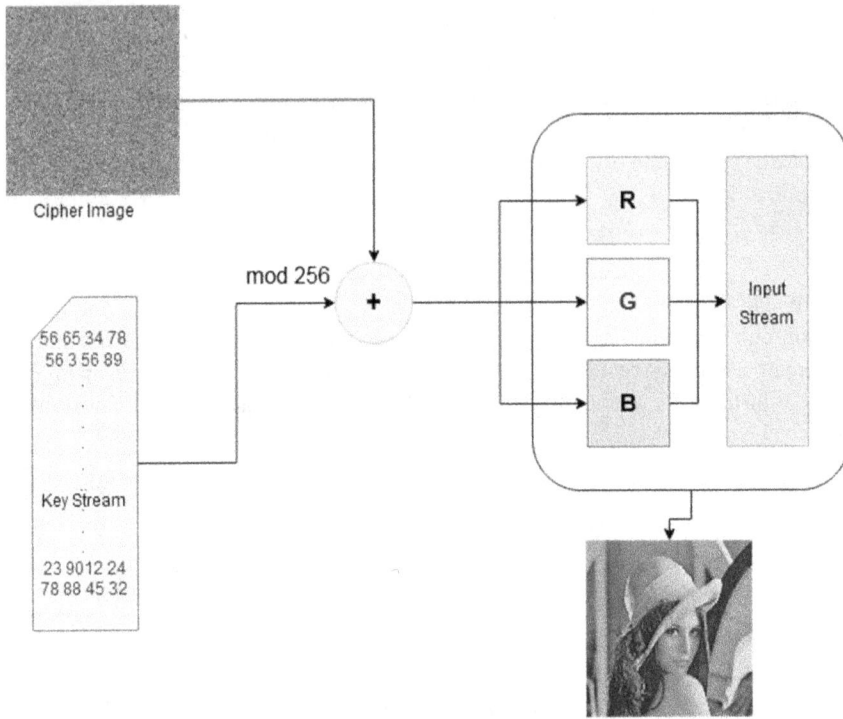

FIGURE 14.7   General process of decryption of an image.

14.3.3.2   Decryption Unit

The decryption unit takes the keystream file from the file, as shown in Figure 14.7, and forms a modulus 256 to the input to create a cipher. The information must be a ciphertext, as shown in Figure 14.7. At last, we split the channels correspondingly to recreate the original image. Equations 14.13 and 14.14 explain the operations mathematically.

$$C_i = (P_i + K_i)\mod256 \tag{14.13}$$

$$P_i = (C_i - K_i)\mod256 \tag{14.14}$$

The process of decryption is highlighted in Figure 14.7. Notice the cipher image is complete gibberish, showing that the message is entirely distributed along with the image.

14.4   IMPLEMENTATION

This section distills the detailed information used to implement our proposed system. It contains traces of algorithms, pseudocodes, and architectures of the entire process.

14.4.1   MTCNN

MTCNN is a framework industrialized to explain mutual face detection and face alignment (Sannidhan et al. 2021). The method entails three phases of convolutional networks that can

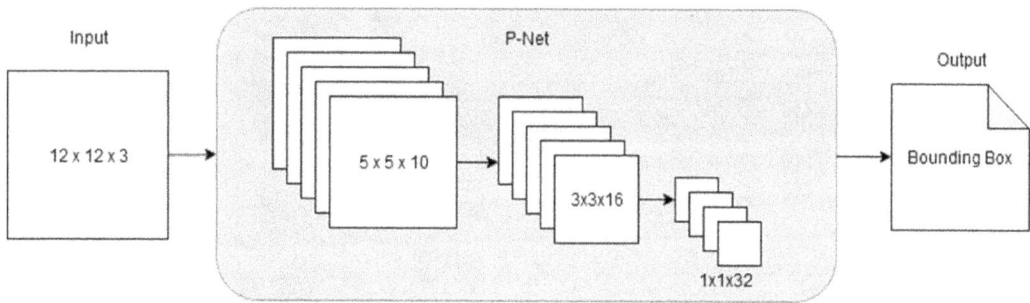

FIGURE 14.8 The architecture of P-net.

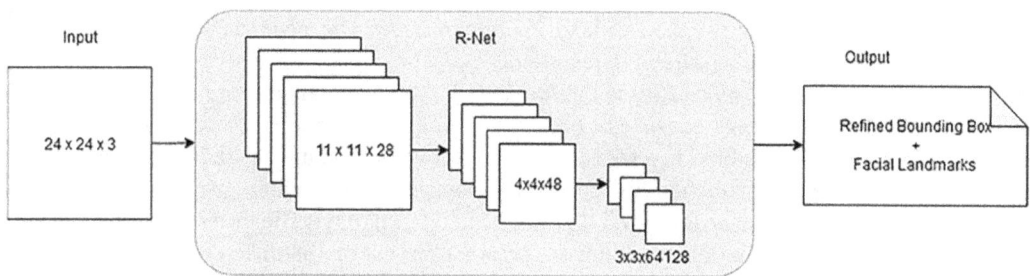

FIGURE 14.9 Architecture of R-net.

distinguish human faces and basic landmark position such as nose, mouth, and eyes. The three sages are P-net, R-net, and O-net. The detailed orientation of them is shown under the following subsections (Martis & Balasubramani 2019).

14.4.1.1 P-Net

P-net is a full convolutional network that does not employ any dense part for classification. The term P-net means Proposed-net, meaning it isolates regions of interest and their bounding boxes. The P-net architecture consists of three convolutional layers of 3×3 connected one to another to break down the entire image into tiny bounding boxes. Figure 14.8 shows the architecture of the P-net.

14.4.1.2 R-Net

The job of the R-net is to refine the rough edges construed by the P-Net. This network is a CNN-based network due to the involvement of a dense layer at the end. It performs calibration of the bounding boxes and also helps in merging overlapping candidates: the R-net outputs a refined bounding box and a simple one-dimensional ten-element facial landmark vector. Figure 14.9 portrays the architecture of R-Net.

14.4.1.3 O-Net

Termed as the Output layer, the O-net is the final layer of the MTCNN. The architecture of the O-net is very much similar to that of the R-net. The only difference is that it varies with the dimension of the inputs. The major contribution of this network is to output precise landmark locations like eyes, mouth, and nose. The more specific facial landmarks are right and left eyes, a single nose, and right and left mouth corner. The architecture of the O-net is shown in Figure 14.10.

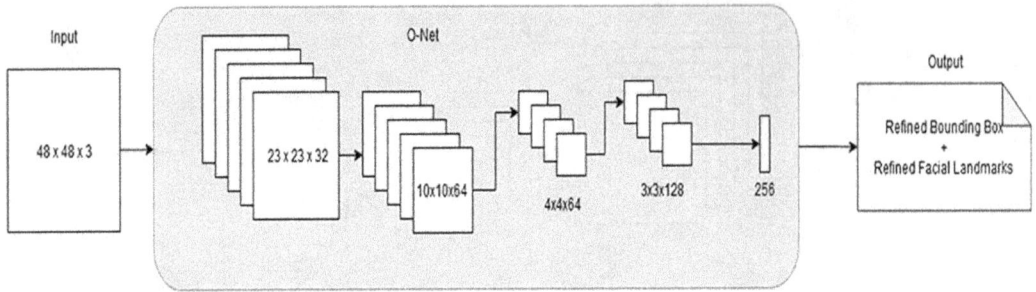

FIGURE 14.10 The architecture of O-net.

14.4.2 FACENET

Our Facenet architecture is tweaked to take 2D human faces and generate feature embeddings across 512 dimensions. The highlighting feature of Facenet is due to the availability of pre-trained models in it. We train human faces using the framework provided and generate 512 embeddings of every person. Our training set consisted of 188 facial images taken from the CUHK dataset. All generated embeddings are stored in a faces database portrayed in Figure 14.10. Facenet uses a standard triplet loss generation function that compares an anchor image with a genuinely positive and false negative called anchor positive and anchors negative. The image features generated in huge dimensions are compared among all photos to create a clear cut of 188 different embeddings from the CUHK dataset. Similar images will always give similar embeddings since the network is highly attuned to an extrapolation based only on the trained set. Equations 14.15 and 14.16 highlight the gist of Facenet triplet loss working (He et al. 2020).

$$E_c = \sum_{\text{facial database}} T_L + L_2 \tag{14.15}$$

$$T_L = \left[f(A) - f(P)\right]^2 + \alpha - \left[f(A) - f(N)\right]^2 \tag{14.16}$$

Here, E_c determines the embeddings generated from an image, T_L is the total loss function determined from the absolute value of anchor, positive, and negative images denoted by $f(A)$, $f(P)$, and $f(N)$, respectively. α forms to be the tuning factor determining the learning rate of the neural network. The value of α ranges between 0 and 1. L_2 is the regularization function kept at a constant of 0.1. Table 14.1 highlights the architecture of our Facenet network.

TABLE 14.1

Table Describing the Architecture of Our Facenet

Neural Layer	General Description
Preprocessing layer (input)	Three convolutional layers normalized and max pool on a 3×3 level.
Depth concat layer 1	Twelve convolutional layers are divided among two depth concat and maxpool layers.
Max pooling layer	One maxpooling layer of 3×3.
Depth concat layer 2	30 convolutional layers divided among five depth concat and maxpooling layers.
Max pooling layer	One maxpooling layer of 3×3.
Depth concat layer 3	Twelve convolutional layers are divided among two Depth concat and maxpool layers.
Average pooling layer	One averaging layer of 7×7.
Fully connected layer (final)	Layer that connects all neurons together.

ALGORITHM 14.1
Algorithm Illustrating the Entire Process of Encryption

Input: (Image from camera, Image)
Output: (Encryption Image)

1   **for each** image in datastore f, **do**

2     $I\left(x_{\text{org}}, y_{\text{org}}\right) \leftarrow$ Capture sample from a camera in Pixel Format.

3     $\left(W_{\text{org}}, H_{\text{org}}\right) \leftarrow$ Length and Width of Image Format

    Boundingbox = MTCNN$\left[I\left(x_{\text{org}}, y_{\text{org}}\right)\right]$

4     $\text{Image}_{\text{face}} = \text{Crop}\left[I\left(x_{\text{org}}, y_{\text{org}}\right), \text{Boundingbox}\right]$

5     $\langle \text{em}_1 | \ldots | \text{em}_{512}\rangle = \text{Facenet}\left[\text{Image}_{\text{face}}\right]$

6     $\langle \text{em}_1 | \ldots | \text{em}_{512}\rangle = 256 \times \begin{bmatrix} \text{em}_1 \\ \vdots \\ \text{em}_{512} \end{bmatrix}$

7     $\langle e_1 | \ldots | e_{64}\rangle = T\langle \text{em}_1 | \ldots | \text{em}_{512}\rangle$

    for each embedding e_i, **do**

8     $e_i = \langle \text{em}_1 | \ldots | \text{em}_8\rangle \oplus_{256}$ **and so on**

9   **end for**

10   **end for**

11   **for each** embedding in groups of four, **do**

12     $N_n = \left(N_0 \oplus_{256} N_2\right)$

13     $N_3 = N_n$

14     $N_2 = N_3$

15     $N_1 = N_2$

16     $N_0 = N_1$

17   **store and continue until** there is no match in subsequences

18   **end for**

19   **cipher=** Encrypt$\left(\text{image}, \text{suibsequence}\right)$

14.4.3   LFSR

Our LFSR is divided into four single sequence registers, which uses a block of four embeddings obtained from Facenet. Taking the block mentioned above into account, we generate a total of 128 LFSR sequences. Algorithm 14.1 shows the pseudocode used in our system.

14.5   RESULTS

We have implemented our neural network on an NVIDIA Tesla P100 1000 core 16GB RAM having a total computational capability of 13.3 TFLOPS (Tera Floating Point Operations Per Second). We have taken embeddings from an Indian faces dataset containing 25,000 labeled faces. We have also stored our neural network weights on Keras platform-based h5 file system because this training sequence can be used to extract facial features. Table 14.2 shows a sample of the input and the output obtained from our designed network that generates embedding sequences.

TABLE 14.2
Sample Output from Our Neural Network

Network	Sample Output
MTCNN architecture	
Facenet architecture	
Output	[554, 611, 669, 730, 792, 855, 916, 973, 1,009, 1,028,…]

14.5.1 VISUAL PRESENTATION OF THE ENCRYPTION AND DECRYPTION SEQUENCE

In this section, we show the encryption image and the encrypted image. We also present the decrypted version of the image for comparison. We have performed this test on a standard Lenna image of size 256×256. Table 14.3 summarizes the results.

Our LFSR system took a total of four blocks to generate subsequences. This step was repeated until there was no repetition of subsequences. Table 14.4 shows a sample output from the sequences obtained.

From Table 14.4, it is evident that there is a high chance of getting 771 subsequences, thereby creating a massive difference in the encryption scenario. This sequence of subsequences also generates more significant confusion among cryptanalysts, thereby creating an infeasible hacking strategy. Figure 14.11 shows a graph generated from the sequences.

TABLE 14.3
Sample Output from our Encryption and Decryption Approach

Step	Sample Output
Input	
Encryption	
Decryption	

TABLE 14.4

Subsequences Generated from Four Blocks of Sequences

Seed Values	Number of Subsequences Obtained
199, 61, 100, 23	771
61, 181, 23, 205	771
0, 0, 126, 108	385
0, 0, 0, 108 192	192
133, 209, 118, 213	771
193, 148, 142, 72	771
60, 135, 251, 95	771
180, 252, 109, 216	771
32, 216, 140, 30	771
111, 97, 82, 72	771
97, 221, 72, 215	771
221, 251, 215, 15	771
251, 7, 15, 20	771
7, 12, 20, 4	771
12, 47, 4, 81	771
47, 43, 81, 94	771
43, 105, 94, 176	771
105, 141, 176, 231	771
141, 172, 231, 17	771
172, 210, 17, 74	771
210, 242, 74, 127	771
242, 30, 127, 196	771
30, 53, 196, 221	771
53, 65, 221, 220	771

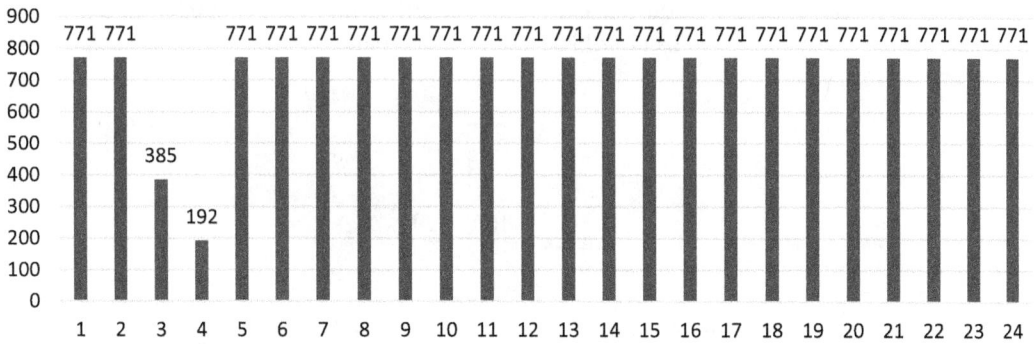

FIGURE 14.11 Graph showing the number of subsequences generated from the Facenet facial features.

14.5.2 Tests on Subsequence Generated

To prove our system's competence and validate the strength of the key generated, we evaluate it with specific test parameters such as Chi-square, runs up-down, and encryption–decryption performance. The details of these tests are particularized below.

TABLE 14.5

Chi-Square Test Based on Various Intervals

Sample Count	Interval Count	Chi-Square Value Calculated	Expected Standard Value
100	100	91.0	126.2
200	100	120.0	125.2
300	50	49.23	57.8
500	10	18.25	20.3

14.5.2.1　Chi-Square Test

We use Chi-Square as a test parameter to experiment with the arbitrariness influence of the engendered key. The scientific agreement is presented in Equation 14.17 which designates the practice of accomplishing the Chi-Square analysis on a random sample of size n with a set of k class intervals.

$$\text{ChiSquare}_0^2 = \sum_{n=1}^{k} \frac{\left(\text{Observed}_i - \text{Expected}_i\right)^2}{\text{Expected}_i} \tag{14.17}$$

From Equation 14.17, parameters Observed_i is an experimental occurrence in the class interval and Expected_i is probable incidence under a similar interval. We have calculated the Chi-Square value using 500 samples with an interval length of hundred each. Table 14.5 summarizes the randomness result obtained in terms of Chi-square formulation.

Comparing the last two columns presented in Table 14.5, it is evident that our experimental results have surpassed the expected results in terms of Chi-Square testing. It also verifies that we have accomplished the randomness criterion.

14.5.2.2　Run Up-Down Test

We perform this test to associate the independence criterion of random numbers. This test envisages the running dependence between the chosen set of numbers. Higher test values indicate no dependency between the engendered sequences and strengthen the random number's ability. The formulae shown in Equations 14.18 and 14.19 describe the computation variance and the mean value of the run test, respectively

$$\text{mean}_a = \frac{2N-1}{3} \tag{14.18}$$

$$\text{Variance} = \frac{16N-29}{90} \tag{14.19}$$

Here, N shows the number of samples used to determine the values. If the examples are extensive, the value used to present the standard curve mean and variance is shown in Equations 14.20 and 14.21.

$$\text{mean} = \frac{a - N_a}{\sigma_a} \tag{14.20}$$

$$\text{Stdev} = \frac{a - \left(\dfrac{2N-1}{S}\right)}{\sqrt{\dfrac{(16N-29)}{90}}} \tag{14.21}$$

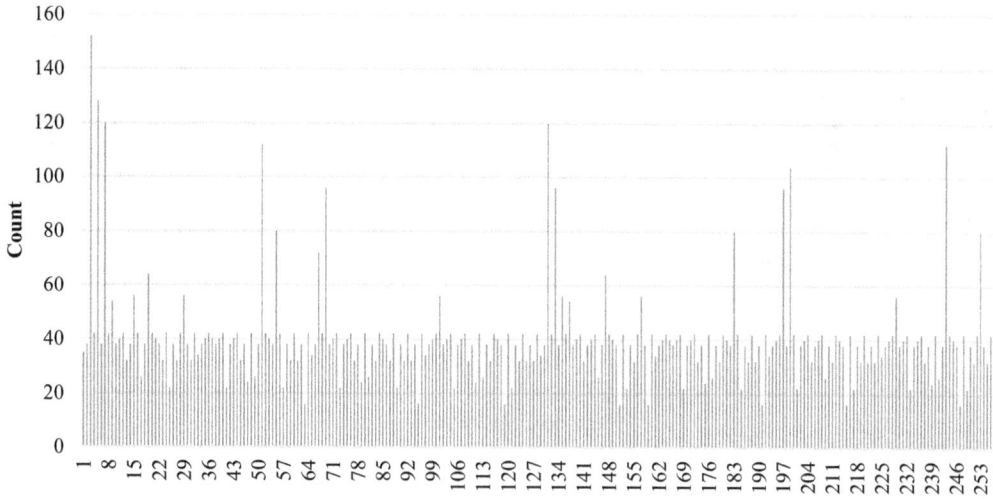

FIGURE 14.12   Histogram analysis of the run-up-down test.

For our test, we took 10,370 samples that raised a total running length value of 6,815, having a mean of 6,920 and a variance of 1,842.133. We also compared this with our average curve statistics that fetched a mean of −1.9936 assuming a 1% failure rate based on a two-tailed hypothesis. These values mentioned above-engendered project the true randomness of the series. Figure 14.12 demonstrates this effect concerning histogram analysis.

14.5.2.3   Performance Analysis of Encryption and Decryption

Figure 14.13 presents the histogram levels of the encryption and decryption images along the pixels ignoring the overheads. Figures 14.13a and b show every pixel's encrypted and decrypted values. There is a massive difference between the two figures, which directly approves the strength of encryption and decryption.

14.6   CONCLUSION

Our proposed system accomplished the defined objectives of the research. The main intention of the work concentrated on generating random keys of enormous length through implementing a novel random key generator. We successfully developed a novel LFSR pseudo-random key generator to generate subsequences from facial features. We have also incorporated a Neural Network that extracts 512 features called facial embeddings to extract many unique feature points from the facial images. By this, we have ensured that the LFSR uses seed values that are also random to further generate huge subsequences for a group of four seed values. Statistical tests like Chi-square and runs test were conducted to check the randomness factor of the generated subsequences. The outcomes of the tests proved that the subsequences satisfy the property of randomness. Further, we successfully generated a random key with an extended length of 50,763 keys that is comparatively 771 times larger than the original seed values. Finally, the stream cipher system is validated by utilizing generated keys via encryption and decryption operations.

14.6.1   LIMITATIONS AND FUTURE SCOPE OF THE SYSTEM

Even though the proposed system attained all the expected objectives, the following are the few identified limitations of our system

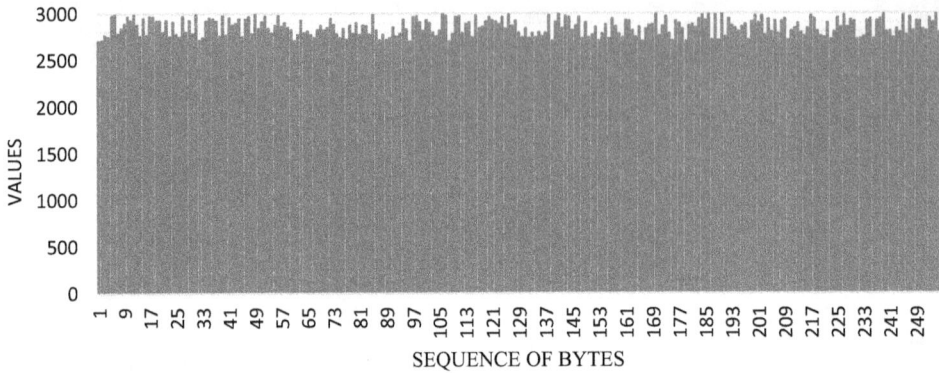

(a) Encryption Sequence

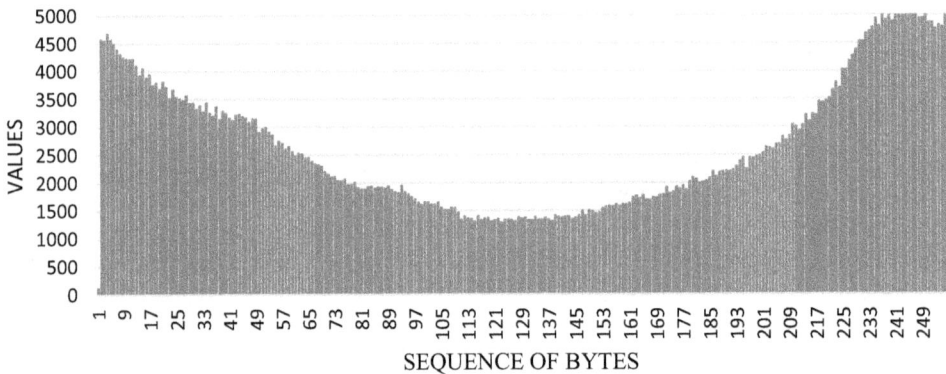

(b) Decryption Sequence of Bytes

FIGURE 14.13   Histogram analysis of the (a) encryption and (b) decryption sequence.

1. Incorporating neural networks required a high amount of computation and storage, thereby reducing the system's performance in terms of time and space.
2. There is no authentication scheme implemented to ensure data robustness.
3. The system does not support multiple key strategies for a stream cipher.

Concerning the system's limitations portrayed, we have considered the following implications as future scope: (1) a quantum/parallel computing approach to enhance computation performance, (2) a classification method to classify the facial images to deal with spatial complexity, and (3) a class-specific file to store the consistent facial embeddings corresponding to the face. To avoid data corruption and ensure data robustness, the proposed system can be improved by utilizing a Stegnographical system or hashing methodology. Finally, to overcome the pitfall of using multiple key cryptosystems, the system can be installed with centralized or distributed key sharing architecture.

REFERENCES

Abed, L., Clarke, N., Ghita, B., & Alruban, A. (2018). Securing Cloud Storage by Transparent Biometric Cryptography. Lecture Notes in Computer Science (Including Subseries Lecture Notes in Artificial Intelligence and Lecture Notes in Bioinformatics), 11359 LNCS, 97–108. https://doi.org/10.1007/978-3-030-12942-2_9.

Blackledge, J., Bezobrazov, S., & Tobin, P. (2015). Cryptography using artificial intelligence. *Proceedings of the International Joint Conference on Neural Networks*, 2015-September. https://doi.org/10.1109/IJCNN.2015.7280536.

Cotrina, G., Peinado, A., & Ortiz, A. (2021). Gaussian pseudorandom number generator using linear feedback shift registers in extended fields. *Mathematics*, 9(5), 556. https://doi.org/10.3390/MATH9050556.

Dhanda, S. S., Singh, B., & Jindal, P. (2020). Lightweight cryptography: A solution to secure IoT. *Wireless Personal Communications*, 112(3), 1947–1980. https://doi.org/10.1007/S11277–020–07134–3.

Dutta, S., Kar, A., Mahanti, N. C., & Chatterji, B. N. (2008). Network Security Using Biometric and Cryptography. Lecture Notes in Computer Science (Including Subseries Lecture Notes in Artificial Intelligence and Lecture Notes in Bioinformatics), 5259 LNCS, 38–44. https://doi.org/10.1007/978-3-540-88458-3_4.

El-Zoghabi, A., Yassin, A. H., & Hussien, H. H. (2013). Survey report on cryptography based on neural network. *International Journal of Emerging Technology and Advanced Engineering*, 3(12), 456–462.

Hao, F., Anderson, R., & Daugman, J. (2006). Combining crypto with biometrics effectively. *IEEE Transactions on Computers*, 55(9), 1081–1088. https://doi.org/10.1109/TC.2006.138.

He, M., Zhang, J., Shan, S., Kan, M., & Chen, X. (2020). Deformable face net for pose invariant face recognition. *Pattern Recognition*, 100, 107113. https://doi.org/10.1016/J.PATCOG.2019.107113.

Hu, Z., Gnatyuk, S., Okhrimenko, T., Tynymbayev, S., & Iavich, M. (2020). High-speed and secure PRNG for cryptographic applications. *International Journal of Computer Network and Information Security*, 12(3), 1–10. https://doi.org/10.5815/IJCNIS.2020.03.01.

Kalsi, S., Kaur, H., & Chang, V. (2017). DNA cryptography and deep learning using genetic algorithm with NW algorithm for key generation. *Journal of Medical Systems*, 42(1). https://doi.org/10.1007/S10916–017–0851–Z.

Korać, D., Damjanović, B., & Simić, D. (2021). Correction to: A model of digital identity for better information security in e-learning systems. *The Journal of Supercomputing*, 1–1. https://doi.org/10.1007/S11227–021–04006–W.

Karimovich, G. S., & Turakulovich, K. Z. (2016, November). Biometric cryptosystems: Open issues and challenges. In *2016 International Conference on Information Science and Communications Technologies (ICISCT)* (pp. 1–3). IEEE. Tashkent, Uzbekistan.

Maghrebi, H., Portigliatti, T., & Prouff, E. (2016). Breaking Cryptographic Implementations Using Deep Learning Techniques. Lecture Notes in Computer Science (Including Subseries Lecture Notes in Artificial Intelligence and Lecture Notes in Bioinformatics), 10076 LNCS, 3–26. https://doi.org/10.1007/978-3-319-49445-6_1.

Martis, J. E., & Balasubramani, R. (2019). Reckoning of emotions through recognition of posture features, 15(2), 230–254. https://doi.org/10.1080/19361610.2019.1645530.

Martis, J. E., Sudeepa, K. B., Sannidhan, M. S., & Bhandary, A. (2020). A rapid automated process for organizing bacterial cluster segments using deep neural networks. *Proceedings of the 3rd International Conference on Smart Systems and Inventive Technology*, ICSSIT 2020, 963–968. https://doi.org/10.1109/ICSSIT48917.2020.9214173.

Pallavi, S., Sannidhan, M. S., & Bhandary, A. (2021). Retrieval of facial sketches using linguistic descriptors: An approach based on hierarchical classification of facial attributes. *Advances in Intelligent Systems and Computing*, 1133, 1131–1149. https://doi.org/10.1007/978-981-15-3514-7_84.

Panchal, G., Samanta, D., & Barman, S. (2017). Biometric-based cryptography for digital content protection without any key storage. *Undefined*, 78(19), 26979–27000. https://doi.org/10.1007/S11042–017–4528–X.

Pushpalatha, V., Sudeepa, K. B., & Mahendra, H. N. (2021). Pseudo random number generation based on genetic algorithm application. *Advances in Intelligent Systems and Computing*, 1133, 793–808. https://doi.org/10.1007/978-981-15-3514-7_59.

Quinga-Socasi, F., Zhinin-Vera, L., & Chang, O. (2020). A deep learning approach for symmetric-key cryptography system. *Advances in Intelligent Systems and Computing*, 1288, 539–552. https://doi.org/10.1007/978-3-030-63128-4_41.

Sannidhan, M. S., Ananth Prabhu, G., Robbins, D. E., & Shasky, C. (2019). Evaluating the performance of face sketch generation using generative adversarial networks. *Pattern Recognition Letters*, 128, 452–458. https://doi.org/10.1016/J.PATREC.2019.10.010.

Sannidhan, M. S., Prabhu, G. A., Chaitra, K. M., & Mohanty, J. R. (2021). Performance enhancement of generative adversarial network for photograph–sketch identification. *Soft Computing*, 1–18. https://doi.org/10.1007/S00500–021–05700–W.

Sannidhan, M. S., Sudeepa, K. B., Martis, J. E., & Bhandary, A. (2020). A novel key generation approach based on facial image features for stream cipher system. *Proceedings of the 3rd International Conference on Smart Systems and Inventive Technology*, ICSSIT 2020, 956–962. https://doi.org/10.1109/ICSSIT48917.2020.9214095.

Sudeepa, K. B., Aithal, G., Rajinikanth, V., & Satapathy, S. C. (2020). Genetic algorithm based key sequence generation for cipher system. *Pattern Recognition Letters*, 133, 341–348. https://doi.org/10.1016/J. PATREC.2020.03.015

Sudeepa, K. B., Raju, K., Sannidhan, M. S., & Bhandary, A. (2017). Maximum period word oriented non-binary key sequence generation and its application in image encryption/decryption. *Proceedings of IEEE International Conference on Emerging Technological Trends in Computing, Communications and Electrical Engineering, ICETT 2016*. https://doi.org/10.1109/ICETT.2016.7873654.

Wang, S. P. (2021). *Computer Architecture and Organization*. https://doi.org/10.1007/978–981–16–5662–0.

Xi, K., & Hu, J. (2010). *Bio-Cryptography. Handbook of Information and Communication Security*, 129–157. https://doi.org/10.1007/978-3-642-04117-4_7.

Yu, W., & Chen, J. (2018). Deep learning-assisted and combined attack: A novel side-channel attack. *Electronics Letters*, 54(19), 1114–1116. https://doi.org/10.1049/EL.2018.5411/CITE/REFWORKS.

Zhao, X., Lin, S., Chen, X., Ou, C., & Liao, C. (2020). Application of face image detection based on deep learning in privacy security of intelligent cloud platform. *Multimedia Tools and Applications*, 79(23–24), 16707–16718. https://doi.org/10.1007/S11042–019–08014–0.

Zolotar, O. O., Zaitsev, M. M., Topolnitskyi, V. V., Bieliakov, K. I., & Koropatnik, I. M. (2021). Prospects and current status of defence information security in Ukraine. *Linguistics and Culture Review*, 5(S3), 513–524. https://doi.org/10.21744/LINGCURE.V5NS3.1545.

15 Quantum Computing and Its Real-World Applications

Pawan Mishra, Ravi Kamal Pandey, and Pooja
University of Allahabad

CONTENTS

15.1 INTRODUCTION

Quantum Computing is one of the most novel and less-explored area of research. It is based on the quantum processing theory of quantum mechanics that explains the behavior and real nature of energy and particles at the quantum level [1]. It was first ever introduced by the German Physicist Max Planck at the German Physical Society in the beginning of the 19th century. Quantum mechanics allows for a possibility of a superposition state between two stable states of a system. Before we go deeper into quantum computation tasks, let's try to understand what is quantum superposition with the simplest example [2]. Assume a coin flipping scenario with a fair coin. At the instance the coin is flipping in the air, we cannot determine whether it will land head or tail. Therefore, we can say that the coin is in superposition of both head and tail. Similarly, so quantum mechanics tell us there must be an infinite number of unknown superposition states. Although physically developing the quantum computer like our today's classical computer is still not possible, their simulation can be done by some open-source software techniques like cloud computing, combinations of Integrated Circuits (ICs), circuits, transistors, etc. The central focus of this chapter is to help beginners to understand the quantum computing phenomenon and find the best suitable use of this disruptive technology in cyber securities, handling cloud computing and optimizing the mathematical problems as well as it will be compared with respect to their previous state of art concepts.

15.2 QUANTUM COMPUTING

The fundamental turnaround is the basic understanding of the events happening at the atomic scale that started at the beginning of the 20th century. The difficulty in explaining

DOI: 10.1201/9781003267812-15

the distribution in black body radiation led Max Planck to adopt a new line of thought, the exchange of energy between atomic radiators and the surrounding is not continuous but rather discrete [3,4]. Energy was being carried in the form of small indivisible packets (quantum) with associated energy $h\nu$, where ν refers to the frequency of atomic radiator and h is Planck's constant. This assumption helped Planck to obtain the correct explanation for the spectral energy distribution of black body radiation. Almost five years later, Einstein [5] provided an explanation of the photoelectric effect, previously reported by Hertz [6] in 1887, by adopting Planck's quantum hypothesis. Lewis in 1926 [7,8] coined the word "photon" for these quanta of light particles. Erwin Schrödinger [9] found an explanation for discrete energy levels of atoms through his famous wave equation,

$$ih\frac{\partial}{\partial t}\psi(x,\,t) = \frac{-\hbar^2}{2m}\nabla^2\psi(x,\,t) + V(x)\psi(x,\,t) \tag{15.1}$$

Schrödinger's equation is a linear partial differential equation, which has an important implication that any linear combination of its solution is also an allowed solution. This is called the super-position principle in quantum mechanics. Following this, Werner Heisenberg [10] formulated the uncertainty principle, which states that canonically conjugate variables in quantum mechanics like position and momentum of a particle cannot be measured precisely and simultaneously. Paul Dirac unified quantum mechanics with special relativity to obtain relativistic wave function of electrons [11]. During this time, Jon Von Neumann developed the mathematical foundation of quantum mechanics as the theory of linear operators in complex Hilbert space, in the form we use it today [12]. The field of quantum information theory (QIT) was developed mainly in the last three decades of the 20th century by using the principles of quantum mechanics to the existing classical information theories.

The aim of QIT is to understand the quantum nature of information and to formulate, manipulate and process the quantum information to achieve important and useful ends using physical systems, which operate on quantum mechanical principles. The field came into existence with the work of Stephen Wiesner who put forward the concept of conjugate coding, which was a cryptographic tool to protect confidential data using the property of quantum superposition [13]. James Park then demonstrated the no-go theorem in 1970 (which was later derived by Wooters and Zurek in 1982) that expressed the fact that the measurement on a quantum system will destroy its superstition, and thus a general unknown quantum state cannot be cloned [14]. Later, Holevo showed that a computing machine that uses finite qubits can store more data compared to the same number of classical bits [15]. Paul Benioff gave the first quantum mechanical model of a computer, describing the Schrödinger equation possibility of Turing machine [16]. Feynman [8] realized that in order to simulate a quantum system one would require a machine that works on the quantum mechanical principles. In general, it is quite difficult for a classical computer to simulate a quantum system as the dimensionality of Hilbert space required to describe it increases exponentially with the number of particles.

David Deutsch described the first universal quantum computer, which is able to simulate any other quantum computer with a polynomial time delay [17]. Bennet and Brassard introduce the famous BB84 protocol for quantum key distribution using the polarization state of single photon [18]. Ekert showed how the long-range EPR correlations between entangled particles can be used to obtain secure communication [19]. Bennet et al. [20] gave the first idea of teleporting the unknown quantum state of a system using quantum entanglement as a resource. Following these, numerous quantum information processing techniques have been developed using these fundamental protocols. The field, since then, is ever growing as we shall see in the subsequent section.

15.2.1 Key Points of Quantum Theory

In order to appreciate the advantage of quantum computing over classical computing, we briefly develop the basic idea of quantum mechanics with the help of the following postulates:

i. **The postulate of state space**

The state of any quantum system is completely described by a vector in complex vector space known as the Hilbert space. One of the simplest quantum systems is a qubit, which is an element of 2-D Hilbert space requiring two orthonormal basis vectors $\left(\{|0\rangle, |1\rangle\}\right)$ for its representation.

An arbitrary qubit can then be written as the linear combination of these basis states,

$$|\psi\rangle = \alpha|0\rangle + \beta|1\rangle$$

with the only constraint, $|\alpha|^2 + |\beta|^2 = 1$, due to normalization of $|\psi\rangle$.

ii. **The postulate of state evolution**

The evolution of a closed quantum system is described by unitary transformation on the state of the system. The unitary evolution ensures the conservation of total probability density in a closed quantum system. This postulate is no more than the consequence of the wave equation given by Brook Taylor. In quantum computing, the evolution of qubit is done via quantum gates, which essentially is a unitary transformation.

iii. **The postulate of measurement**

The measurement of a quantum system consists of a set of measurement operators, O_m, which acts on the state of the system to be measured. Once the measurement is made, the quantum state jumps into one of the eigenstates of the observable being measured. For example, in the case of a qubit representing the polarization state of a single photon (with horizontal and vertical polarization state represented by $|H\rangle$ and $|V\rangle$, respectively), a general polarization state, $|\psi\rangle = a|H\rangle + b|V\rangle$ would give us result $|H\rangle$ with probability $|a|^2$ and $|V\rangle$ with probability $|b|^2$. In special case when the measurement operators are orthogonal, we obtain orthogonal projectors also known as the Von Neumann measurement.

15.2.2 Qubit, Superposition, and Entanglement

If we consider a classical two-level system that can exist in states 0 or 1, specification of state of this system is the smallest piece of information called as a bit or a classical bit (c-bit). In quantum mechanics, a system can exist in two orthogonal states $|0\rangle$ or $|1\rangle$ and can also exist in any one of the infinite superposed states, $|\psi\rangle = a|0\rangle + b|0\rangle$, where a and b are two complex numbers having the only restriction, $|a|^2 + |b|^2 = 1$, because of normalization of $|\psi\rangle$. Information about the quantum state of the two-level system is called a qubit. Sometimes, the quantum two-level system is also referred to as a qubit. If we consider a pure quantum state of two qubits, the most general is

$$|\psi\rangle = a|00\rangle + b|01\rangle + c|10\rangle + d|11\rangle$$

There can be situations where $|\psi i\rangle$ can be written as a product of states of two qubits, i.e.,

$$|\psi\rangle = \left(\alpha|0\rangle + \beta|1\rangle\right) \otimes \left(\gamma|0\rangle + \delta|1\rangle\right)$$

This gives

$$|\psi\rangle = \alpha\gamma|00\rangle + \alpha\delta|01\rangle + \beta\gamma|10\rangle + \beta\delta|11\rangle$$

or possibility of expressing a, b, c, d in the forms

$$a = \alpha\gamma,\ b = \alpha\delta,\ c = \beta\gamma \text{ and } d = \beta\delta$$

If the above condition holds, we call the state $|\psi\rangle$ is separable and if $|\psi\rangle$ cannot be separated by factorization it is called entangled. The two qubits that exist in a pure separable state show no correlation as for measurement of operators $\widehat{O_1}$ and $\widehat{O_2}$ for two qubit-1 and qubit-2 existing in a separable state $|\psi\rangle = |\psi_1\rangle \otimes |\psi_2\rangle$ of qubits gives

$$\left\langle \psi|\widehat{O_1}\widehat{O_2}|\psi \right\rangle = \left\langle \psi_1|\widehat{O_1}\middle|\ \psi_1 \right\rangle \left\langle \psi_1|\widehat{O_2}\middle|\psi_1 \right\rangle$$

Measurement of O_1 in $|\psi_1\rangle$ is completely independent of measurement of O_2 in $|\psi_2\rangle$.

This is, however, not the case when the composite state $|\psi i\rangle$ of the two qubits is inseparable. Consider, e.g., two qubits existing in the state

$$|\psi\rangle = \left[\sqrt{2}\right]^{-1}\left(|00\rangle + |11\rangle\right)$$

Probability of finding any one of two qubits in state $|0\rangle$ is 1/2 and in the state $|1\rangle$ is also 1/2. But if measurement is made on any one of qubit, the state of the other is determined. Thus, if the first qubit is found to exist in state $|0\rangle$ the second will also exist in the same state. This property is called entanglement, which is a quantum mechanical phenomenon. Quantum entanglement is a useful resource for many important tasks such as remote state preparation, quantum teleportation and super dense coding.

15.2.3 SUPREMACY OF QUANTUM COMPUTING OVER CLASSICAL COMPUTER

A piece of recent news had flown over the world about achieving a milestone by Google that they have successfully achieved quantum supremacy over the classical processor, and they had first-ever physically proved the quantum supremacy. Their research article has been published by Frank et al. [21]. The quantum computer outperforms very efficiently compared to our classical computation machine, so the researchers called this millstone to achieve supremacy over the classical supercomputer. They have used 2^{53} qubit processor devices "Sycamore" to perform the simulation of the problem. And they claim that their task has been performed within a few minutes, if the same task would be processed with today's most advanced state of art supercomputer, then it would take thousands of years and if we give it to a classical computer to perform then it might take millions of years even not sufficient to provide the result.

According to Martinis, quantum supremacy has long been seen as a watershed moment since it indicates that quantum computers can surpass traditional computers [22]. Although the benefit has only been demonstrated for a single scenario, it demonstrates to physicists that quantum mechanics works as expected when applied to a real-world complex mathematical problem. There are many relatable comments that have been made by several scientists like Michelle Simmons (a quantum physicist at the University of New South Wales in Sydney, Australia).

Recently, IBM Corporation had argued on Google's claim on achieving quantum supremacy. "Although the calculation Google has chosen for validating the outputs of a quantum random-number

TABLE 15.1

Classical Computing vs Quantum Computing [1]

Classical Computer	Quantum Computer
1. Classical computers are large-scale integrated multi-purpose computers.	1. QC is mainly a high-speed parallel computer based on the physical phenomenon of quantum mechanics.
2. It uses the classical bits to perform any task.	2. It uses qubits or quantum bits to perform any task.
3. It performs traditional tasks like data processing, Internet suffering and much more daily life task.	3. It performs complex tasks like large-scale optimization, big data processing and solving hard mathematical problems, etc.
4. Performance of the classical computer depends on the number of bits processing per second.	4. Quantum computer depends on qubits, i.e., 0's, 1's and superposition state. (superposition state allows bits to represent a 1 or 0 at the same time).
5. Classical computers are very large and robust in nature.	5. Quantum computers are very large, complex and fragile.
6. Classical computers work with a low error rate and can be operated at room temperature.	6. Quantum computer has a very high error rate and needs to be kept in extremely cold temperatures.
7. Circuit behavior is managed by a classical physics phenomenon.	7. The circuit behavior is explicitly managed by a quantum mechanics phenomenon.
8. Circuits of classical computers are implemented by macroscopic technologies such as Complementary Metal-Oxide Semiconductor (CMOS).	8. Quantum circuits use microscopic technologies like NMR (nuclear magnetic resonance).

generator has limited practical applications" [23]. IBM claims that complete simulations of the same task can be performed with much greater accuracy on an existing supercomputer system in 2.5 days. Honestly, I would like to say, it's still pretty impressive to be able to solve a problem within a few minutes that would take days on a supercomputer. In fact, many approaches have been concurrently taken under-ways to build a physically working quantum computer by scientists and researchers.

15.2.4 COMPUTER COMPUTING VS CLASSICAL COMPUTING

The differences between Classical Computing and Quantum Computing are listed in Table 15.1.

15.2.5 RUMORS AND REALITIES ABOUT QUANTUM COMPUTING

There are many rumors that exist in our environment, which puts the question mark over the working mechanism of quantum computers. Mostly, people poke up it as a mythical computational device that if built will magically solve almost everything you can imagine even within a nap of time. But in reality, the performance of quantum computers has been tested over limited and specific problems. In this section, we are trying to clarify these existing mythical thoughts.

Rumor 1: Quantum computers perform operations much faster than classical computers?

Reality: Ofcourse, quantum computers are much faster than our classical computer, but there is some limitation that also comes with fame like quantum computers might be performed only problem-oriented tasks. Somehow classical computers also have unique abilities to handle specific tasks that cannot be matched by even powerful quantum computers. So, with the concluding remark, we can say that both technologies have their importance and dominance at some level of problem-solving skills.

Rumor 2: Quantum computers perform all the possible computational paths at the same time and frequently achieve the best solutions?

Reality: Yeah, in 1994, Peter Shor devised a well-known quantum algorithm that factors the integers exponentially and increases the data processing speed more than known classical algorithms. It means they have more potential to process exponentially vast amount of data as compared to classical computers. We know that quantum computer uses qubits to perform any task, so it provides more opportunities to achieve its goal. Therefore, it explores the possibilities of solutions. Qubits can be prepared by superposition of exponential number states. Then, it (Shor's Quantum algorithm) should compute all possible inputs at the same time. This phenomenon might vastly exceed the processing speed of quantum computers.

Rumor 3: Quantum computers can smoothly solve complex NP-complete and NP-hard problems?

Reality: Currently, the two most known and notable complexity problems are "P" and "NP". Here, P represents problems that can be solved within the polynomial time by a classical computer. NP problems are problems that cannot be solved in polynomial time by classical computers, but the problem can be computed and verified by any classical computer.

A Quantum computing algorithm named Shor's algorithm can solve the NP-hard problem by factoring a large number of "N" in polynomial time. The Inverse Quantum Fourier Transform process on the quantum circuit library might solve this problem within the polynomial amount of time.

15.3 HAND-HELD APPLICATIONS OF QUANTUM COMPUTING

This section will build the foundation that allows a comprehensive analysis of Quantum computing in cyber security, cloud service and evolutionary computing. The significance of quantum computational technology has been discussed for cyber security.

This section shows the role of Quantum computing in cloud service and has compared the challenges of cloud computing in the post-quantum world. And trying to relate, how this disruptive technology (quantum computing) will change the world of cloud computing. Furthermore, the study has covered the quantum-inspired evolutionary computing, which shows ultimate changes in optimization after the quantum-inspired phenomenon induced in evolutionary computing and discusses its tremendous effects, which will change the field of optimization in the global scenario (Figure 15.1).

15.3.1 QUANTUM COMPUTING IN CYBER SECURITY

Cyber security is new age technology that comes with many challenges for developers and end users [24]. The developers try hard to make a robust system with a safe and secure environment because even a single loophole might be breaching the entire security. In this immersing world, there are

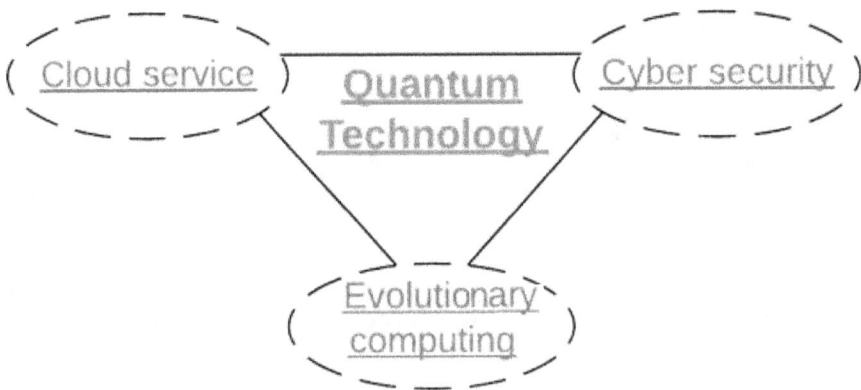

FIGURE 15.1 ER diagram of quantum computation with its applications.

several powerful systems with high computing features are available that might also pose a challenging task in front of developers [25]. With rapid development in technologies, a new dimension of threats has occurred in front of cyber security specialists [26]. It requires risk evaluations and makes the necessary scientific changes in currently available cyber security algorithms, which are not enough to handle the modern intense of threats; this may have time to perform significant changes in existing traditional mechanism, complex asymmetric mathematical methods of data encryptions in cyber security [27].

Cyber security is one of the most interesting study areas of the computer science field, so it provides a vast area of research, and respectively yet to get more explored [28]. Let's talk about the reasons behind the frequent development of cyber security. It was presented in 1970 by Bob Thomas, as research tasks, named as The Advanced Research Projects Agency Network (ARPANET) [29,30]. This project was focused to make a computer program, which has the ability to get transmitted over the ARPANET network, and it was supposed to provide a secure networking environment for the computers. This program was known as CREEPER, which shows a message ("I'm the CREEPER: CATCH ME IF YOU CAN") on the computer screen. The desired outcome of this project was, the computer program exploits the security and shows the possibility of data manipulations, deletion in those computer environments and it had sought the attention of larger private farms and governmental organizations toward the major loopholes in computer networking. Finally, then the academia and researchers put their attention and efforts into this problem and that might be overcome with secure cyber security management [31].

Cyber security is one of the core programs of any computer system or network environment; sometimes, it has been used for unethical acts like money extortion, intentional data manipulation and deletions, or may have asked for some other disastrous demands by cyber-attackers [32,33]. They perform such unethical things either in a group or alone and they could be a random person or your known ones. In the past history, in 1986 Russians with the help of German hackers had made a cyber-attack on the American military and breached the networks to get sensitive information of military intelligence. At that time, there were more than 400 computers hacked including a network of central commands center pentagon. It was a huge blunder and could be disastrous for America. Within a few years of these attacks, the USA made an immediate cyber response team in 1988 for toggling future cyber-attacks. At present, the USA became one of the most cyber-secured countries [34–36].

At the initial phase of secure cyber computing, the network system was very less secure and probably accessed by even a simple computer program, so at that time hacking of network was very common. The modern age technological development has also put hard challenges in front of cyber network developers. This exponential advancement helps unethical persons.

After several simultaneous cyber-attacks, a group of academia and scientist have proposed certain standard protocols that provide a safe environment to the cyber network. We have discussed some state of art protocols and traditional algorithms that are used to make a robust computer network environment, which is very secure in nature and not easy to get hacked by an unethical personality.

Quantum Cyber Security, the growth of large quantum computers, and the additional computing power it will bring, could have disastrous consequences for cyber security [37–39].

Completely securing classical protocols with respect to quantum-equipped technology, which adversaries are possible; however, it requires more care that might be beyond a perfect choice of cryptosystems. Quantum technology may also have an effective effect on cyber safety. Quantum gadgets with the current advanced technological era may be used to ensure and enhance safety where classically not possible to accomplish the tasks, such as secret key expansion in the robust security environment. Since quantum computer systems will become a crucial part of our network communications and computations, the need for current trends is to increase sensible methods to use quantum computer systems that ensure safety and provide a secure environment for classical computing.

15.3.2 QUANTUM COMPUTING IN CLOUD COMPUTING

Cloud Computing is the concept that provides facilities of large date storage, virtual servers, accessing the application software, database networking and many more. It uses the servers computing phenomenon to manage these facilities [40]. Cloud Computing is a field of computing wherein the computing power is provided as a utility. This paves way for a shift in computing, i.e., rather than computing being a product that is purchased; it would be delivered to the consumers over the internet from far-off large-scale data centers. The important point is that the customers will avail of all these features without actually knowing how the cloud works. These data centers in which the data is actually stored and processed can be termed as the "clouds", this is how cloud computing gets its name. The implementation of this technology allows the cost of computation, application hosting and data storage to be reduced significantly. There are three service models in cloud computing. These are Software as a Service (Saas) [41], Platform as a Service (PaaS) and Infrastructure as a Service (IaaS) [42–45]. Web applications are distributed through the browser to thousands of SaaS cloud computing consumers in the model. The development environment is supplied as a service in PaaS cloud computing.

The infrastructure, such as storage, is given as a service to consumers via the internet in the IaaS form of cloud computing. In addition, there are four deployment methods in cloud computing, public cloud, private cloud, hybrid cloud and community cloud are some of the options. The cloud services are supplied to the general public in a public cloud. The cloud service is made available to a certain organization or group of people in a private cloud. It is not accessible to the general population. Multiple clouds are combined in a hybrid cloud (public, private or community). Individual clouds retain their individuality, but are tied together as a group. The cloud service in a community cloud is arranged to serve a common function of one or more organizations. With such cloud computing applications, it's also known as "sky computing", due to the various independent clouds. As the field of cloud computing progresses, it faces a major dilemma that stems from the very foundation of its success. This is a matter of safety. Cloud computing will not be truly secure until this problem is resolved. However, the dilemma is how to protect the cloud utilizing current methodologies in order to build trust in cloud computing technology among enterprises and individuals. The solution to this question can be found in yet another branch of computing that is about to make the leap from the lab to the real world. This quantum computing next-generation technology is drastically used in these days to remove the security problem; quantum computing may play an important role due to its inherent characteristics that we are going to discuss it in upcoming sections.

Quantum Cloud Computing, to be in reality the quantum cloud pair will be another technology. It would bring "quantum computing as a service" to the domain of cloud computing for the first time [46]. It would provide such significant technological improvement in the realm of computing that today's computers would appear to be obsolete. In the near future, this technological duo would give numerous benefits. It would have the essential benefits of next-generation powerful quantum computing, as well as cloud computing significant qualities like cost reduction, resource pooling, cloud storage, cloud processing, easier maintenance, enhanced accessibility, flexibility and mobility. This technology would bring about unthinkable changes in the technology we use every day, and it would be available through a variety of touch points.

The most significant benefit would be the elimination of security concerns associated with cloud computing [47,48]. Quantum Cryptography, an intrinsic property of quantum computing, would be used to provide this security. Quantum cryptography is a technology that uses quantum mechanics to provide information security. Information security is predicated on the inviolability of quantum physics' rules. Quantum computers would use extremely strong encryption techniques to safeguard data stored in the cloud, which would take a hacker's machine millions of years to breach. This would aid in the development of consumer trust in cloud technologies [49,50]. In addition, the quantum cloud duo would put massive processing power in the hands of customers. This means that users will be able to use the internet to access quantum computing's super-fast technologies from anywhere. Such a technology would pave the way for the next generation of computers, which would make computing more portable and perhaps lead to a massive increase in the number of smart-phones on the market. This is because the application-supported infrastructure would be provided by end users [51,52].

Quantum Integrated Cloud Technology is housed on a cutting-edge, shared cloud infrastructure in line with current trends, but at a faster rate than ever before, allowing for high-quality applications. It would also result in mobile gadgets that are slimmer and more powerful. Furthermore, quantum computing role in cloud computing will serve as a foundation for super-fast database operations on the cloud. Lov Grover developed a quantum algorithm in 1996 [1] that allowed him to explore a comparable quantum database with N terms in O (. J N) steps. In traditional calculation, finding the required entry would take $N/2$ steps. Grover's technique would require…J (10,000 = 100) steps for a database of 10,000 names, rather than 5,000 for a database of 10,000 names. The method works by establishing a superposition of all 10,000 entries, with each entry having the same chance of appearing in response to the system's measurement [53]. A quantum computer could cut the search time from hundreds of years to minutes by applying Grover's technique. As a result, it would be tremendously beneficial in terms of offering quick database operations in cloud storage. This would be yet another benefit of this dual-technology approach. Furthermore, such remarkable technology will enable enterprises of all sizes, large and small, to accept and implement cloud technology. It would open the door for a wave of global trust in cloud computing [54,55].

Importance of Quantum-based Cloud Services: Cloud-based quantum computer is exponen-tially increasing area of research by combining both technologies and helps to explore the new era of research. Quantum computing could aid in the following ways,

1. By decoding the complex structure of the chemical compounds.
2. Managing risk in cloud services, providing safeguard facilities from external threats.
3. Discovery of new medications, by efficiently analyzing the big data set and providing the most prominent composition of chemical.
4. Data might be transmitted over the internet with efficiently safer encryption, due to such ability of quantum-based cloud computing that handles more complex numbers.
5. Financial trading provides a robust environment to the trading giants.
6. Supply chain optimization is some of the other cloud service applications, where the vari-ous concurrent problems might be handled over the same environment.

15.3.3 QUANTUM COMPUTING IN EVOLUTIONARY COMPUTING

Evolutionary algorithm is mainly influenced by the Charles Darwin's nature-inspired hypothesis, "Survival of fittest" (Letter it became the theory, 1859), where he said that the only species will be survived in the environment, which have followed the adaptation strategies (fitness) of nature [56]. The species or the candidate solution that has diversity in nature will defiantly go for the next gen-eration of evolution and vice-versa might be possible. The best suitable example of Darwin's this epic theory is the evolution with giraffe, where the neck of giraffe had got long due to some reason (probably, the lack of food over the ground, grass and tinny trees) of evolution. What is evolutionary computing? [57], it mainly comes under the fields of Artificial Intelligence (AI), specifically in the sub-fields of AI like Computational Intelligence, Soft Computing. The working mechanism of evo-lutionary computing is to make the system more reliable so that it might survive in the most difficult scenarios like competitive, qualitative and reliable challenges. Moreover, to handle the dynami-cally changing continuous environment, so here the biological operators (mutations, reproduction, recombination, natural selection strategies over all "survival of the fittest" theme) might provide the needful environment (to generate, evaluate and select the best candidate solutions, which have more ability to survive) for the evolutionary algorithm. This algorithm is one of the most famous

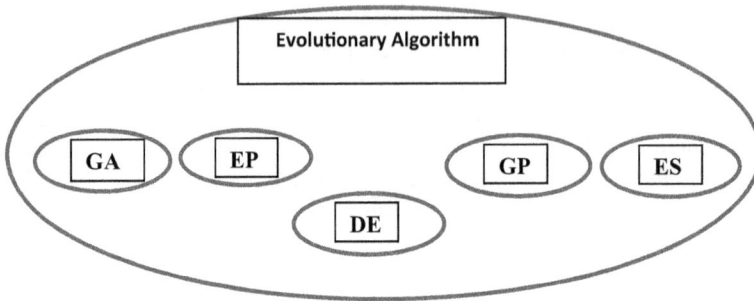

FIGURE 15.2   The classification of evolutionary algorithms.

nature-inspired stochastic algorithms for optimization. There are mainly four algorithms that come under evolutionary algorithm as shown in Figure 15.2 [58–60].

1. Genetic Algorithm (GA) [61,62] **2.** Evolutionary Programming (EP) [63]
3. Genetic Programming (GP) [64] **4.** Evolutionary Strategies (ES) [65]
5. Differential Evolution (DE) [66–70]

Each evolutionary algorithm has several variants that already exist in the study. Here, we are going to focus on the differential evolutionary algorithm and its novel hybridization with the Quantum-Inspired Differential evolutionary algorithm. In this study, we have discussed the impact of the quantum computation mechanism in the DE algorithm.

Standard Differential Evolutionary Algorithm: The concept of DE was first introduced by Storn R and Price K in 1996 individually [66,67]. But the officially tested DE had come in 1997 with joint efforts of both scientists. Differential evolutionary algorithm is an efficient and one of the powerful population-based search techniques for solving complex mathematical optimization problems over a continuous search space [68–70]. The main objective of DEs is based on the difference of randomly sampled pairs of solutions in the population [71–75].

Algorithm of DE

STEP 1. Create Initial population (Np*D), where Np=no of population, $D=$ Dimension
STEP 2. Evaluate each individual population
       $\rightarrow$ **2.1** Mutation
       $\rightarrow$ **2.2** Crossover (Recombination)
       $\rightarrow$ **2.3** Selection
STEP 3. Best candidate solution.

Quantum-Inspired Evolutionary Computing: Quantum computation is an emerging field that incorporates quantum mechanics and information science. Many academics have been curious about quantum computation since Deutsch presented the "Deutsch–Jozsa algorithm" in 1985 [1–3]. The quantum evolutionary algorithm (QEA) is a probabilistic and stochastic optimization algorithm based on the quantum computation theory.

The chromosomes or random vectors are encoded by qubits and updated through quantum rotation gates, which facilitates the evolutionary search in QIEA [76,77].

Once the quantum mechanism is incorporated with the evolution computing field is likely known as a quantum-inspired evolutionary algorithm. Both technologies are very recently introduced by scientists. However, various research articles have already been published by academia but many researchers are still trying to explore it more as more with hope of finding the most suitable techniques in optimization of different real-world problems. In this section, various past and recent developments have been cited and trying to create an understandable environment for new age researchers, who are looking for collective and concise knowledge and the frequent research studies of this emerging field. Some QEAs have recently been proposed for a variety of combinatorial optimization problems, including the traveling salesman problem [78–88], the knap-sack problem [89] and 8-queen problem [90]. The QEA is rapidly gaining popularity and exploring the next level of research paradigm.

Quantum-Inspired DE Algorithm: The aim behind the development of quantum-inspired DE algorithm is to increase the optimization of the existing DE algorithm. Therefore, this study has focused on the understanding of the differential evolutionary algorithm along with the incorporation of a novel quantum-inspired mechanism in the same algorithm, so that it really put a drastic change over the performance of the DE algorithm [91,92].

The quantum-inspired mechanism has been first ever introduced for nature-inspired algorithm in 2010 by W. Fang and team, they had proposed the quantum-behaved particle swarm optimization algorithm. In the present study, the quantum-inspired evolutionary strategies and qubits rotation in block sphere have been discussed. The individual candidate is encoded in qubit format within the block sphere and the corresponding rotation angle of these individual candidates is achieved by evolutionary strategies.

Algorithm for Quantum-Inspired DE Algorithm:

STEP 1. Begin

STEP 2. Initialization: Generate random real-valued population and perform the quantum observation and fitness evaluations.

STEP 3. Mutation: Calculate mutated population through adaptive mutation strategy based on quantum observation and fitness evaluations.

STEP 4. Crossover (Recombination): Use self-adaptive crossover mechanism for crossover population and perform quantum observation and fitness evaluations.

STEP 5. Selection: Perform binary observation trail on each individual population and check whether it is better than current individuals or not.

    5.1 if: Yes then: Select the better trail individual and quantum trail individual.
        5.1.1 Store the best solutions.
    5.2 else: Determine the quantum rotation angle.
        5.2.1 Update the quantum individual with quantum rotation gate.
    5.3 Update the iteration (**iter+1**).
    5.4 Update till maximum iteration(iter<=maxi_iter).

STEP 6: End

Why Quantum-Inspired DE Algorithm? Quantum-inspired DE algorithm is a novel approach for solving the optimization problems. This algorithm combined the DE operations and superposition principles of quantum computing. This algorithm uses mutation, crossover and selection operations of the DE algorithm. Furthermore, it contained the principles of quantum computing like quantum gates and superposition states. It provides the global optimal solution by establishing a great balance between the exploitation and exploration mechanism of DE.

On behalf of the vast literature review, we can say that the quantum-inspired DE is very much able to enhance diversity and convergence performance for most of the real-world engineering and complex mathematical problems. In addition, this novel mechanism provides effective and efficient in finding the optimal solutions for high-dimensional situations.

15.4 DISCUSSION AND CONCLUSION

Quantum computing, a new age technology that uses the laws of quantum mechanics to achieve exponentially higher performance in certain types of computations, offers opportunities for major breakthroughs in all areas. After going through the vast literature review, we can say that the quantum computing has radically changed the way in which we proceed and simulate the problems of many fields. With the unbeatable processing power of quantum computers, we can simulate digital versions of chemistry, physics and engineering problems. This "digital twinning" enables a number of opportunities to test out theories and predict real actions of expected results.

Obviously, quantum computers can do things that classical computers are not even sufficient to handle it. This phenomenon we call quantum supremacy. However, even the tech giants are desperate to make that happen.

REFERENCES

1. Everett, H. "Relative state" formulation of quantum mechanics. *American Physical Society*, 454–462, 1957.
2. Chuang, IL, Yoshihisa, Y. Simple quantum computer. *A Physical Review*, 52(5), 3489, 1995.
3. Planck, M. On an improvement of Wien's equation for the spectrum. *Annalen der PhysikAnn. Physik*, 1, 719–721, 1900.
4. Planck, M. On the law of distribution of energy in the normal spectrum. *Annalen der physik*, 4(553), 1, 1901.
5. Einstein, A. Uber einem die erzeugung und verwandlung des lichtes betref- ̈ fenden heuristischen gesichtspunkt. *Annalen der physik*, 4, 1905.
6. Hertz, H. Ueber einen einfluss des ultravioletten lichtes auf die electrische entladung. *Annalen der Physik*, 267(8), 983–1000, 1887.
7. Lewis, GN. The conservation of photons. *Nature*, 118(2981), 874–875, 1926.
8. Feynman, RP. Simulating physics with computers. *International Journal of Theoretical Physics*, 21(6), 1982.
9. Schrödinger, E An undulatory theory of the mechanics of atoms and molecules. *Physical Review*, 28, 1049, 1926.
10. Heisenberg, W. *The Physical Principles of the Quantum Theory*. Courier Corporation. Dover: New York, NY, 1930.)
11. Dirac, P. On the theory of quantum mechanics. In *Proceedings of the Royal Society of London. Series A, Containing Papers of a Mathematical and Physical Character*, vol. 112, pp. 661–677, 1926.
12. Von Neumann, J. *Mathematical Foundations of Quantum Mechanics*. Princeton University Press, 2018.
13. Weisner, SJ. Conjugate coding. *SIGACT News*, 15(1), 78–88, 1983.
14. Park, J. The concept of transition in quantum mechanics. *Foundations of Physics*, 1, 23–33, 1970.
15. Holevo, AS. Bounds for the quantity of information transmitted by a quantum communication channel. *Problemy Peredachi Informatsii*, 9(3), 3–11, 1973.
16. Benioff, P. The computer as a physical system: A microscopic quantum mechanical Hamiltonian model of computers as represented by Turing machines. *Journal of Statistical Physics*, 22, 563–591, 1980.
17. Deutsch, D. Quantum theory, the Church–Turing principle and the universal quantum computer. *Proceedings of the Royal Society of London. A. Mathematical and Physical Sciences*, 400, 97–117, 1985.
18. Bennett, C, Brassard, G Quantum cryptography: Public key distribution and coin tossing. ArXiv Preprint ArXiv:2003.06557. (2020).
19. Ekert, A. Quantum cryptography based on Bell's theorem. *Physical Review Letters*, 67, 661, 1991.
20. Bennett, C, Brassard, G, Crépeau, C, Jozsa, R, Peres, A, Wootters, W Teleporting an unknown quantum state via dual classical and Einstein-Podolsky-Rosen channels. *Physical Review Letters*, 70, 1895, 1993.
21. Frank, A, Kunal, A, Ryan, B, et al. Quantum supremacy using a programmable superconducting processor. *Nature*, 574, 505–510, 2019.
22. Courtland, R. Google aims for quantum computing supremacy. *IEEE Spectrum*, 54(6), 9–10, 2017.
23. Edwin, P, John, G, Dmitri, M, Jay, G, On: "quantum supremacy". *IBM Research Blog*, 2019. https://www.ibm.com/blogs/research/2019/10/on-quantumsupremacy/.
24. Zeng, E, Mare, S, Roesner, F. End user security and privacy concerns with smart homes. In *Thirteenth Symposium on Usable Privacy and Security*, pp. 65–80, 2017.

25. Dillon, T, Wu, C, Chang, E. Cloud computing: Issues and challenges. In *IEEE International Conference on Advanced Information Networking and Applications*, pp. 27–33, 2010.
26. Zhang, Q, Cheng, L, Boutaba, R. Cloud computing: State-of-the-art and research challenges *Journal of Internet Services and Applications, Springer Open*, 1, 7–18, 2010.
27. Takabi, H, Joshi, JBD, Ahn, G. Security and privacy challenges in cloud computting environments. *IEEE Security & Privacy*, 8(6), 24–31, 2010.
28. Torngren, M, Sellgren, U. Complexity challenges in development of cyber physical systems. In *Principles of Modeling*. Springer, pp. 478–503, 2018.
29. Abbate, JE. *From ARPANET to INTERNET: A History of ARPA Sponsored Computer Networks.* University of Pennsylvania, pp. 1966–1988, 1994.
30. Nye, JS, et al. *The Regime Complex for Managing Global Cyber Activities.* University of Pennsylvania, vol. 1, 2014.
31. Jang-Jaccard, J, Surya, N. A survey of emerging threats in cyber security. *Journal of Computer and System Sciences*, Elsevier, 80(5), 973–993, 2014.
32. Pfleeger, CP, Shari Lawrence, P. *Analyzing Computer Security: A Threat/Vulnerability/Countermeasure Approach.* Prentice Hall Professional, 2012.
33. Shinder, DL, Michael, C. *Scene of the Cybercrime.* Elsevier, 2008.
34. Baezner, M, Robin, P. *Cyber-Conflict between the United States of America and Russia.* ETH Zurich, 2017.
35. Stiennon, R. A short history of cyber warfare. In *Cyber Warfare*. Routledge, pp. 7–32, 2015.
36. Rid, T. Cyber war will not take place. *Journal of strategic studies*, 35(1), 5–32, 2012.
37. De Wolf, R. The potential impact of quantum computers on society. *Ethics and Information Technology*, (4), 271–276, 2017.
38. Keplinger, K. Is quantum computing becoming relevant to cyber-security? *Network Security*, 16–19, 2018.
39. Wallden, P, Kashefi, E. Cyber security in the quantum era *Communications of the ACM*, 62(4), 120, 2019.
40. Hayes, B. 2008. Cloud computing. *Communications of the ACM*, 51(7), 9–11, 2008. https://doi.org/10.1145/1364782.1364786.
41. Satyanarayana, S. Cloud computing: SAAS. *Computer Sciences and Tele Communications*, 4, 76–79, 2012.
42. Mohammed, CM, Zeebaree, SRM. Sufficient comparison among cloud computing services: IaaS, PaaS, and SaaS: A review. *International Journal of Science and Business*, 5(2), 17–30, 2021.
43. Bhardwaj, S, Jain, L, Jain, S Cloud computing: A study of infrastructure as a service (IAAS). *International Journal of Engineering and Information Technology*, 2(1), 60–63, 2010.
44. Goyal, S. Software as a service, platform as a service, infrastructure as a service – A review. *International Journal of Computer Science & Network Solutions*, 1(3), 53–67, 2013.
45. Zhou, M, et al. Services in the cloud computing era: A survey. In *2010 4th International Universal Communication Symposium*. IEEE, 2010.
46. Möller, M, Vuik, C. On the impact of quantum computing technology on future developments in high-performance scientific computing. *Ethics and Information Technology*, 19(4), 253–269, 2017.
47. Singh, H, Sachdev, A. The quantum way of cloud computing. In *2014 International Conference on Reliability Optimization and Information Technology (ICROIT)*. IEEE, 2014.
48. Grodzinsky, FS, Wolf, MJ, Miller, KW. Quantum computing and cloud computing: Humans trusting humans via machines. In *2011 IEEE International Symposium on Technology and Society (ISTAS)*. IEEE, 2011.
49. Ahmad, S, Mehfuz, S, Beg, J. Empirical analysis of security enabled quantum computing for cloud environment. In *Quantum and Blockchain for Modern Computing Systems: Vision and Advancements*. Springer: Cham, pp. 103–125, 2022.
50. Rahaman, M, Islam, MM. A review on progress and problems of quantum computing as a service (QCAAS) in the perspective of cloud computing. *Global Journal of Computer Science and Technology*, 2015.
51. Zhou, L, et al. Quantum technique for access control in cloud computing II: Encryption and key distribution. *Journal of Network and Computer Applications*, 103, 178–184, 2018.
52. Kaiiali, M, Sezer, S, Khalid, A. Cloud computing in the quantum era. In *2019 IEEE Conference on Communications and Network Security (CNS)*. IEEE, 2019.
53. Murali, G, Sivaram Prasad, R. CloudQKDP: Quantum key distribution protocol for cloud computing. In *2016 International Conference on Information Communication and Embedded Systems (ICICES)*. IEEE, 2016.

54. Mardirossian, N, et al. Novel algorithms and high-performance cloud computing enable efficient fully quantum mechanical protein-ligand scoring. arXiv preprint arXiv:2004.08725 (2020).

55. Pandya, M. Securing Cloud-The Quantum Way. arXiv preprint arXiv:1512.02196 (2015).

56. Darwin, C. *Charles Darwin's Natural Selection: Being the Second Part of His Big Species Book Written from 1856 to 1858*. Cambridge University Press, 1987.

57. Eiben, AE, Smith, JE. *Introduction to Evolutionary Computing*. Springer: Berlin, vol. 53, 2003.

58. Eiben, AE, Schoenauer, M. Evolutionary computing. *Information Processing Letters*, 82(1), 1–6, 2002.

59. Radcliffe, NJ, Surry, PD. Fundamental limitations on search algorithms: Evolutionary computing in perspective. *Computer Science Today*, 275–291, 1995.

60. Ghosh, A, Tsutsui, S, eds. *Advances in Evolutionary Computing: Theory and Applications*. Springer Science & Business Media, 2012.

61. Whitley, D. A genetic algorithm tutorial. *Statistics and Computing*, 4(2), 65–85, 1994.

62. Mirjalili, S. Genetic algorithm. In *Evolutionary Algorithms and Neural Networks*. Springer: Cham, pp. 43–55, 2019.

63. Yao, X, Liu, Y, Lin, G. Evolutionary programming made faster. *IEEE Transactions on Evolutionary Computation*, 3(2), 82–102, 1999.

64. Langdon, WB, Poli, R. *Foundations of Genetic Programming*. Springer Science & Business Media, 2013.

65. Beyer, H-G. *The Theory of Evolution Strategies*. Springer Science & Business Media, 2001.

66. Price, KV. Differential evolution: A fast and simple numerical optimizer. In *Proceedings of North American Fuzzy Information Processing*, pp. 524–527, 1996. doi: 10.1109/NAFIPS.1996.534790.

67. Storn, R, Price, K. Differential evolution – A simple and efficient heuristic for global optimization over continuous spaces. *Journal of Global Optimization*, 11, 341–359, 1997.

68. Price, KV. Differential evolution. In *Handbook of Optimization*. Springer: Berlin Heidelberg, pp. 187–214, 2013.

69. Back, T. *Evolutionary Algorithms in Theory and Practice: Evolution Strategies, Evolutionary Programming, Genetic Algorithms*. Oxford University Press, 1996.

70. Eiben, AE, Hinterding, R, Michalewicz, Z. Parameter control in evolutionary algorithms. *IEEE Transactions on Evolutionary Computation*, 3, 124–141, 1999.

71. Karafotias, G, Hoogendoorn, M, Eiben, AE. Parameter control in evolutionary algorithms: Trends and challenges. *IEEE Transactions on Evolutionary Computation*, 19, 167–187, 2015.

72. Chao Liu, C, Zhao, Q, Yan, B, Gao, Y. A new hypervolume-based differential evolution algorithm for many-objective optimization. *Rairo: Operations Research*, 51(4), 1301–1315, 2017.

73. Lynn, N, Ali, MZ, Suganthan, PN. Population topologies for particle swarm optimization and differential evolution. *Swarm and Evolutionary Computation*, 39, 24–35, 2018.

74. Ali, MZ, Awad, NH, Suganthan, PN. Multi-population differential evolution with balanced ensemble of mutation strategies for large-scale global optimization. *Applied Soft Computing*, 33, 304–327, 2015.

75. Zhong, X, Cheng, P. An improved differential evolution algorithm based on dual-strategy. *Mathematical Problems in Engineering*, 1–14, 2020.

76. Deng, W, et al. An improved quantum-inspired differential evolution algorithm for deep belief network. *IEEE Transactions on Instrumentation and Measurement*, 69(10), 7319–7327, 2020.

77. Draa, A, et al. A quantum-inspired differential evolution algorithm for solving the N-queens problem. *Neural Networks*, 1(2), 21–27, 2011.

78. Hota, AR, Pat, A. An adaptive quantum-inspired differential evolution algorithm for 0–1 knapsack problem. In *2010 Second World Congress on Nature and Biologically Inspired Computing (NaBIC)*. IEEE, 2010.

79. Han, K-H, Kim, J-H. Quantum-inspired evolutionary algorithm for a class of combinatorial optimization. *IEEE Transactions on Evolutionary Computation*, 6(6), 580–593, 2002.

80. Fiasché, M. A quantum-inspired evolutionary algorithm for optimization numerical problems. In *International Conference on Neural Information Processing*. Springer: Berlin, Heidelberg, 2012.

81. da Cruz, AVAbs, Vellasco, MMBR, Pacheco, MAC. *Quantum-Inspired Evolutionary Algorithms Applied to Numerical Optimization Problems*. IEEE Congress on Evolutionary Computation. IEEE, 2010.

82. Zhang, G. Quantum-inspired evolutionary algorithms: A survey and empirical study. *Journal of Heuristics*, 17(3), 303–351, 2011.

83. Talbi, H, Draa, A. A new real-coded quantum-inspired evolutionary algorithm for continuous optimization. *Applied Soft Computing*, 61, 765–791, 2017.

84. Su, H, Yang, Y. Quantum-inspired differential evolution for binary optimization. In *2008 Fourth International Conference on Natural Computation*. IEEE, vol. 1, 2008.

85. Zouache, D, Moussaoui, A. Quantum-inspired differential evolution with particle swarm optimization for knapsack problem. *Journal of Information Science and Engineering*, 31(5), 1757–1773, 2015.
86. Wang, Y, Wang, W. Quantum-inspired differential evolution with Grey Wolf optimizer for 0–1 Knapsack Problem. *Mathematics*, 9, 1233, 2021. https://doi.org/10.3390/math9111233.
87. Feng, XY, et al. Quantum-inspired evolutionary algorithm for travelling salesman problem. In *Computational Methods*. Springer: Dordrecht, pp. 1363–1367, 2006.
88. Papalitsas, C, Kastampolidou, K, Andronikos, T. Nature and quantum-inspired procedures–a short literature review. *GeNeDis*, 129–133, 2020.
89. Han, K-H, et al. Parallel quantum-inspired genetic algorithm for combinatorial optimization problem. In *Proceedings of the 2001 Congress on Evolutionary Computation (IEEE Cat. No. 01TH8546)*. IEEE, vol. 2, 2001.
90. Draa, A, et al. A quantum-inspired differential evolution algorithm for solving the N-queens problem. *Neural Networks*, 1(2), 21–27, 2011.
91. Pat, A, Hota, AR, Singh, A. Quantum-inspired differential evolution on bloch coordinates of qubits. In *International Conference on Advances in Computing, Communication and Control*. Springer: Berlin, Heidelberg, 2011.
92. Jiao, B, Gu, X, Xu, G. An improved quantum differential algorithm for stochastic flow shop scheduling problem. In *Proceedings of IEEE International Conference on Control and Automation*, pp. 1235–1240, 2009.

16 Encrypted Network Traffic Classification and Application Identification Employing Deep Learning

Jyoti Mishra and Mahendra Tiwari
University of Allahabad

CONTENTS

16.1 INTRODUCTION

In modern day's network communication systems, network traffic classification is a critical task [1]. Tahaei et al.'s interest in traffic categorization algorithms to address various issues in Internet of Things (IoT) applications have increased due to the fast expansion of IoT devices and the diversity of IoT traffic patterns. Because of the significant increase in excessive traffic needs, it is critical to distinguish distinct types of network applications and manage them correctly. As a result, proper traffic categorization must be a need for improved network management operations such as providing appropriate Quality-of-Service, Anomaly Detection, and pricing, among others. As a result, network traffic categorization has piqued the interest of researchers and commercial network managers alike [2–4].

To understand the importance of network traffic categorization, consider the asymmetric structure of today's network access lines built on the premise that users install more data than they upload. However, the increasing use of symmetric-demand applications like peer-to-peer (P2P) programs, voice over Internet protocol (VoIP), and video conferencing have changed users' requirements, prompting them to stray from the previous premise. As a result, specific application-level expertise is required to assign adequate assets to such applications to deliver a satisfactory customer experience. Furthermore, the expansion of modern apps and interactions among multiple components via online platforms has dramatically increased the complexity and prejudice of this community, resulting in severe annoyance. The following sections go through some of the most

DOI: 10.1201/9781003267812-16

critical challenges in network traffic classification. Initially, the growing demand for client data security and privacy has resulted in a significant increase in the amount of encrypted traffic on the Internet today [4]. The encryption procedure turns the original data into a pseudo-random-like pattern, making it difficult to decrypt. As a result, the encrypted data contains several discriminative styles for determining network traffic. As a result, accurately identifying encrypted communication has become a difficult challenge in today's networks [2]. It is, nevertheless, significant enough to mention that several of the suggested network traffic categorization approaches, such as payload inspection, statistical methods, and Machine Learning, need the extraction of patterns or features with the assistance of an expert. This technique is time-consuming, error-prone, and costly. Finally, because of excessive bandwidth consumption and copyright infringements, certain Internet service providers (ISPs) obstruct File transfer programs [5].

To minimize traffic control systems, these applications use protocol integration methodologies and approaches [6]. Identifying these traits and functions is one of the most difficult challenges in network traffic categorization. The categorization of network traffic has aroused a lot of curiosity [7–10]. However, a significant proportion of them have focused on identifying a protocol family, known as traffic characterization (e.g., streaming, P2P, and chatting), rather than a single application, known as application identification (e.g., Hangouts, Bit Torrent, and Spotify) [11]. On the other hand, this chapter presents a technique for characterizing and identifying network traffic based on recent advances in machine learning, specifically deep learning [12,13]. The following are the potential advantages of this suggested technique that make it preferable to other classification schemes:

- This method eliminates the need for a network traffic expert to extract network traffic-related operations. Furthermore, the time-consuming process of discovering and retrieving different characteristics has been avoided in favour of this strategy.
- This method is used to detect traffic at both granular levels (traffic characterization and application identification) delivering cutting-edge outcomes as compared to previous efforts performed on the same dataset [9,14].
- This method is capable of classifying among the hardest classes of applications, termed as P2P [10]. These applications use the latest port obfuscation techniques to avoid ISPs' control measures by embedding their information in well-known payload protocols and using randomized terminals.

16.2 LITERATURE REVIEW

Several other related studies in the past for network traffic classification are given below. The techniques used in these studies are broadly categorized into three major classes:

Port-Based Approach: The oldest or even well scheme for such a task is traffic classification by port number [2]. Packets' information inside the transfer control protocol (TCP) or user datagram protocol (UDP) headers employs by port-based classifiers to find the port number that is presumed to be connected with a specific application. When you've already received the associated port number, you can move on to the next phase of characterized network traffic, which is analysed to the Internet Assigned Numbers Authority (IANA) TCP/UDP port numbers. The technology is quite simple; also, encryption techniques have no impact on port numbers. This approach is commonly used in firewalls and access control lists [15] because of the rapid extraction process. The easiest and most efficient way to distinguish network traffic is port-based classification. However, the accuracy of this method has declined significantly due to the widespread use of other approaches such as port encryption, network address translation (NAT), protocol encapsulation, random port allocations, and port forwarding.

Port-based classification algorithms can only classify 30%–70% of contemporary Internet traffic, according to certain authors [10,16]. To classify modern network traffic, increasingly complicated traffic classification measures are employed.

Techniques of Payload Inspection: These techniques are based on analysing data contained in the application layer payload of packets [11]. As signatures for each protocol, most payload assessment methods, commonly called deep packet inspection (DPI), use predetermined patterns such as regular expressions [17,18]. The patterns that are generated are then used to differentiate between protocols. One of the most significant drawbacks of this approach is the necessity to update patterns whenever a new protocol is published, as well as user privacy concerns.

Sherry et al. [19] suggested an alternate DPI system that may be used to detect encrypted payloads without attempting to decrypt them. Sherry et al. [19] suggested an alternate DPI system that may be used to detect encrypted payloads without attempting to decrypt them. As a result, the user's privacy concern is resolved, despite the fact that it can only manage HTTP Secure (HTTPS) traffic. To overcome this issue of current traffic categorization approaches, which do not have any device of real-time information for encrypted data, a software-defined network (SDN) that is dependent on payload Inspection techniques is introduced [20]. Setiawan et al. used Home Gateway for Congestion (SDNHGC) to emphasize monitoring network traffic in real-time and core network resource distribution that distributes power all over the connected network, resulting in an advanced home network with better end-to-end network traffic monitoring.

Machine-Learning Approach and Statistical Methods: A few of these techniques, specifically statistical analysis methods, depend on the erroneous presumption that each application might have specific characteristics within the underlying traffic. The characteristics of each application are closely distinctive to it. Each statistical method does have its relatively particular statistical model and functions. Crotti et al. [21], proposed protocol fingerprints based on packet interarrival time and standardized thresholds on a probability density function (PDF). For the acquisition of protocols, including HTTP is Hyper Text Transfer Protocol, Post Office Protocol 3, call it as POP3 Protocol, and Simple Mail Transfer Protocol, is SMTP in short, they scored up to 91% accuracy. Wang et al. [22] looked at PDF of packet size in related research. With correctness of up to 87%, their technique can identify a wider range of protocols, that includes TELNET, file transfer protocol (FTP), secure shell (SSH), and Internet Message Access Protocol.

To categorize traffic, a large number of machine learning methods are employed. Dong [23] has proposed a new support vector machine (SVM) algorithm called cost-sensitive SVM to resolve the imbalance problem in network traffic identification. Afuwape et al. [24] evaluated the performance of secured network traffic classification employing machine learning techniques. To improve the performance in Virtual Private Network (VPN) Traffic as measured by Precision, Recall, and F1-score employing Ensemble Classifiers, different Machine learning algorithms are considered for the classification and detection of VPN traffic. Auld et al. [25] introduced a widely used Bayesian neural network that attained 99% accuracy in classifying the best-known P2P protocols, BitTorrent, which also includes Kazaa and Gnutella. Using a kernel density estimation method and a well-known Naive Bayes classifier, Moore et al. [26] acquired 96% of correctness over the applications of the same domain. For traffic determination, artificial neural networks (ANNs) have been considered [27,28]. Furthermore, ANN methodology has been shown to outperform Naive Bayes algorithms. Machine learning techniques are used in two of the most important ideas based on the "ISCX VPN-nonVPN" dataset of Network traffic. Gil et al. employed the k-nearest neighbour (k-NN) and decision tree algorithm C4.5 for classifying the network traffic employing time-related parameters like bytes flow per second, flow duration, and interarrival time of forward as well as backward. Using the C4.5 technique, they were able to characterize six key traffic-based: web surfing, streaming, email, file transfer, chat, and VoIP. On the same dataset, which was routed across VPN, they got roughly 88% recall using the C4.5 algorithm. Using the k-NN method [14], Yamansavascilar et al. manually picked 111 flow characteristics [29] and attained 94% accuracy for 14 types of applications. The primary disadvantage of all these methods is that the feature selection and feature extraction processes are essentially performed with the assistance of a professional. As a consequence, the above-mentioned methods are time-consuming, costly, and prone to human error. Furthermore, as mentioned [14], when utilizing k-NN classifiers for prediction, it is well recognized that the execution time of these algorithms is a considerable concern. Wangc [30] documented just a study on the basis of deep learning theories before this work, to the best of our knowledge. They classified specific network traffic using stacked auto-encoders for many protocols such as HTTP, SMTP, and others. However, they did not mention the dataset they utilized in their technical report. Furthermore, their strategy's approach, implementation details, and a detailed explanation of their results are all missing.

The survey of Chen et al. [31] reflects that there are further hidden threats to cyber security, such as information leaks and hostile cyber assaults. This study describes and interprets the Cyber Security of Smart Cities using deep learning techniques, as well as relevant advancements on IoT security in advanced cities. This survey of numerous deep learning models, specifically Boltzmann machines, Recurrent Neural Networks (RNNs), Deep Belief Networks, Adversarial Generative Networks, and Convolutional Neural Networks (CNNs), highlighted applications of cyber security, including use cases in smart cities on the basis of deep learning technologies. The DISTILLER classifier is the result of a created novel multimodal system (by learning both intra- and inter-modality dependencies) multitask deep learning technique for network traffic classification [32].

16.3 DEEP LEARNING AND CNN

16.3.1 DEEP LEARNING

Deep learning is the subpart of machine learning, a subset of artificial intelligence. On the other hand, machine learning is a subfield of artificial intelligence. Artificial intelligence is a catch-all term for methods that enable machines/computers to mimic the intelligent behaviour of humans. On either end, machine learning refers to a set of data-driven techniques that allow systems to learn without being explicitly taught. Deep learning is a type of machine learning classification algorithm that is influenced by human neural activity. By continuously analysing data with a specific logical structure, deep learning algorithms strive to generate identical results as humans by continuously analysing data with a specific logical structure. The Neural Network's (NNs) model relies on the anatomy of the nervous system.

Neurons can be trained to carry the same tasks on data that our nervous systems do when searching for patterns and categorizing various types of data. Each layer of NNs can be regarded as a type of filter that operates from coarse to fine, increasing the likelihood of correctly detecting and producing a result. When we acquire new information, our brain attempts to compare it to previously acquired objects in a comparable manner. Deep Neural Networks [33] use the same approach. Ferrag et al. [34] provided an overview of deep learning algorithms for detecting cyber security intrusions. Deep Neural Networks, RNN, Deep Belief Neural Networks, Restricted Boltzmann and Deep Boltzmann machines, Deep auto-encoders, and CNNs are overall studied deep learning models. Izadi et al. [35] used three deep learning networks, namely, CNN, Deep Belief Network, and Multi-layer Perceptron for classifying network traffic.

16.3.2 CONVOLUTIONAL NEURAL NETWORKS

CNNs are deep learning systems that extract features from input data by layering convolutional processes. Convolutional networks are influenced by the sensory/perceptual mechanism in living organisms [36]. A CNN's basic building element is the convolutional layer, which is depicted as follows:

Let's suppose a convolutional layer with a filter ω of size m×m and N×N square neuron layer as input. The output layer is denoted by zl, which is of size $(N-m+1)\times(N-m+1)$ and is calculated as follows:

$$z_{ij}^l = f\left(\sum_{a=0}^{m-1}\sum_{b=0}^{m-1} \omega_{ab} z_{(i+a)(j+b)}^{l-1}\right)$$

To learn more complicated features from the input, a nonlinear function f, like the rectified linear unit (RELU), is frequently imposed over the convolutional output, as shown above. A pooling layer is also used in some applications. The primary reason for using a pooling layer is to combine several features of low level in a neighbourhood to achieve invariance locally, which also assists in lowering the network's computational cost during the training and testing phases.

16.4 MATERIAL AND METHODS

In this work, one of the handiest deep learning techniques in this work is used, specifically CNN, for both "application identification" and "traffic characterization" tasks. Before training the NNs, the network traffic data must be put together to be properly fed into NNs. A preprocessing step over the dataset was carried out to this end. The overall structure of this method is demonstrated in Figure 16.1. A pre-trained NN is utilized to anticipate the traffic class that the packet belongs to, and a pre-trained NN is utilized, similar to the sort of identification of application, classification, and network traffic characterization employed in the test phase. The preprocessing phase's dataset, implementation, layout information, and the structure of suggested NNs are defined in Figure 16.1.

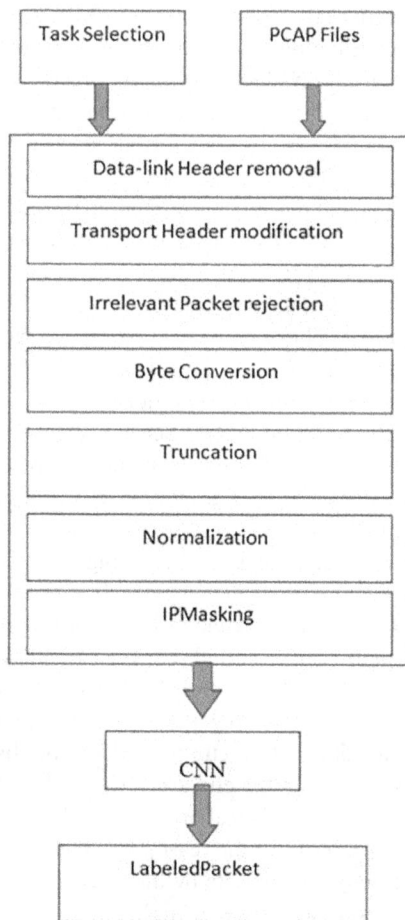

FIGURE 16.1 Preprocessing steps of proffered neural network structure [37].

16.5   DATASET

The "ISCX VPN-nonVPN" traffic dataset is used in this study, which contains collected traffic from a variety of applications in pcap layout files. The captured packets are divided into individual pcap files in this dataset, which are categorized as per the need of the application that generates the packets (such as Skype and Hangouts) and the precise tasks that were involved in the application throughout the capture session such as voice call, file transfer, chatting, and video conferencing.

This study uses the "ISCX VPN-nonVPN" traffic dataset, which contains pcap layout files containing traffic from various applications. In addition, this collection includes traffic recorded by the Tor software application. This network traffic is most likely created whenever the Tor Browser is used. Also, it contains labels like Twitter, Google, and Facebook. An open-source (free) software called "Tor" is a tool that allows users to communicate anonymously. Tor routes user traffic over its unrestricted, underlying global network comprising servers run by volunteers. Tor was developed to protect users from "traffic analysis," a type of Internet spying. Tor constructs a circuit of encrypted connections via a network of relays for establishing a non-public network pathway in such a manner that an entity relay never knows a data packet's whole route.

16.6   PREPROCESSING

Initially, the "ISCX VPN-nonVPN" dataset is acquired over the data link layer. As a result, it consists of the Ethernet header. The data link layer header carries information about its physical connectivity. That is crucial for frame forwarding over the network system. However, inadequate information is worthless for activities like application identification and traffic characterization. In such a way, the Ethernet header is first eliminated from the preprocessing steps. The header period of transport layer segments, such as TCP and UDP, varies. The former usually adheres 20-byte period header, while the latter has an 8-byte header. The transport layer segments are made uniform so that zeros are injected at the ends of UDP segment headers allow them to be identical in length to TCP headers.

Network packets are further converted from bits to bytes, allowing us to minimize the NNs' feed size of the input. Because the dataset is taken using a real-world simulation, it carries a few extra packets that need to be eliminated. The datasets consist of a few TCP segments along with the ACK, SYN, or FIN flags which are set to 1. Thus, it ensures that there is no payload. These segments are required for the three-way handshaking protocol while creating or terminating a link. However, they contain no data about the application that originated them and may be securely discarded. In addition, the file comprises a few Domains Name Service segments. To resolve Hostnames and while converting URLs to IP addresses, TCP segments are often used, but these segments don't apply to both. Consequently, network traffic categorization, as well as Identification of Applications, might be neglected from the dataset.

The length of packets varies across the dataset, while using NNs requires a fixed-length input. As a result, a fixed-length termination or no padding is inescapable. The packet length statistics are carefully inspected to discover the finite size for truncation. According to preliminary research, about 96% of packets have a payload length significantly lower than 1,480 bytes. With a consistent remark, it is estimated that the maximum of the PC networks is restricted via the Maximum Transmission Unit of length to 1,500 bytes. Therefore, the IP header is preserved, with each primary packet being 1,480 bytes long, which leads to a 1,500-byte vector to feed into our hypothetical NNs. Packets with an IP payload of fewer than 1,480 bytes at the end are zero-padded. To reap the highest efficiency, each packet byte is split evenly by 255; the highest price is for a byte, resulting in all the input values lying within the range of [0, 1]. Moreover, for the reason that there may be the possibility of the NN trying to perceive the classification of the packets by their actual IP addresses is that the dataset is gathered within a fixed range of hosts and servers, it has been discovered that over-fitting could occur if the NNs try to learn the classification of the packets by its actual IP addresses. In this case, it has been ensured that the NN does not perform classification using irrelevant or redundant features.

16.6.1 Labelling Dataset

As noted, before, the dataset's pcap documents are categorized as per the applications and operations in which they have been involved. Furthermore, the labels for application identification or traffic characterization activities must be redefined regarding every activity. All pcap documents categorized as a specific application that have been gathered all through a nonVPN session are then integrated right to form a single file for application determination. In the end, there are 17 different labels as indicated in Table 16.1. To characterize the network traffic, the recorded traffic of distinctive programs engaged in similar tasks is pooled into a single pcap file, regardless of whether they were using a VPN. Table 16.2 shows this result in a dataset with 12 classes. When looking at Tables 16.1 and 16.2, it's clear that the dataset is drastically unbalanced, and also it can be viewed that the number of samples changes dramatically between multiple classes. It is well understood that such a discrepancy in training records results in poor class performance.

Sampling is an essential and effective method to triumph over this issue. Therefore, the proposed NNs are learned by using the under-sampling approach, in which the samples of main significant classes (major classes/classes with higher samples) are neglected randomly until the classes are fairly balanced.

16.6.2 Model Architecture

Lotfollahi et al. 2019 [37] depict a simplified example of the developed model based on one-dimensional CNN. To achieve good performance with this model, a grid search has been performed within a subspace of the hyper-parameter space. Moreover, to hinder the over-fitting problem, the dropout strategy is used where two-dimensional sensor is compressed into a vector space of one-dimensional, then connected into a three-layered community of fully linked neurons as indicated in Figure 16.2. For classification, a SoftMax classifier is incorporated and the best values observed for the hyper-parameters [37] are shown in Table 16.3.

TABLE 16.1
Various Labels for Application Identification through a Non-VPN Session [37]

Application	Size
AIMChat	5K
Email	28K
YouTube	251K
FTFS	7,872K
Gmail	12K
Netflix	299K
ICQ	7K
Hangouts	3,766K
SCP	448K
SFTP	418K
Skype	2,872K
Spotify	40K
Torrent	70K
Tor	202K
Voipbuster	842K
Vimeo	146K
Facebook(fb)	2,502K

TABLE 16.2

Dataset Used for Traffic Characterization [37]

Class Name	Size
Chat	82K
Email	28K
FileTransfer	210K
Streaming	1,139K
Torrent	70K
VoIP	5,120K
VPN:Chat	50K
VPN:Email	251K
VPN:FileTransfer	13K
VPN:Streaming	479K
VPN:Torrent	269K
VPN:VoIP	753K

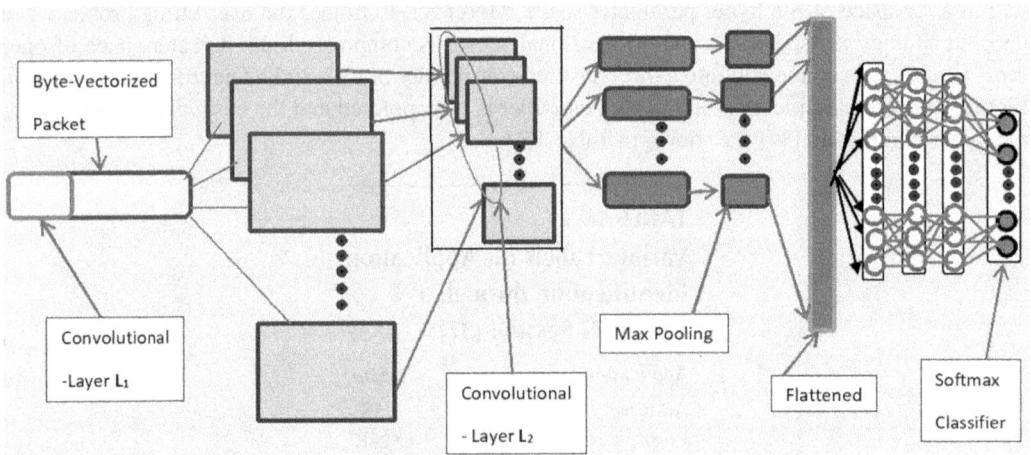

FIGURE 16.2 Proposed architecture depending on one-dimensional CNN.

TABLE 16.3

Obtained Values for Hyper Parameters

Tasks	C1 Filter			C2 Filter		
	Sizes	No.	Strides	Sizes	No.	Stride
Traff. class.	5	200	3	4	200	3
App. iden.	4	200	3	5	200	1

16.7   EXPERIMENTAL RESULTS AND DISCUSSION

To implement the NN, Python libraries such as Pyspark, Numpy, and Panda are used. The proffered model is trained and tested using an independent test set extracted from the dataset. The dataset is categorized randomly into three different sets: the initial set, which incorporates 64% of samples, is employed for training as well as managing the adjustment of weights and biases; the next stage, which contains 16% of samples, is employed for validation in the course of the training phase; eventually, the last one, consists of 20% of data factors, is used for testing the model. In addition, "an early stopping technique" was used to overcome the over-fitting problem. When the cost of the loss function on the validation set continues to remain nearly unchanged for several epochs, the training procedure is stopped in this approach, consequently preventing the network from over-fitting on the training data. The Batch Normalization approach was employed in the proposed model to accelerate the learning phase. To implement the suggested one-dimensional CNN, Adamas was used as an optimizer, and the categorical cross-entropy was used as a loss function. Here, the network is stretched for 300 epochs. Also, all layers have had ReLU as an activation function.

To examine the overall performance, Recall (Rc) and Precision (Pr) metrics have been used. The above metrics are defined mathematically as follows:

$$Rc = \frac{TP}{TP+FN}$$

$$Pr = \frac{TP}{TP+FP}$$

Here, TP is True Positive, FP is False Positive, and FN is False Negative.

Tables 16.4 and 16.5 demonstrate the achieved performance for both Application Identification and Network Classification, respectively.

It is evident from Figure 16.3 that deep learning is an effective technique for both Network Classification and Application Identification.

TABLE 16.4

Performance Result of Application Identification

	Label	Rc	Pr
0	AIMChat	0.48050	0.930946
1	Email	0.185192	0.647936
2	Facebook	0.833078	0.901853
3	FTFS	0.998785	0.995186
4	Gmail	0.162633	0.910223
5	Hangouts	0.923520	0.929447
6	ICQ	0.093160	0.752518
7	Netflix	0.950540	0.981588
8	SCP	0.789973	0.925415
9	SFTP	0.998043	0.964007
10	Skype	0.971092	0.763326
11	Spotify	0.387922	0.966516
12	Torrent	0.271715	0.978052
13	Tor	0.991099	0.997363
14	Vimeo	0.963770	0.980871
15	Voipbuster	0.994875	0.986675
16	YouTube	0.940438	0.953821

TABLE 16.5

Performance Result of Network Classification

	Label	Rc	Pr
0	Chat	0.558824	0.140479
1	Email	0.132695	0.167066
2	FileTransfer	0.790650	0.998449
3	Streaming	0.966900	0.710631
4	Torrent	0.128468	0.993157
5	VoIP	0.861641	0.616195
6	VPN:Chat	0.333166	0.902410
7	VPN: Email	0.995494	0.844921
8	VPN:FileTransfer	0.171821	0.995940
9	VPN:Streaming	0.998482	0.994873
10	VPN:Torrent	0.990486	0.995405
11	VPN:VoIP	0.885191	0.487451

Although there are some areas where these methods are not as effective as it is related to some other areas, there is always scope for further improvement. Therefore, experimentation should be undertaken using different methods and algorithms to get the desired results.

16.8   CONCLUSION

In this work, a deep learning method (using a CNN) is proposed to tackle the problem of network traffic classification. A lot of work has been done on this problem, but most of it was not nearly as efficient in solving it. This work differs from previous works, including application identification and traffic classification. By doing so, we make the task of data analysis a much easier task. This method successfully classifies the Tor and VPN traffic, a significant achievement in this field. We can enhance the performance and results with better solutions for the imbalance problems by employing various other machine learning algorithms. In future, we will work on the joint optimization techniques of deep learning and extend it by considering different advanced deep learning layers such as inception, attention, residual, and semi-supervised multitask learning.

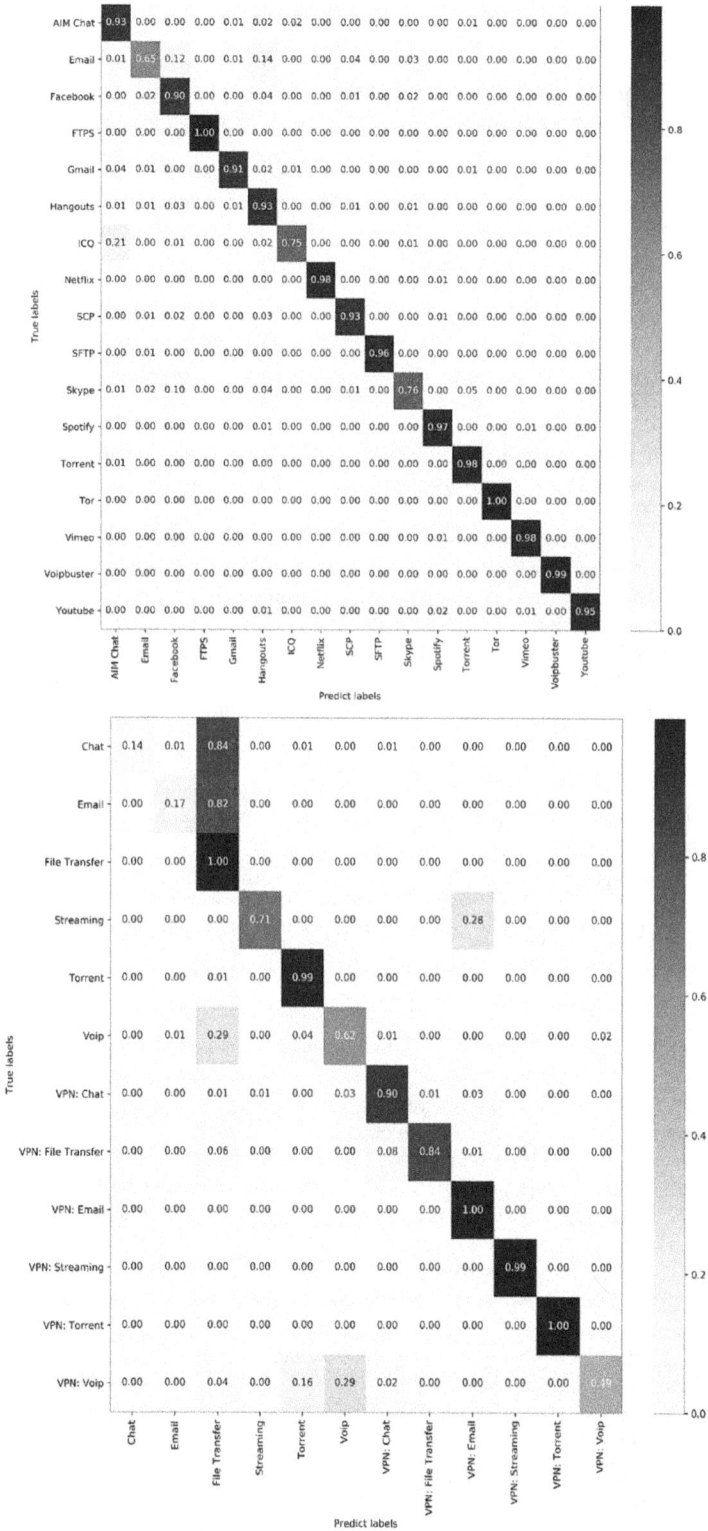

FIGURE 16.3   Confusion matrix of application identification and network classification.

REFERENCES

1. Tahaei H, Afifi F, Asemi A, Zaki F, Anuar N (2020) The rise of traffic classification in IoT networks: A survey. *Journal of Network and Computer Applications* 154:102538.
2. Dainotti A, Pescape A, Claffy KC (2012) Issues and future directions in traffic classification. *IEEE Network* 26(1).
3. Finsterbusch M, Richter C, Rocha E, Muller JA, Hanssgen K (2014) A survey of payload-based traffic classification approaches. *IEEE Communications Surveys & Tutorials* 16(2):1135–1156.
4. Velan P, et al. (2015) A survey of methods for encrypted traffic classification and analysis. *International Journal of Network Management* 25(5):355–374.
5. Lv J, Zhu C, Tang S, Yang C (2014) Deepflow: Hiding anonymous communication traffic in p2p streaming networks. *Wuhan University Journal of Natural Sciences* 19(5):417–425.
6. Alshammari R, Nur Zincir-Heywood A (2011) Can encrypted traffic be identified without port numbers, IP addresses and payload inspection? *Computer Networks* 55(6):1326–1350.
7. Kohout J, Pevn´y T (2018) Network traffic fingerprinting based on approximated kernel two-sample test. *IEEE Transactions on Information Forensics and Security* 13(3). doi: 10.1109/TIFS.2017.2768018.
8. Perera P, et al. A comparison of supervised machine learning algorithms for classification of communications network traffic. In *International Conference on Neural Information Processing*. Springer, Cham, 2017.
9. Draper-Gil G, et al. Characterization of encrypted and vpn traffic using time-related. In *Proceedings of the 2nd International Conference on Information Systems Security and Privacy (ICISSP)*, 2016.
10. Moore AW, Papagiannaki K Toward the accurate identification of network applications. In *PAM*. Springer, vol 5, pp 41–54, 2005.
11. Khalife J, Hajjar A, Diaz-Verdejo J (2014) A multilevel taxonomy and requirements for an optimal traffic classification model. *International Journal of Network Management* 24(2):101–120.
12. Bengio Y (2009) Learning deep architectures for AI. *Foundations and Trends in Machine Learning* 2(1):1–127.
13. LeCun Y, Bengio Y, Hinton G (2015) Deep learning. *Nature* 521(7553):436–444.
14. Yamansavascilar B, et al. Application identification via network traffic classification. In *2017 International Conference on Computing, Networking and Communications (ICNC)*. IEEE, 2017.
15. Qi Y, et al. (2009) Packet classification algorithms: From theory to practice. *IEEE INFOCOM*. IEEE.
16. Madhukar A, Williamson C A longitudinal study of P2P traffic classification. In *14th IEEE International Symposium on Modeling, Analysis, and Simulation*. IEEE, 2006.
17. Yeganeh SH, Eftekhar M, Ganjali Y, Keralapura R, Nucci A Cute: Traffic classification using terms. In *Computer Communications and Networks (ICCCN), 2012 21st International Conference on, IEEE*, pp 1–9, 2012.
18. Sen S, Spatscheck O, Wang D Accurate, scalable in-network identification of p2p traffic using application signatures. In *Proceedings of the 13th International Conference on World Wide Web*. ACM, New York, NY, pp 512–521, 2004.
19. Sherry J, et al. Blindbox: Deep packet inspection over encrypted traffic. In *Proceedings of the 2015 ACM Conference on Special Interest Group on Data Communication*, 2015.
20. Setiawan R, et al. (2021) Encrypted network traffic classification and resource allocation with deep learning in software defined network. *Wireless Personal Communications*:1–17.
21. Crotti M, et al. (2007) Traffic classification through simple statistical fingerprinting. *ACM SIGCOMM Computer Communication Review* 37(1):5–16.
22. Wang X, Parish DJ Optimised multi-stage TCP traffic classifier based on packet size distributions. In *Communication Theory, Reliability, and Quality of Service (CTRQ), 2010 Third International Conference on, IEEE*, pp 98–103, 2010.
23. Dong S (2021) Multi class SVM algorithm with active learning for network traffic classification. *Expert Systems with Applications* 176:114885.
24. Afuwape AA, et al. (2021) Performance evaluation of secured network traffic classification using a machine learning approach. *Computer Standards & Interfaces* 78:103545.
25. Auld T, Moore AW, Gull SF (2007) Bayesian neural networks for internet traffic classification. *IEEE Transactions on Neural Networks* 18(1):223–239.
26. Moore AW, Zuev D Internet traffic classification using Bayesian analysis techniques. In *ACM SIGMETRICS Performance Evaluation Review*. ACM, vol. 33, pp 50–60, 2005.
27. Sun R, Yang B, Peng L, Chen Z, Zhang L, Jing S Traffic classification using probabilistic neural networks. In *Natural computation (ICNC), 2010 Sixth International Conference on, IEEE*, vol 4, pp 1914–1919, 2010.

28. Ting H, Yong W, Xiaoling T Network traffic classification based on kernel self-organizing maps. In *Intelligent Computing and Integrated Systems (ICISS), 2010 International Conference on, IEEE*, pp 310–314, 2010.

29. Moore A, Zuev D, Crogan M *Discriminators for Use in Flow-Based Classification*, 2013.

30. Wang Z (2015) The applications of deep learning on traffic identification. *BlackHat* 24(11):1–10.

31. Chen D, Wawrzynski P, Lv Z (2021) Cyber security in smart cities: A review of deep learning-based applications and case studies. *Sustainable Cities and Society* 66:102655.

32. Aceto G, Ciuonzo D, Montieri A, Pescapé A (2021) DISTILLER: Encrypted traffic classification via multimodal multitask deep learning. *Journal of Network and Computer Applications* 183:102985.

33. https://towardsdatascience.com/what-is-deep-learning-and-how-does-it-work-2ce44bb692ac.

34. Ferrag M, Maglaras L, Moschoyiannis S, Janicke H (2020) Deep learning for cyber security intrusion detection: Approaches, datasets, and comparative study. *Journal of Information Security and Applications* 50:102419.

35. Izadi S, Ahmadi M, Rajabzadeh A (2022) Network traffic classification using deep learning networks and Bayesian data fusion. *Journal of Network and Systems Management* 30(2):1–21.

36. Hubel DH, Wiesel TN (1968) Receptive fields and functional architecture of monkey striate cortex. *The Journal of Physiology* 195(1):215–243.

37. Lotfollahi M, Jafari Siavoshani M, Shirali Hossein Zade R, Saberian M (2019) Deep packet: A novel approach for encrypted traffic classification using deep learning. *Soft Computing* 24(3):1999–2012. doi: 10.1007/s00500-019-04030-2.

Index

Note: **Bold** page numbers refer to tables and *italic* page numbers refer to figures.

Taylor & Francis Group
an **informa** business

Taylor & Francis eBooks

www.taylorfrancis.com

A single destination for eBooks from Taylor & Francis
with increased functionality and an improved user
experience to meet the needs of our customers.

90,000+ eBooks of award-winning academic content in
Humanities, Social Science, Science, Technology, Engineering,
and Medical written by a global network of editors and authors.

TAYLOR & FRANCIS EBOOKS OFFERS:

A streamlined
experience for
our library
customers

A single point
of discovery
for all of our
eBook content

Improved
search and
discovery of
content at both
book and
chapter level

REQUEST A FREE TRIAL
support@taylorfrancis.com

Routledge
Taylor & Francis Group

CRC Press
Taylor & Francis Group

For Product Safety Concerns and Information please contact our EU
representative GPSR@taylorandfrancis.com
Taylor & Francis Verlag GmbH, Kaufingerstraße 24, 80331 München, Germany

www.ingramcontent.com/pod-product-compliance
Lightning Source LLC
Chambersburg PA
CBHW061347210326
41598CB00035B/5908

* 9 7 8 1 0 3 2 2 1 3 2 1 7 *